4th International Conference on Manufacturing Optimization, Industrial and Material Engineering (MOIME 2016)

AA004657

IOP Conference Series: Materials Science and Engineering
Volume 131

Bali, Indonesia
19 – 20 March 2016

Editors:

Ford Lumban Gaol
Jeff Webb
Jun Ding

ISBN: 978-1-5108-2450-8
ISSN: 1757-8981

Printed from e-media with permission by:

Curran Associates, Inc.
57 Morehouse Lane
Red Hook, NY 12571

Some format issues inherent in the e-media version may also appear in this print version.

Printed by Curran Associates, Inc. (2016)

For permission requests, please contact the Institute of Physics
at the address below.

Institute of Physics
Dirac House, Temple Back
Bristol BS1 6BE UK

Phone: 44 1 17 929 7481
Fax: 44 1 17 920 0979

techtracking@iop.org

Additional copies of this publication are available from:

Curran Associates, Inc.
57 Morehouse Lane
Red Hook, NY 12571 USA
Phone: 845-758-0400
Fax: 845-758-2633
Email: curran@proceedings.com
Web: www.proceedings.com

Table of contents

Volume 131

4th International Conference on Manufacturing, Optimization, Industrial and Material Engineering (MOIME 2016)
19–20 March 2016, Bali, Indonesia

Accepted papers received: 16 May 2016
Published online: 6 June 2016

Preface

Papers

IOP Conf. Series: Materials Science and Engineering **131** (2016) 011001 doi:10.1088/1757-899X/131/1/011001

The 4th International Conference on Manufacturing, Optimization, Industrial and Material Engineering 2016 (MOIME 2016) was held in Bali Rani Hotel, Bali, Indonesia, from 19–20 March 2016.

The MOIME 2016 conference aims to bring together researchers, engineers and scientists from around the world who work in the areas of interest. MOIME 2016 is focused on promoting interaction between the theoretical, experimental, and applied communities, so that a high level exchange is achieved in new and emerging areas within material engineering, industrial engineering and all areas that relate to optimization.

We would like to express our sincere gratitude to all in the Technical Program Committee who have reviewed the papers and developed a very interesting conference program, as well as to the invited and plenary speakers.

This year, we received 117 papers and after rigorous review, 25 papers were accepted.

The participants came from seven countries and there were six parallel sessions and three keynote speakers.

It is an honour to present this volume *of IOP Conference Series: Materials Science and Engineering (MSE)* and we deeply thank the authors for their enthusiastic and high-grade contributions.

Finally, we would like to thank the conference chairmen, the members of the steering committee, the organizing committee, the organizing secretariat and the financial support from the conference sponsors that allowed MOIME 2016 to be a success.

the Editors of the MOIME 2016 Proceedings

Dr. Ford Lumban Gaol,
Jeff Webb, Ph.D
Prof. Jun DING, Ph.D

IOP Conf. Series: Materials Science and Engineering **131** (2016) 011002 doi:10.1088/1757-899X/131/1/011002

Keynote Speakers

1. Dr. Pragya Shandilya
Motilal Nehru National Institute of Technology, Allahabad

Title of Talk : The design of experiments (or experimental design)

Abstract : The design of experiments (or experimental design) is the design of any task that aims to describe or explain the variation of information under conditions that are hypothesized to reflect the variation. The term is generally associated with true experiments in which the design introduces conditions that directly affect the variation, but may also refer to the design of quasi-experiments, in which natural conditions that influence the variation are selected for observation. The lecture will examine how to design experiments, carry them out, and analyze the data they yield. Various designs will be discussed with their respective differences, advantages, and disadvantages. In particular, factorial and fractional factorial designs will be discussed in more detail. These are designs in which two or more factors are varied simultaneously; the experimenter wishes to study not only the effect of each factor, but also how the effect of one factor changes as the levels of other factors change. The latter is generally referred to as an interaction effect among factors. The lecture will be elementary in terms of mathematics.

Throughout the lecture we emphasize applications, using real examples from the area of manufacturing. With the use of examples of manufacturing processes, the lecture will show how to optimize the machining parameters among number of experiments.

Short Bio :

Dr. Pragya Shandilya
Assistant Professor
Mechanical Engineering Department
Motilal Nehru National Institute of Technology, Allahabad

EDUCATION

Ph.D 2012 IIT Roorkee, India
M.Tech. (Design & Production) 2008 College of Tech., Pant Nagar

RESEARCH INTERESTS

- Advanced machining process
- Hybrid machining process
- Composite materials
- Application of soft computing techniques to manufacturing process
- Manufacturing process parameters optimization

PUBLICATIONS: 50+ (book chapters, international journals and conferences).

2. Prof Ding Jun

Department of Materials Science & Engineering, National University of Singapore, Singapore 119260

Title of Talk : In-vivo measurement of magnetic nanoparticle concentration for cancer treatment by hyperthermia.

Abstract : Recently, high-performance agents for magnetic hyperthermia have been obtained. Excellent intratumoral results have been achieved. However, intravenous injection is much more ideal for cancer treatment. For this purpose, we have prepared Gd_2O_3 and Gd-doped Fe_3O_4 nanoparticles. For ultra-small Gd_2O_3 nanoparticles (Fig. 1), an improved longitudinal relaxivity r1 of 12.1 mM1 s1 at 7 T has been achieved, which is around three times that of the commercial contrast agent Magnevist@. The results have shown that our Gd_2O_3 nanoparticles with a particle size of 3 nm are an excellent MRI positive contrast enhancement agent (Fig. 2). Our Gd-doped Fe_3O_4 nanoparticles have shown a long blood circulation time. A significant number of magnetic particles have been successfully delivered into the tumor. Our research has shown that these particles can be qualitatively monitored using MRI. Magnetic hyperthermia has been

IOP Conf. Series: Materials Science and Engineering **131** (2016) 011002 doi:10.1088/1757-899X/131/1/011002

obtained. This implies a novel method for in-vivo measurement of magnetic nanoparticles for magnetic hyperthermia treatment.

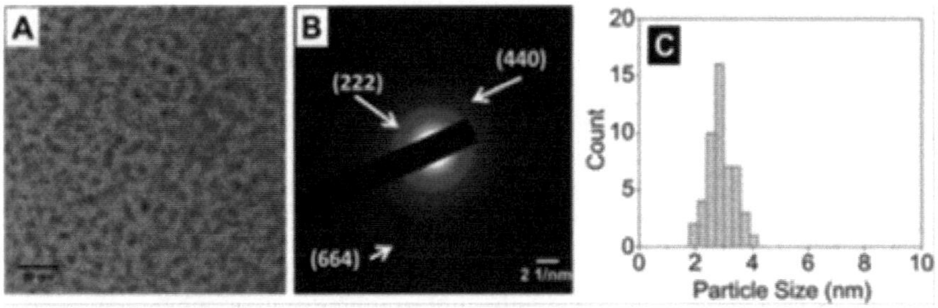

Fig. 1. (A) TEM image of Gd2O3-OA, (B) SAED pattern and (C) Statistics of the measured particle size of (A).

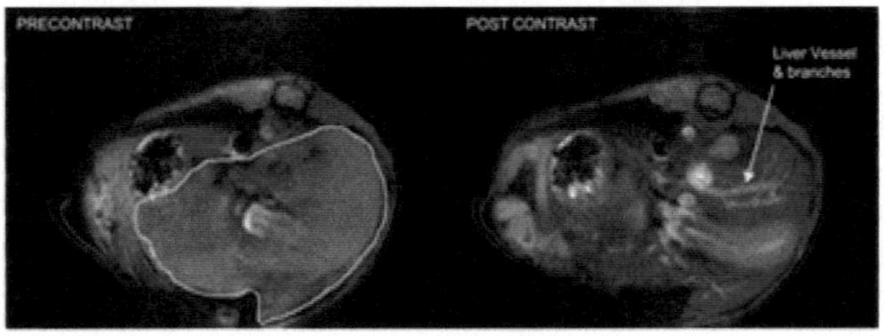

Fig. 2. T1-weighted liver images obtained before injection (left); 12 min after injection (right) of Gd2O3-PVP.

Short Bio : Dr Jun Ding obtained his PhD from Ruhr-University Bochum, Germany, in 1990. After working as Research Fellow and Senior Research Fellow at the University of Western Australia for six years, Dr Ding joined the National University of Singapore (NUS) as a Faculty. Now, he is a Professor at the Department of Materials Science & Engineering (NUS) and Visiting Professor at NIMTE, Chinese Academy of Science. Dr Ding has been working in magnetic materials for ~25 years. He has published over 300 papers with > 7000 citations and an H-index of 47. His current research work focuses on nanomagnetics and magnetic nanoparticles for bio-medical applications.

3. Assoc Prof. Jeffrey Webb, Ph.D

Department of Mechanical, Materials and Manufacturing Engineering, The University of Nottingham Malaysia Campus.

Title of Talk : Some Insights into Quantum Cascade Lasers using a Multi-Level Rate Equation Model

Abstract : Quantum cascade lasers (QCLs) have many interesting applications including terahertz wave generation. Here a multi-level model of the cascade of wells in a QCL will be discussed and it will be shown how some useful quantities important for the characterisation of QCLs can be calculated in a relatively straightforward manner by employing a set of uncoupled rate equations. Graphical illustrations of the dependence between various parameter pairs can also be constructed from solutions of the rate equations, giving further insight into the operation of QCLs.

Short Bio : Jeffrey F. Webb received his PhD in Physics at the University of Strathclyde, Scotland, UK. He is an Associate Professor in the Department of Mechanical, Materials and Manufacturing Engineering, The University of Nottingham Malaysia Campus. His research interests include nanoscale ferroelectric materials, the theory of graphene, and quantum cascade laser modelling. He is the author of many research papers published in international journals and international conference proceedings, and has contributed several chapters to books on ferroelectrics and nanotechnology.

IOP Conf. Series: Materials Science and Engineering **131** (2016) 011003 doi:10.1088/1757-899X/131/1/011003

Peer review statement

All papers published in this volume of *IOP Conference Series: Materials Science and Engineering* have been peer reviewed through processes administered by the proceedings Editors. Reviews were conducted by expert referees to the professional and scientific standards expected of a proceedings journal published by IOP Publishing.

IOP Conf. Series: Materials Science and Engineering **131** (2016) 012001 doi:10.1088/1757-899X/131/1/012001

Optimal physicians schedule in an Intensive Care Unit

L Hidri
Department of Industrial Engineering, College of Engineering, King Saud University, PO Box 800 Riyadh 11421, Saudi Arabia.
E-mail: lhidri@ksu.edu.sa

M Labidi[1]
Department of Industrial Engineering, College of Engineering, King Saud University, PO Box 800 Riyadh 11421, Saudi Arabia.
E-mail: mlabidi@ksu.edu.sa

Abstract. In this paper, we consider a case study for the problem of physicians scheduling in an Intensive Care Unit (ICU). The objective is to minimize the total overtime under complex constraints. The considered ICU is composed of three buildings and the physicians are divided accordingly into six teams. The workload is assigned to each team under a set of constraints. The studied problem is composed of two simultaneous phases: composing teams and assigning the workload to each one of them. This constitutes an additional major hardness compared to the two phase's process: composing teams and after that assigning the workload. The physicians schedule in this ICU is used to be done manually each month. In this work, the studied physician scheduling problem is formulated as an integer linear program and solved optimally using state of the art software. The preliminary experimental results show that 50% of the overtime can be saved.

1. Introduction

The personnel scheduling problems consist on the assignment of some tasks to a set of workers under a number of constraints that can be conflicting in some circumstances. These constraints make the actual scheduling problem harder than the simple assignment problem. From practical point of view, the preparation of the personnel schedule is a hard task that requires taking in consideration a plenty of rules such as: workers preferences, assignment of the night shifts, weekends, restrictions on the number of consecutive shifts, days off and vacation periods. The personnel scheduling problems are of practical interest since they are met in several working environments. In the other hand, these scheduling problems are of theoretical interest because of their hardness. Subsequently, during the last decades these scheduling problems have received a lot of attention from researchers community where a satisfactory literature have been provided and several approaches have been proposed ([1, 2, 3 and 4]).

In the health care sector, the personnel scheduling problems present a particular interest since several services (emergency rooms, intensive care unit) should be ensured continuously twenty four hours per day, seven days a week to preserve human life. Thus, several authors have addressed the personnel scheduling problems and most of them treated the case of nurse scheduling ([5, 6, 7, and 8]). In addition, there are several differences between the nurses scheduling and the physicians scheduling in terms of preferences, requirements, specialties…

[1] To whom any correspondence should be addressed.

Some works on physicians scheduling problem have been proposed in literature. The Authors in ([9, 10, 11 and 12]) tackled the emergency rooms scheduling problem, and ([13 and 14]) treated the same scheduling problem in operating rooms. More recent works ([15, 16 and 17]) considered the physician scheduling problem and provided a master scheduling for the hospital (operating rooms, emergency rooms….).

In this work, a case study in an Intensive Care Unit in a local hospital is considered. In this ICU the proposed physicians schedule is carried out manually each month under several particular restrictions and rules. The performing of such a schedule is a time consuming one with critical distribution from physician's side. To overcome all the manual schedule drawbacks an integer linear programming model is proposed and preliminary experiments are conducted and compared with the existing ones.

The paper is organized as follows: in Section 2, we give a brief description of the studied ICU. The mathematical formulation is presented in section 3. In addition, an experimental analysis of the proposed integer linear program is presented in section 4. Finally, we conclude and we present some directions for future research.

2. Description of the studied ICU

The considered ICU is composed of three buildings according to the provided service. These buildings are:

1. **Building 1: The general ICU building** which is the main building for the ICU department.
2. **Building 2: Maternity & burn ICU building** which concern the maternity and burn cases.
3. **Building 3: Mobile ICU for accidents & emergency,** where a team of physicians is located and intended to intervene outside in cases of accidents, in order to treat severe cases.

The available set of physicians is divided into six teams, where each one is assigned to a building. The required number of physicians, for each one of these buildings depends on the building itself and on the season (summer, fall, winter, spring). The determination of such a number is based on the previous collected data. The workload per day is composed of two shifts (day and night shifts), with duration of 12 hours for each one. The week starts Sunday until Thursday and the weekend is for Friday and Saturday. In addition, the physicians schedule is performed each month manually under the following constraints.

- Each physician is assigned to one team.
- The regular workload for each physician is 208 hours per month.
- The night shift for all the buildings will be served by one team only.
- All the team members are assigned to the same location at the same time.
- Building 1 is the main ICU building which serves mostly the severe cases. So, for the comfort of the patient, the same team should serve for the whole week days (Sunday – Thursday).

- The work in building 1 is intense. So, each physician shouldn't work two consecutive weeks at that building.
- Each physician should have two consecutive days as a week vacation.
- At the weekend, the team assigned for the day shift in building 1 will also be assigned for the night shift (24 hours working).
- The team working on Saturday or Friday at building 1 must have a day-off before and a day-off after.
- If the physician worked at the weekends in buildings 2 or 3, he should work Friday and Saturday.
- If the physician worked at the night shift today, he shouldn't work tomorrow morning.

3. Mathematical formulation

In this section, the physician scheduling problem is formulated as an integer linear programming. For that aim we introduce the following notations and definitions.

N: The number of available physicians during a month.

I: set of physicians. $I = \{1, 2...N\}$

J: set of shifts. $J = \{1, 2\}$

K: set of buildings. $J = \{1, 2, 3\}$

L: set of days. $L = \{1, 2... 28\}$

M: set of Team. $M = \{A, B, D, G, E, H\}$.

R_{kj}: the minimum required number of physicians in building k for the shift j.

The considered decision variables are:

$$x_{ijklm} = \begin{cases} 1 \text{ if physician } i \text{ is assigned to shift } j \text{ on building } k \text{ for day } l \text{ and team } m \\ 0 \text{ otherwise} \end{cases}$$

$$y_{im} = \begin{cases} 1 \text{ if physician } i \text{ is assigned to team } m \\ 0 \text{ otherwise} \end{cases}$$

$$U_{il} = \begin{cases} 1 \text{ if physician } i \text{ has two consecutive days off (within a week) starting from day } l \\ 0 \text{ otherwise} \end{cases}$$

$$v_{ihm} = \begin{cases} 1 \text{ if physicians } i \text{ and } h \text{ belong to the team } m \\ 0 \text{ otherwise} \end{cases}$$

The objective function presenting the total over time is given by the following expression:

$$Z=\sum_{i=1}^{N}\sum_{j=1}^{2}\sum_{k=1}^{3}\sum_{l=1}^{28}\sum_{m=1}^{6}12x_{ijklm}-\sum_{i=1}^{N}\sum_{l=1}^{28}\sum_{m=1}^{6}(12x_{i22lm}+12x_{i23lm})-208N$$

where,

208 N is the minumum load for the physicians(each one with *208* hours per month).

$-\sum_{i=1}^{N}\sum_{l=1}^{28}\sum_{m=1}^{6}(12x_{i22lm}+12x_{i23lm})$: we substract this expression since the night shift for all the buildings will be served by only one team.

$\sum_{i=1}^{N}\sum_{j=1}^{2}\sum_{k=1}^{3}\sum_{l=1}^{28}\sum_{m=1}^{6}12x_{ijklm}$: the total loadwork with repetition presented in the latter expression.

1- Each physician *i* is assigned exactly to one team:

$$\sum_{m=1}^{6}y_{im}=1 \qquad \forall i=1,..,N$$

2- The number of team members is from three to six members:

$$3\leq\sum_{i=1}^{N}y_{im}\leq 6 \qquad \forall m=1,..,6$$

3- The relation between the decision variables *x* and *y*:

$$x_{ijklm}\leq y_{im} \qquad \forall i=1,..,N,\forall j=1,..,2,\forall k=1,..,3,\forall l=1,..,28,\forall m=1,..,6$$

4- The regular workload for each physician is *208* hours per month:

$$\sum_{j=1}^{2}\sum_{k=1}^{3}\sum_{l=1}^{28}\sum_{m=1}^{6}12x_{ijklm}-\sum_{l=1}^{28}\sum_{m=1}^{6}(12x_{i22lm}+12x_{i23lm})\geq 208 \qquad \forall i=1,..,N$$

5- Each physician *i* should work one shift per day at building 2 and 3:

$$\sum_{j=1}^{2}x_{ijklm}\leq 1 \qquad \forall i=1,..,N,\forall k=1,..,3,\forall l=1,..,28,\forall m=1,..,6$$

6- Each physician *i* should work one shift per day at building 1: (Sunday – Thursday):

$$\sum_{j=1}^{2}x_{ij1lm}\leq 1 \qquad \forall i=1,..,N,\forall k=1,\forall l=1,..,28/\{6,7,13,14,20,21,27,28\},\forall m=1,..,6$$

7- The night shift for all the buildings will be served by one team only.

$$x_{i21lm}=x_{i22lm}=x_{i23lm} \qquad \forall i=1,..,N,\forall l=1,..,28,\forall m=1,..,6$$

8- Building 1 is the main ICU buildings which served mostly the severe cases. So, for the comfort of the patient, the same team should serve for the whole week days (Sunday – Thursday):

$$x_{i11lm}=x_{i11lm} \qquad \forall i=1,..,N,\forall\, 2\leq l\leq 5,\forall m=1,..,6$$

$$x_{i118m} = x_{i11lm} \qquad \forall i = 1,.., N, \forall 9 \le l \le 18, \forall m = 1,..,6$$

$$x_{i11(15)m} = x_{i11lm} \qquad \forall i = 1,.., N, \forall 16 \le l \le 19, \forall m = 1,..,6$$

$$x_{i11(22)m} = x_{i11lm} \qquad \forall i = 1,.., N, \forall 23 \le l \le 26, \forall m = 1,..,6$$

9- The work in building 1 is intense. So, each physician shouldn't work two consecutive weeks at this building:

$$x_{i11lm} + x_{i11(7+l)lm} \le 1 \qquad \forall i = 1,.., N, \forall 1 \le l \le 5, \forall m = 1,..,6$$

$$x_{i118m} + x_{i11(14+l)lm} \le 1 \qquad \forall i = 1,.., N, \forall 1 \le l \le 5, \forall m = 1,..,6$$

$$x_{i11(15)m} + x_{i11(21+l)lm} \le 1 \qquad \forall i = 1,.., N, \forall 1 \le l \le 5, \forall m = 1,..,6$$

10- The required number of physician per day should be more than R_{kj} :

$$\sum_{j=1}^{2} x_{ijklm} \ge R_{kj} \qquad \forall i = 1,.., N, \forall k = 1,..,3, \forall l = 1,..,28, \forall m = 1,..,6$$

11- Each physician should have two consecutive days as a week vacation:

$$\sum_{l=1}^{6} U_{il} \ge 1 \qquad \forall i = 1,.., N$$

$$\sum_{l=8}^{13} U_{il} \ge 1 \qquad \forall i = 1,.., N$$

$$\sum_{l=15}^{20} U_{il} \ge 1 \qquad \forall i = 1,.., N$$

$$\sum_{l=22}^{27} U_{il} \ge 1 \qquad \forall i = 1,.., N$$

12- The relation between the variables U and x:

$$x_{i1klm} + x_{i2(s)lm} + x_{i1t(l+1)m} + x_{i2w(l+1)m} \le 4(1 - U_{il})$$

$$\forall i = 1,.., N, \forall k = 1,...,3, \forall l = 1,..,28/\{7,14,21,28\}$$

$$\forall m = 1,..,6, \forall t = 1,...,3, \forall s = 1,...,3, \forall w = 1,...,3.$$

13- At the weekend, the team assigned for the day shift in building 1 will also be assigned for the night shift (24 hours working):

$$x_{i11lm} = x_{i21lm} \qquad \forall i = 1,..,35, \forall l = \{6,7,13,14,20,21,27,28\}, \forall m = 1,..,6$$

14- The team working Saturday or Friday at building 1 must have day-off before and day-off after:

$$x_{ijk(l-1)m} \le 1 - x_{i11lm} \qquad \forall i = 1,.., N, \forall j = 1,..,2, \forall k = 1,..,3, \forall l = \{6,13,20,27\}, \forall m = 1,..,6$$

$$x_{ijk(l+1)m} \le 1 - x_{i11lm} \qquad \forall i = 1,.., N, \forall j = 1,..,2, \forall k = 1,..,3, \forall l = \{7,14,21\}, \forall m = 1,..,6$$

$$x_{i11(27)m} \le 1 - x_{i11(28)m} \qquad \forall i = 1,.., N, \forall m = 1,..,6$$

15- If the physician worked at the weekends in buildings 2 or 3, he should work Friday and Saturday:

$$x_{i1klm} = x_{i1k(l+1)m} \qquad \forall i = 1,.., N, \forall k = 2,..,3, \forall l = \{6,13,20,27\}, \forall m = 1,..,6$$

16- If the physician worked at the night shift today he shouldn't work tomorrow morning:

$$x_{i2klm} + x_{i1s(l+l)m} \leq 1 \qquad \forall i = 1,.., N, \forall s = 1,..,3, \forall k = 1,..,3, \forall l = 1,...,27, \forall m = 1,..,6$$

17- A physician has to be assigned in only one location for the morning shift:

$$x_{i11lm} + x_{i12lm} + x_{i13lm} \leq 1 \qquad \forall i = 1,.., N, \forall l = 1,...,28, \forall m = 1,..,6$$

18- All the team members are assigned to the same location at the same time:

$$x_{ijklm} - x_{hjklm} \leq 1 - v_{ihm} \qquad \forall i = 1,.., N, \forall i < h \leq N, \forall j = 1,..,2, \forall k = 1,..,3, \forall l = 1,...,28, \forall m = 1,..,6$$

$$v_{ihm} + 1 \geq y_{im} + y_{hm} \qquad \forall i = 1,.., N, \forall i < h \leq N, \forall m = 1,..,6$$

$$\sum_{m=1}^{6} v_{ihm} \leq 1 \qquad \forall i = 1,.., N, \forall i < h \leq N$$

$$v_{ihm} \leq y_{im} \qquad \forall i = 1,.., N, \forall i < h \leq N$$

$$v_{ihm} \leq y_{hm} \qquad \forall i = 1,.., N, \forall i < h \leq N, \forall m = 1,..,6$$

4. Preliminary computational results

3.1 *Test problems*

The performance of the proposed integer linear programming model has been assessed through experimental tests. These tests focus on the month of September, where the total number of available physicians is *N=18* and the number of the teams is *m=6*. The manual schedule for this month is displayed in the following table (**Table 1**).

Table 1. Results of the physician scheduled for 28 days by the manual method.

Sep	*building 1*		*building 2*		*building 3*	
day	Shift 1	Shift 2	Shift 1	Shift 2	Shift 1	Shift 2
1	H	A	D	A	E	A
2	H	E	B	E	A	E
3	H	G	B	G	A	G
4	H	G	B	G	A	G
5	H	E	B	E	A	E
6	H	H	G	H	A	H
7	B	B	G	B	A	B
8	D	G	A	G	H	G
9	D	G	A	G	H	G
10	D	B	A	B	H	B
11	D	B	A	B	H	B
12	D	G	A	G	H	G
13	D	D	B	D	H	D
14	A	A	B	A	H	A
15	E	B	H	B	D	B
16	E	B	H	B	D	B
17	E	A	H	A	D	A
18	E	A	H	A	D	A
19	E	B	H	B	D	B
20	E	E	A	E	D	E
21	H	H	A	H	D	H
22	G	A	D	A	E	A
23	G	A	D	A	E	A
24	G	H	D	H	E	H
25	G	H	D	H	E	H
26	G	A	D	A	E	A
27	G	G	H	G	E	G
28	D	D	H	D	E	D

3.2 *Numerical results*
We report in Table 2 the obtained results after solving to optimality the proposed integer linear programming model:

Table 2. Physician scheduled for 28 days by the computerized model.

	building 1		*building 2*		*building 3*	
day	Shift 1	Shift 2	Shift 1	Shift 2	Shift 1	Shift 2
1	A	E	D	E	G	E
2	A	G	D	G	B	G
3	A	D	H	D	B	D
4	A	H	B	H	E	H
5	A	B	G	B	E	B
6	H	H	E	H	G	H
7	D	D	E	D	G	D
8	H	G	A	G	E	G
9	H	G	A	G	D	G
10	H	G	D	G	B	G
11	H	A	B	A	E	A
12	H	G	B	G	E	G
13	D	D	B	D	E	D
14	A	A	B	A	E	A
15	D	E	G	E	H	E
16	D	E	G	E	H	E
17	D	A	B	A	H	A
18	D	A	B	A	H	A
19	D	H	G	H	B	H
20	A	A	B	A	G	A
21	E	E	B	E	G	E
22	G	D	B	D	A	D
23	G	B	A	B	E	B
24	G	E	A	E	D	E
25	G	A	H	A	D	A
26	G	A	H	A	B	A
27	E	E	B	E	H	E
28	D	D	B	D	H	D

According to the obtained results (**Table 2**), we observe that all the constraints are satisfied in addition to new ones (compared to the 11 constraints given in section 2 for the manual schedule). Furthermore, the over time is reduced by 50%. This allows balancing the workload, which impacts the performance of the physicians and reduces the financial burden.

5. Conclusion and future directions:
In this paper, we investigated the physicians scheduling problem in an Intensive Care Unit. The schedule used to be performed manually each month. For that aim, we proposed an adequate integer linear programming model. This integer linear program is solved to optimality for one month (September) as preliminary work. The obtained results show the efficiency of the proposed model since the reduction of the over time is 50%. As future directions of this work, we plan to enlarge the experimental tests to other months. In addition, some new objective functions will be proposed and tested.

Acknowledgements
This work is funded by the National Plan for Science, Technology and Innovation (MAARIFAH), King Abdulaziz City for Science and Technology, Kingdom of Saudi Arabia, Awarad Number (12-MAT3123-02).

.

References

[1] Ernst AT, Jiang H, Krishnamoorthy M, Owens B and Sier D 2004, An annotated bibliography of personnel scheduling and rostering, Annals of Operations Research 127, pp. 211–144.
[2] Van J den Bergh, Belien J, De Bruecker P, Demeulemeester E and De Boeck L 2013, Personnel scheduling: A literature review, European Journal of Operations Research 226, pp.367–385.
[3] Brucker P, Qu R and Burke EK 2011, Personnel scheduling: Models and complexity, European Journal of Operations Research 210, pp. 467–473.
[4] Brucker P and Qu R 2014, Network flow models for intraday personnel scheduling problems, Annals of Operations Research 218, pp. 107–114.
[5] Burke EK, De Causmaecker P, Vanden Berghe G and Van Landeghem H 2004, The state of the art of nurse rostering, Journal of Scheduling 7, pp. 441–499.
[6] Cheang B, Li H, Lim A and Rodrigues B 2003, Nurse rostering problems—a bibliographic survey, European Journal of Operations Research 151, pp. 447–460.
[7] Glass C and Knight R 2010, The nurse rostering problem: A critical appraisal of the problem structure, European Journal of Operations Research 202, pp. 379–389.
[8] M'Hallah R and Alkhabbaz A 2013, Scheduling of nurses: A case study of a Kuwaiti health care unit, Operations Research: Health Care 2, pp. 1–19.
[9] Carter MW and Lapierre SD 2001, Scheduling emergency room physicians, Health Care Management Science 4, pp. 347–360.
[10] Beaulieu H, Ferland JA, Gendron B and Michelon P 2000, A mathematical programming approach for scheduling physicians in the emergency room, Health Care Management Science 3, pp. 139–200.
[11] Puente J, Gomez A, Fernandez I and Priore P 2009, Medical doctor rostering problem in a hospital emergency department by means of genetic algorithms, Computers and Industrrial Engineering 56, pp. 1232–1242.
[12] Gendreau M, Ferland J, Gendron B, Hail N, Jaumard B, Lapierre S, Pesant G and Soriano P 2007, Physician scheduling in emergency rooms, in: PATAT06 Lecture Notes in Computer Science, 3867, pp. 53–66.
[13] Roland B, Di Martinelly C, Riane F and Pochet Y 2010, Scheduling an operating theatre under human resource constraints, Computers and Industrrial Engineering 58, pp. 212–220.
[14] Testi A, Tanfani E and Torre G 2007, A three-phase approach for operating theatre schedules, Health Care Management Science 10, pp. 163–172.
[15] Gunawan A and Lau HC 2013, Master physician scheduling problem, Journal of the Operations Research Society 64, pp. 410–425.
[16] Stolletz R and Brunner BO 2012, Fair optimization of fortnightly physician schedules with flexible shifts, European Journal of the Operations Research 219, pp. 622–629.
[17] Bruni R and Detti P 2014, A flexible discrete optimization approach to the physician scheduling problem, Operations Research Health Care, 3, pp. 191-199.

Influence of wood species on properties of injection mould natural flour-HDPE composites

Thanate Ratanawilai Massalan Leeyoa and Yoawanat Tiptong
Department of Industrial Engineering, Faculty of Engineering,
Prince of Songkla University, Hat Yai, Songkhla, Thailand

thanate.r@psu.ac.th

Abstract. Four combinations of wood flour, HDPE, and maleic anhydride (MA) include; (1) rubberwood:HDPE (30:70), (2) rubberwood: HDPE:MA (30:67:3), (3) palm oil:HDPE (30:70), and (4) palm oil:HDPE:MA (30:67:3) were studied. The injection moulding machine was used to produce wood plastic composites (WPCs). Maleic anhydride is an ingredient in bonding agents used to manufacture wood plastic composites. Extrusion molding process was conducted to prefabricate WPCs. Consequently, the effect of temperature and pressure ranging from 180, 190, 200°C and 2300, 2400, 2500 bar on injection molding was evaluated. Mechanical properties were tested including flexural testing and tensile testing according to ASTM D790 and D638, respectively. Hardness testing according to ASTM D2240 and scanning electron microscope (SEM) were also performed. Five replications were done on each test. The result showed that rubberwood:HDPE (30:70) gave a highest strength. The values of ultimate tensile strength, flexural strength, and hardness are 24.9 MPa, 33.3 MPa and 67.2 shore D, respectively. Finally, the uniform distribution of particle in WPCs, examined through SEM was achieved.

1. Introduction

Wood plastic composites (WPCs) have made a name for themselves in extruded decking and fencing boards. WPCs are competitive with calcium carbonate-filled or talc-filled PP in cost, performance, and processing. As they contain up to 50% organic fibers, WPCs offer a material option that is more environmentally friendly than typical petrochemically derived polymers. WPCs have the advantage of lower density resulting to reduce cost and can be beneficial in transportation and other uses on light weight applications. These include automotive, construction, sports, toys, and other consumer goods. WPCs can be made with a variety of plastics, such as polyethylene [1], polypropylene [2], polystyrene [3], and high-density polyethylene (HDPE) [4]. HDPE has a higher strength-to-density ratio and therefore was selected to study. Many natural fibers such as rice hulls, palm fiber waste, or rubberwood flour can utilize to make biocomposites. Many studies noted that maleic anhydride has proven to be having a bridging effect between wood filler and the polymer matrix, resulting in improvement of material mechanical properties when used as compatibilizer [5]. This element is able to improve the WPC flow characteristic. The selection of wood species and coupling agent is very important to suit to the WPC product application and processing methods requirement.

In the process of fabrication, hot press [6], extrusion [7, 8] and injection moulding [9] are among the most popular techniques. However, good candidates for injection molding with WPCs are parts with thick walls and ones that would benefit from excellent rigidity and dimensional stability. However, some injection molders are hesitant to experiment with new materials like wood plastic composites because of inconsistent quality, inconsistent supply, and generally greater difficulty in working with WPCs in comparison with more familiar injection molding materials. However, recent developments in the manufacture of WPC compounds have significantly improved the quality, consistency, and capability of this environmentally friendly material. Good candidates for injection molding with WPCs are parts with thick walls and ones that would benefit from excellent rigidity and dimensional stability. While WPCs are best suited to parts with thick walls, processors can compensate for thinner walls by blending WPC with additional neat polymer. WPCs offer the combined properties of wood and plastic—excellent moisture barrier plus the ability to be screwed or nailed like wood, along with an organic look and feel. In addition, WPCs reduce a molder's exposure to increasing petroleum prices, reduce the energy costs associated with production, and produce an end-product with great structural rigidity, an aesthetically pleasing finish, and new, highly marketable performance capabilities.

This study was focused on two parts; (1) evaluating the effect of wood species which are rubberwood flour (RWF) and palm oil flour (POF) and (2) evaluating the effect of maleic anhydride on the properties of WPCs. Therefore, the objective of this research is to evaluate the effect of wood species and coupling agents on the properties of WPC.

2. Experimental

2.1. Materials
Recycled high density polyethylene, supplied by Withaya Intertrade Co., Ltd. in Samutprakarn, Thailand in the pellet form was used in this study. Maleic anhydride obtained from Sigma-Aldrich Pte Ltd, Singapore, was used as coupling agent. Rubberwood flour from S.T.A Furniture Group Co., Ltd. and palm oil flour from local industry in Songkhla, Thailand were used as lignocellulosic filler. The RWF and POF were sieved through 80 meshes and dried in an oven at 110°C for 8 h. The dried wood flour was stored in a sealed plastic container to prevent the absorption of water vapor. WPCs with 30% of wood flour and 70% of HDPE or 67% and 3% of MAPP were prepared. Each formulation in table 1 was weighed and stirred for 5 min to obtain uniform dispersion.

Table 1. Formulations used in making wood plastic composites.

Sample ID	Mixture formulation (%)			
	Rubberwood	Palm oil	HDPE	MA
r30h70	30	-	70	-
r30h67m3	30	-	67	3
p30h70	-	30	70	-
p30h67m3	-	30	67	3

* r30h67m3 = rubberwood flour(30):HDPE(67):maleic an anhydride(3)
p30h67m3 = palm oil wood flour(30):HDPE(67):maleic an anhydride(3)

2.2. WPCs Processing
The sample preparation was carried out in two stages. In the first stage, RWF and POF along with HDPE were premixed and extruded using the twin-screw extruder model SHJ-36 from En Mach Co., Ltd. In the second stage, The WPC samples were then injection moulded by 50-ton press moulding machine model Arburg Allrounder 320C. The temperature of the mould was ranged from 180 to 200°C with the pressure in the range of 2,300 to 2,500 bar. Test procedures of tensile testing, flexural testing, and hardness testing were conducted according to the ASTM-D 638, 790, and 2240, respectively. High magnification micrographs on the microcellular structure of WPC produced were examined with scanning electron microscope (SEM) model FEI Quanta 400 (Czech Republic) working at 15 kV.

2.3. WPCs testing

Three tests of WPCs were carried out. Firstly, the WPCs specimens of dimensions 19 mm x 165 mm x 3.2 mm were subjected to tensile test as per ASTM-D-638-98, using the Instron Universal Testing Machine (Model 5582). A crosshead speed of 5 mm/min and a gauge length of 50 mm was used for carrying out the test. Secondly, the composite specimens of dimensions 25 mm x 80 mm x 3.2 mm were taken for flexural test, under three point bending, using the same Instron Universal Testing Machine (Model 5582), in accordance with ASTM-D 790-92, at a crosshead speed of 1.3 mm/min and a span length of 50 mm. lastly, the same specimens of dimensions 25 mm x 80 mm x 3.2 mm as for flexural test were evaluated for hardness test, using the Teclock Durometer (Model GS-702N), in accordance with ASTM-D 2240. Each of the above tests was evaluated at room temperature (25°C) for five replications. The mean values and corresponding standard deviations along with the measurement uncertainty values for the experimental data were reported.

3. Results and Discussions

3.1. Mechanical properties

The effect of wood species and different amounts of maleic anhydride on the mechanical properties of wood-HDPE composites was shown in figures 1 and 2. It was observed in figure 1 that the tensile and flexural strengths of the rubberwood-HDPE composites are higher than that of the palm oil wood-HDPE composites. The reason is rubberwood generally has higher tensile stress perpendicular-to-grain (28 kg/cm^2) than that of palm oil wood (10.53 kg/cm^2) [10] and therefore improve the strengths of the composites. In addition, the SEM micrographs of wood flour/HDPE composites in figure 3(a) and (b) showed that the rubberwood flour had less porosity than that of palmoil wood resulting in increasing the strength of the composites. On the other side, the porosity of palmoil wood could be easily filled with polymer as shown in figure 3(b) resulting in poor stress transfer in the composite materials. When adding the maleic anhydride as coupling agent, the tensile and flexural strength of the composites were decreased i.e. from 24.9 to 20.6 MPa for rubberwood composites-tensile strength. Furthermore, the composites with maleic anhydride made from rubberwood flour or palm oil wood has approximately the same values of the tensile and flexural strength i.e. 20.6 and 20.5 MPa for rubberwood and palm oil wood composites-tensile strength. As can be seen in figure 3(c) and (d), the SEM micrograph of the composites made from both rubberwood and palmoil wood could not clearly indentify the difference. This is in good agreement with the mehanical testing result.

The tensile and flexural modulus of the composites as shown in figure 2 decreased with addition of the maleic anhydride. Furthermore, the palm oil wood-HDPE composites appeared to give higher tensile and flexural modulus than that of the rubberwood-HDPE composites. It indicates that the composites made from palm oil wood can be stretched when a load is applied rather than rubberwood.

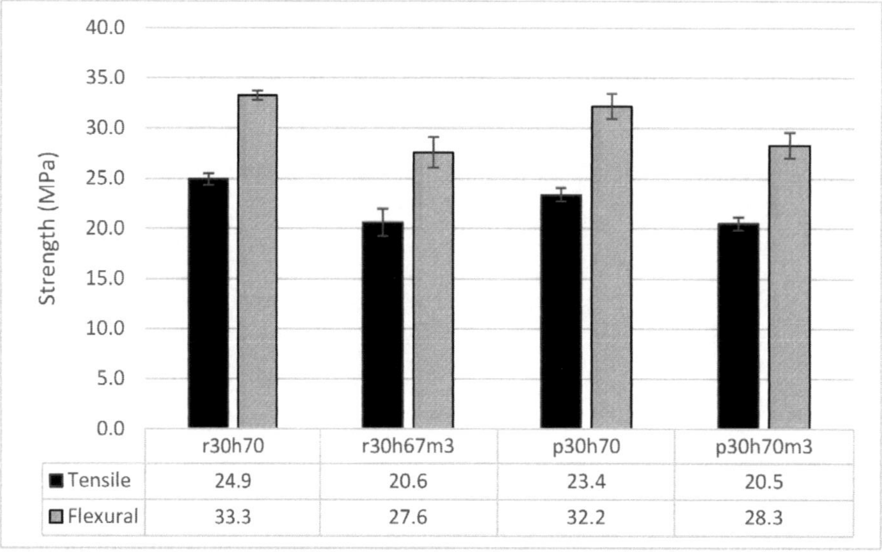

	r30h70	r30h67m3	p30h70	p30h70m3
Tensile	24.9	20.6	23.4	20.5
Flexural	33.3	27.6	32.2	28.3

Figure 1. The strength of wood flour/HDPE composites versus differential wood species and amounts of maleic anhydride

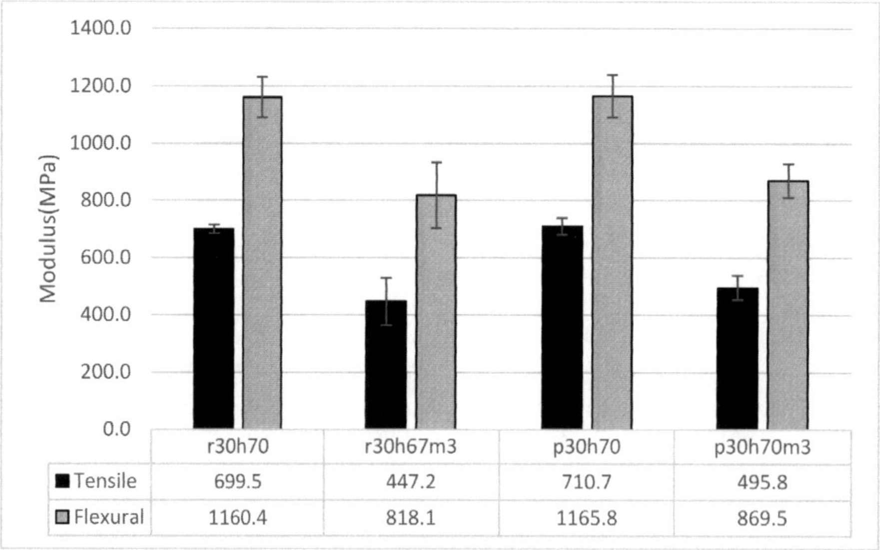

	r30h70	r30h67m3	p30h70	p30h70m3
Tensile	699.5	447.2	710.7	495.8
Flexural	1160.4	818.1	1165.8	869.5

Figure 2. The modulus of wood flour/HDPE composites versus differential wood species and amounts of maleic anhydride

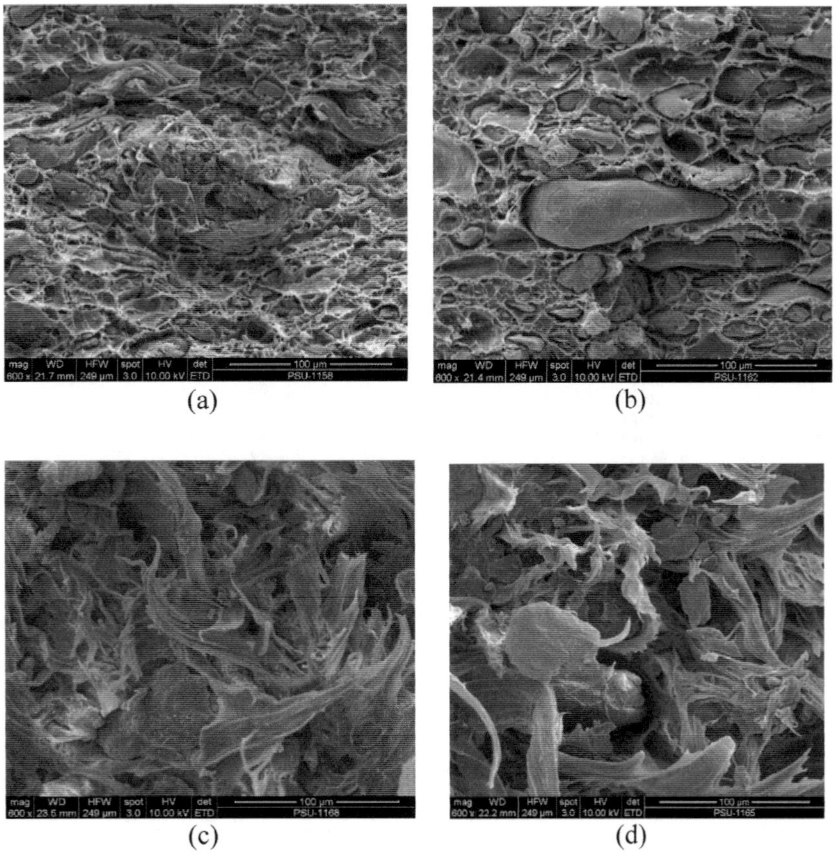

Figure 3. SEM micrographs (600×) of wood flour/HDPE composites containing (a) rubberwood:HDPE (30:70), (b) palm oil wood:HDPE (30:70), (c) rubberwood:HDPE:MA (30:67:3), and (d) palm oil wood:HDPE:MA (30:67:3)

3.2. Hardness properties

The hardness of wood flour/HDPE composites was found in the range of 65.4 to 67.2 shore D as shown in figure 4. The hardness of all composites are not significantly different. This is because all composites were pass through the same injection mould process providing the similar surface property. However, the highest hardness value (67.2 shore D) was obtained from the rubberwood composites without maleic anhydride.

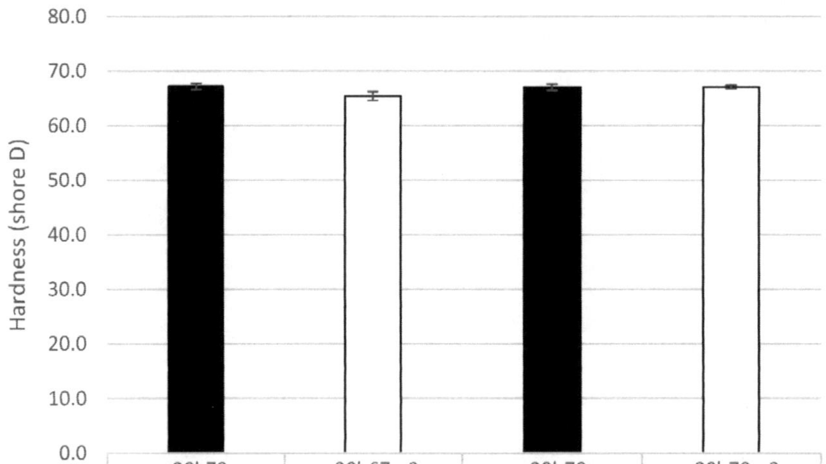

	r30h70	r30h67m3	p30h70	p30h70m3
Hardness	67.2	65.4	67.0	67.1

Figure 4. The hardness of wood flour/HDPE composites versus differential wood species and amounts of maleic anhydride

4. Conclusions

The mechanical properties of composites were affected by the species of the wood flour and the maleic anhydride used as a coupling agent. It was found that wood flour/HDPE composites made from rubberwood provided the better mechanical properties that that of palm oil wood. Morphological evidence was in good agreement with the mechanical results. Howerever, there is unaffected on the hardness from both wood species and addition of maleic anhydride.

Acknowledgements

This work was financially supported by Prince of Songkla University and Faculty of Engineering (ENG-55-2-7-08-0149-S) and the Rubberwood Technology and Management Research Group (ENG-54-27-11-0137-S).

References

[1] Hong H, Liao H, Zhang H, Heb H, Liu T and Jia D 2014 Significant improvement in performance of recycled polyethylene/wood flour composites by synergistic compatibilization at multi-scale interfaces *Composites Part A* **64** p 90–98

[2] Soccalingame L, Bourmaud A, Perrin D, Bénézet JC and Bergeret A 2015 Reprocessing of wood flour reinforced polypropylene composites: Impact of particle size and coupling agent on composite and particle properties *Polym. Degrad. Stab.* **113** p 72–85

[3] Poletto M, Dettenborn J, Zeni M and Zattera AJ 2011 Characterization of composites based on expanded polystyrene wastes and wood flour *Waste Manage. (Oxford)* **31** p 779–784

[4] Sommerhubera PF, Wellinga J and Krauseb A 2015 Substitution potentials of recycled HDPE and wood particles from post-consumer packaging waste in Wood–Plastic Composites *Waste Manage. (Oxford)* **46** p 76–85

[5] Liu W, Wang YJ and Sun Z 2003 Effects of Polyethylene-Grafted Maleic Anhydride (PE-g-MA) on Thermal Properties, Morphology, and Tensile Properties of Low-Density Polyethylene (LDPE) and Corn Starch Blends *J. Appl. Polym. Sci.* **88** p 2904-2911

[6] Ratanawilai T, Thanawattanasirikul N and Homkhiew C 2012 Mechanical and thermal properties of oil palm wood sawdust reinforced post-consumer polyethylene composites *Sci. Asia* **38** p 289-294

[7] Homkhiew C, Ratanawilai T and Thongruang W 2015 Composites from recycled polypropylene and rubberwood flour: Effects of composition on mechanical properties *J. Thermoplast. Compos. Mater.* **28** p 179-194

[8] Yanga TeH, Yanga TsH, Chaoa WC and Leub SY 2015 Characterization of the property changes of extruded wood–plastic composites during year round subtropical weathering. *Constr. Build. Mater.* **88** p 159–168

[9] Schirp A, Mannheim M and Plinke B 2014 Influence of refiner fibre quality and fibre modification treatments on properties of injection-moulded beech wood–plastic composites *Composites Part A* **61** p 245–257

[10] Ratanawilai T, Chumthong T and Kirdkong S 2006 An Investigation on the Mechanical Properties of Trunks of Palm Oil Trees for the Furniture Industry *J. Oil Palm Res.* **Special Issue**, pp. 114-121

Finite element analysis and die design of heading processes of magnesium alloy screws

Y M Hwang[1] and C Y Chang[2]

[1] Professor, National Sun Yat-sen University, Kaohsiung, Taiwan
[2] Graduate student, National Sun Yat-sen University, Kaohsiung, Taiwan
E-mail: ymhwang@mail.nsysu.edu.tw

Abstract. This study is to develop related manufacturing technologies of LZ91 magnesium alloy M6 screws. Firstly, a warm heading process composed of three stages is proposed. The material flow pattern of the billet inside the die is analyzed using the finite element analyses. The effects of the friction factor, die speed and forming temperature on the heading load are discussed. The effects of the stroke at the first stage on the formability at the second stage are also discussed. Finally, warm heading experiments are conducted using a self-designed die set and a lubricant of MoS_2. The experimental values of heading load and product shapes and dimensions are compared with the simulation results to verify the validity of the finite element models and the proposed warm heading procedures.

1. Introduction

Magnesium (Mg) alloys sheets or tubes have been widely used in 3C products, especially in notebooks and flash drives, for lightweight and other excellent material properties, such as electro-magnetic shielding and recycling, etc. However, due to the HCP structure, magnesium alloys have very poor formability at room temperature. As a result, magnesium parts are usually manufactured at elevated temperatures.

Concerning head forging researches, forging die geometries are important factors for obtaining sound products without any defects. Gontarz et al. [1] presented a new forming process of screw spikes. They used the finite volume method (FVM) and finite element method (FEM) to discuss the material flow in the screw spike head forming process. Experiments with lead metal were also conducted to verify the numerical values of heading forces and flow kinematics. Pater et al. [2] developed a new thread rolling technology for sleeper fixing screws by means of two flat wedges provided with special grooves. The numerical simulation results of thread rolling process using finite volumes method (FVM) and finite element method (FEM) were compared with the experimental test results with commercial lead metal. Lin et al. [3] applied CAE (Computer Aided Engineering) combined with finite element simulation to micro screws thread rolling process design to shorten the testing time of a new product design process. Yoshida [4] conducted cold drawing and warm heading on a magnesium alloy to produce a micro screw. So far, most screws are made by carbon steels, stainless steels or aluminum alloys. Few works are involved in the forming process of magnesium screws. The forming technology, such as the influence of the forming temperature, forming speed, etc., has not been established.

2. Finite element simulations of warm heading process

A finite element code "DEFORM 3D" is used to analyze the plastic flow pattern of the magnesium alloy LZ91 during the heading process of a screw. The flow stresses of LZ91 at different strain rates and temperatures obtained by compression tests are shown in Fig. 1. When the temperatures are raised over 250°C, the flow stresses will become smaller and the strain hardening regions will become shorter at each strain rate.

The whole heading process is divided into three stages: (1) shaft extrusion, used to obtain a shaft for the subsequent threading process, (2) head forging, used to obtain outer geometries of the head, and (3) hole forging, used to obtain a hexagonal inner hole. The schematics of these three stages are shown in Fig.2. The left and right halves denote the configurations between the die and billet before and after forming, respectively. In the shaft extrusion process shown in Fig.2(a), a billet with a dimension of 6.5 mm in diameter and 18 mm in length is extruded through a die with an exit diameter of 5.2 mm and a bearing length of 0.5 mm. The detailed dimensions are shown in Fig. 2(b). In the head forging stage shown in Fig.2(c), the bottom of the billet is in contact with by the die and its top part is compressed by an upper die. In the hole forging stage shown in Fig. 2(d), an inner punch die is used to forge out a hole with a hexagonal shape inside the head.

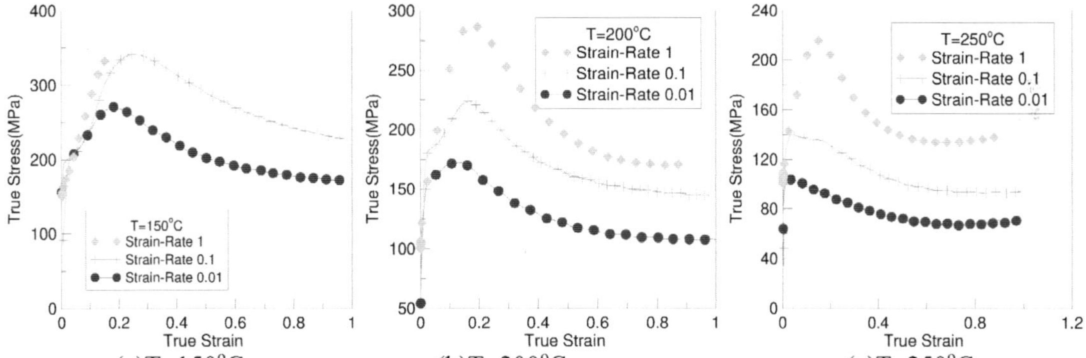

(a)T=150°C (b)T=200°C (c)T=250°C

Fig.1. Flow stresses of LZ91 at different strain rates and temperatures.

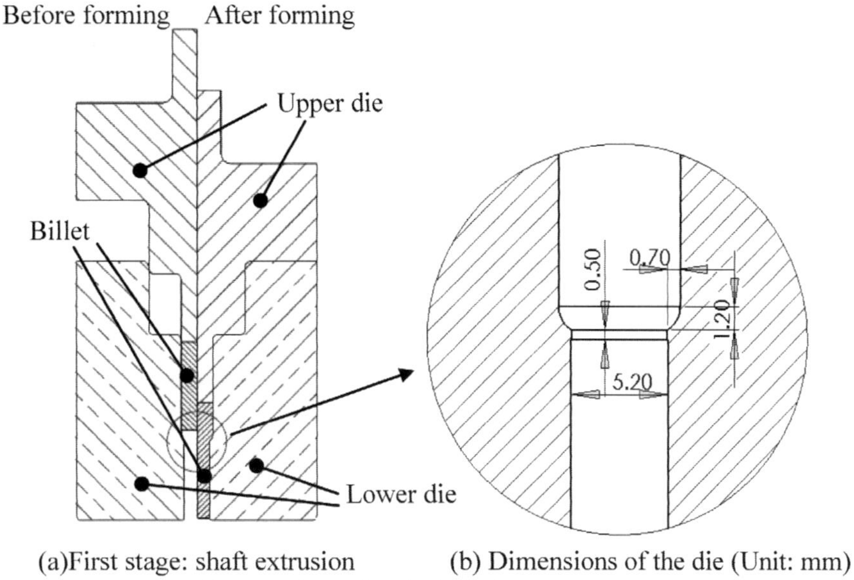

(a)First stage: shaft extrusion (b) Dimensions of the die (Unit: mm)

Fig.2. Schematic of whole heading processes.

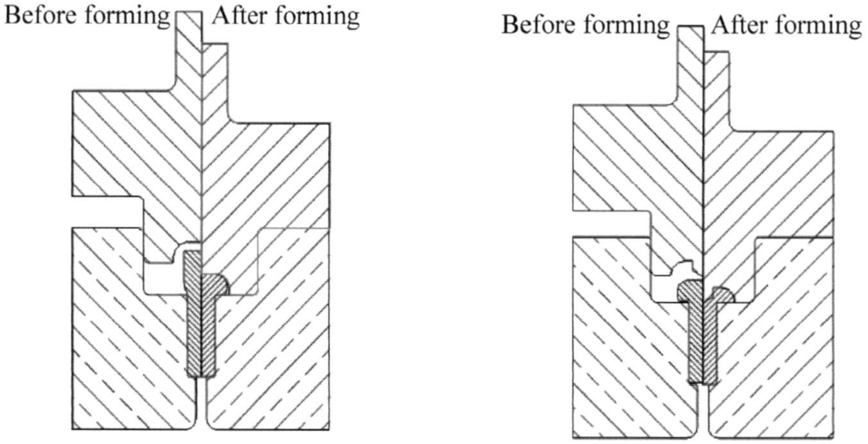

(c) Second stage: head forging (d) Third stage: hole forging

Fig.2. Schematic of whole heading processes.

The expected shape and dimensions of the billet at the initial, intermediate and final stages are shown in Fig.3. During simulations, the dies are assumed to be rigid and the billet is rigid-plastic. The forming temperature is set as 150°C. Due to axis-symmetric, DEFORM 2D is used at the first and second stages, whereas DEFORM 3D is used at the final stage. The relevant forming conditions are summarized in Table 1. Fig. 4 shows the shape and dimensions of the dies used in each stages. Due to symmetry, only the right half parts are shown in the figures.

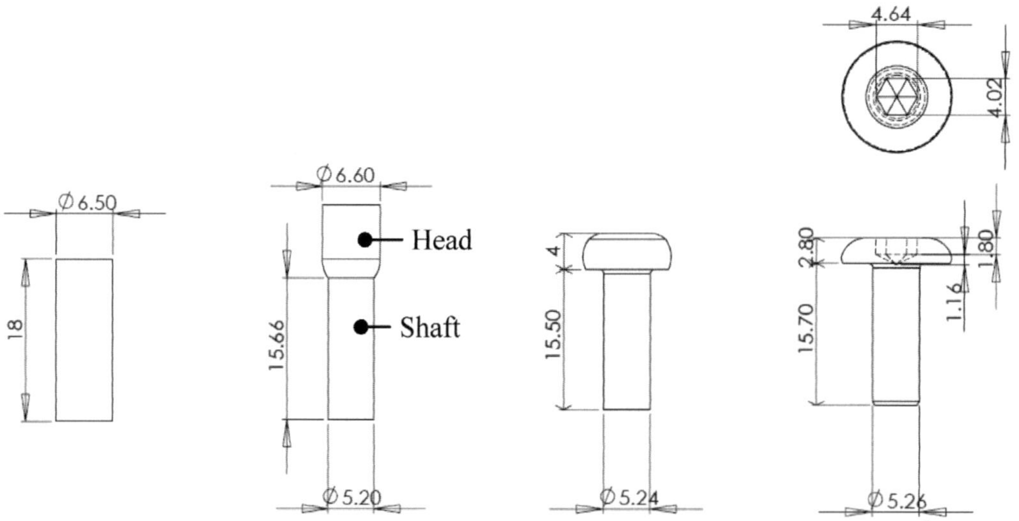

(a)Initial (b) After stage 1 (c) After stage 2 (d) Final

Fig.3. Shapes and dimensions of billet at different stages. (Unit: mm)

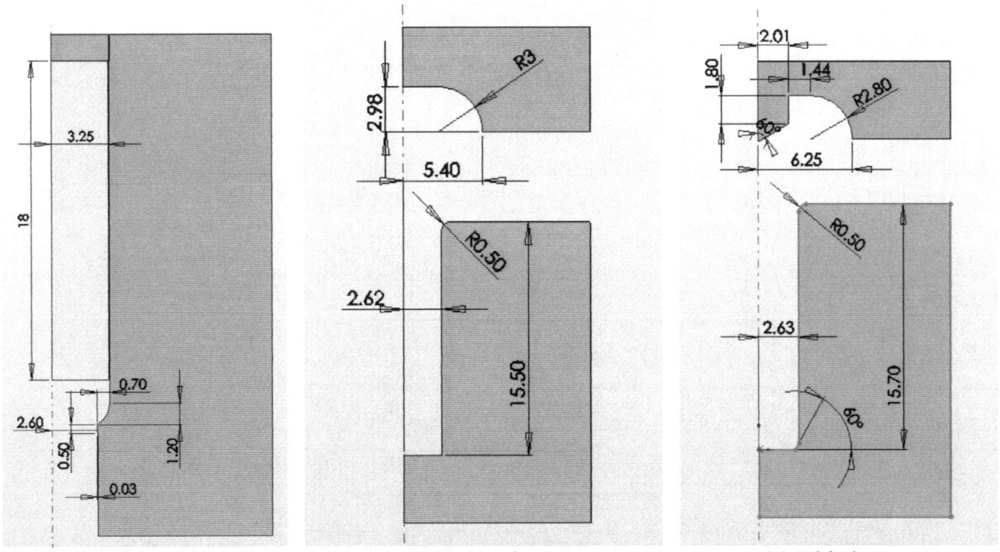

(a) First stage (b) Second stage (c) Third stage

Fig. 4. Shapes and dimensions of dies at different stages. (Unit: mm)

Table 1. Forming conditions of heading process used in finite element analysis.

Software	DEFORM-2D
Material	LZ91
Billet dimensions	$\phi 6.5 \times 18$ [mm]
Speed of upper die	1 [mm/s]
Friction factor, m	0.3
Stroke S_1 (mm)	12~13.5
Forming temperature	150 [°C]

At the second stage of head forging process, the formed product is easily affected by the stroke at the first stage, which means the length of extruded shaft influences significantly the formed results at the second stage. The effects of the strokes of 12~13.5 mm at the first stage on the formed results after the second stage are discussed below.

Figures 5(a) and 5(b) show the mesh configurations before and after the head forging stage, respectively, with a stroke of S_1=12 mm at the first stage. Because the shaft length after the first stage is too short, the neck part contacts the lower die too early, the neck part material flows outward and then accumulates there. Accordingly, a hat-shaped defect occurred after the head forging stage. For a stroke of S_1=13.5 mm at the first stage, the mesh configurations for the initial, intermediate and final states are shown in Figs. 6(a), 6(b) and 6(c), respectively. Because the shaft length after the first stage is too long, the shaft part material flows outward to the upper surface of the lower die as shown in Fig. 6(b). Accordingly, a folding defect occurs at the side wall of the head after the second stage.

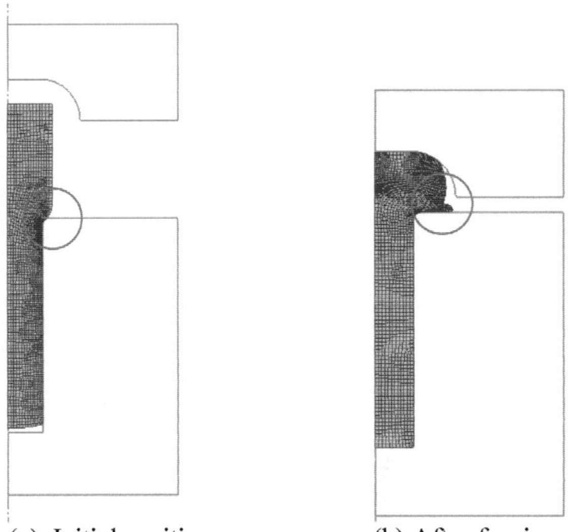

(a) Initial position (b) After forging

Fig. 5 Mesh configurations during the second stage with S_l=12 mm at the first stage.

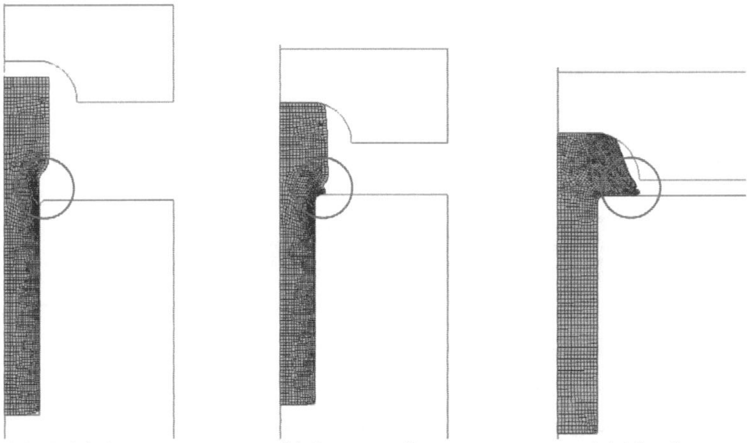

(a) Initial state (b) Intermediate state (c) Final state

Fig. 6 Mesh configurations during the second stage with S_l=13.5 mm at the first stage.

For strokes at the first stage of S_l=12.8~13 mm, the mesh configurations for the initial, intermediate and final states are shown in Figs.7(a), 7(b) and 7(c), respectively. The shaft length after the first stage with S_l=12.8~13 mm is slightly longer, the redundant part material flows outward on the upper surface of the lower die and overlaps with the material from the neck part, which results in a small folding defect at the die corner as shown in Fig. 7(b). This folding defect is not easily found as shown in Fig. 7(c). However, this defect of a small crack usually results in fracture at the head part of the screw after long period usage.

Table 2 shows the mesh configurations after the second stage with different strokes of S_l=12~13.5 mm at the first stage. Clearly, the formed shapes depend largely on the strokes at the first stage. As the strokes are 12.0 ~ 12.4 mm, a hat-shaped defect occurs. As the strokes are 12.5 ~ 12.7 mm, corresponding to shaft lengths of 15.66 ~ 15.97 mm, a sound product without defects can be obtained. As the strokes are 12.8 ~ 13 mm, a not easily seen small folding defect occurs around the die corner. As the strokes are 13 ~ 13.5 mm, an obvious folding defect occurs at the side wall and the die cavity cannot be filled up completely after the second stage.

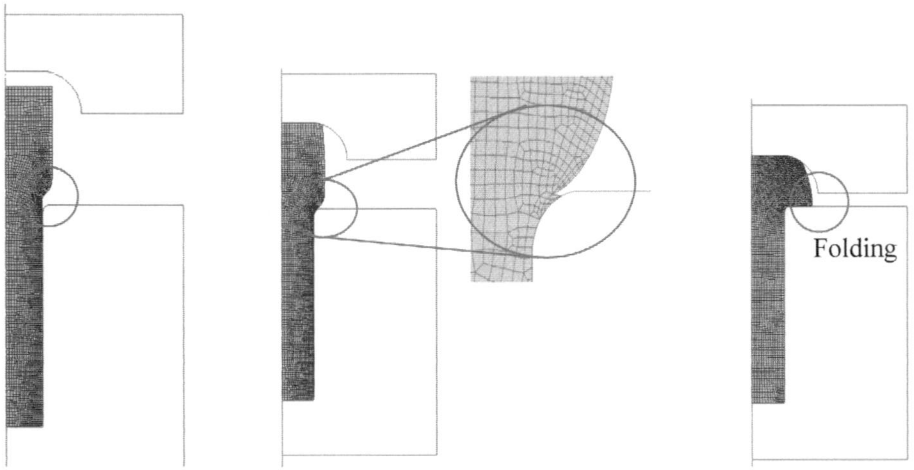

Fig. 7 Mesh configurations for the second stage with S_1=12.8~13 mm at first stage.

Table 2 Mesh configurations after the second stage with S_1=12~13.5 mm at the first stage.

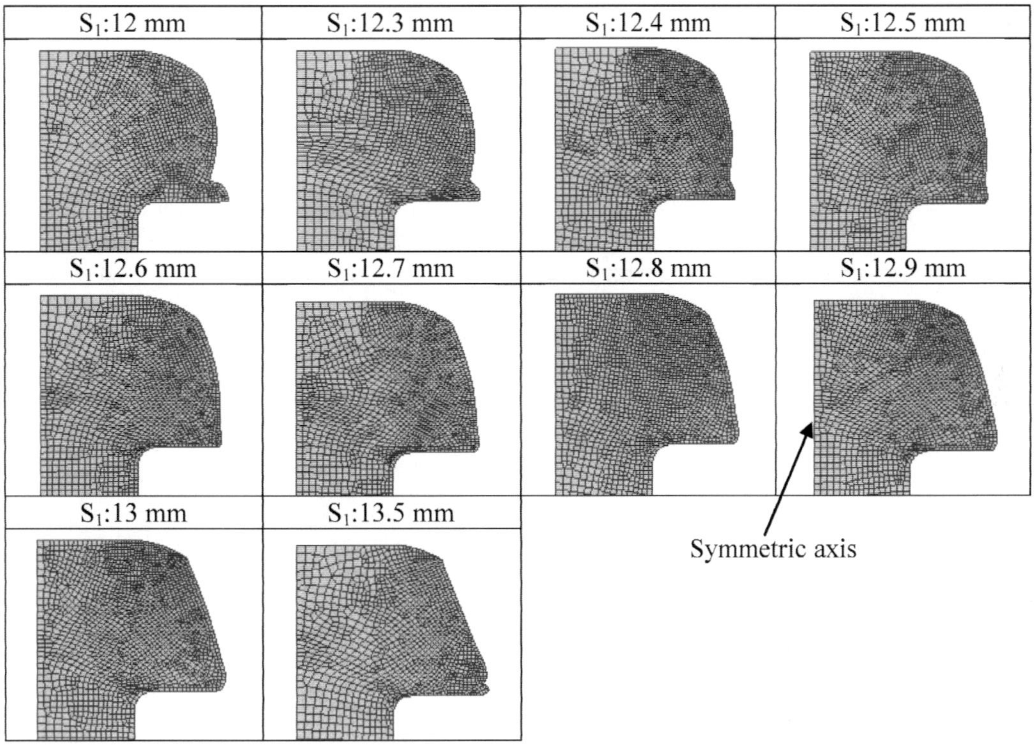

3. Simulation results and discussion

The loads of the upper die for the three stages of the heading processes are shown in Fig. 8. The arc ratios of the die L/H defined in Fig. 2(b) are 2, 2.5, 3 and 3.5. At the first stage, as stroke S=1 mm, the front end of the billet moves to the entrance of the extrusion die. As S=3 mm, the front end of the billet reaches the die exit. As S >3 mm, the billet flows out from the die bearing part and a steady state is reached. At the second stage, the force increases monotonously with the stroke. At the third stage, as $0 < S < 1.5$ mm, the mandrel moves downward and the force increases due to the friction between the outer die and the billet. As 1.5 < S < 3.5 mm, the free outer surface of the billet is not in contact with the die, accordingly the force decreases slightly. As the mandrel approaches to the bottom of the inner hole, the outer surface of the billet begins to fill up the cavity of the outer die and then the force increases dramatically. The minimum heading load was obtained as the die arc ratio L/H=3. The maximum load at the first stage at a arc ratio of 2 is about 1.7 tons, whereas the maximum load at the second stage of heading process is about 2 tons. At a stroke of about 3.5 mm at the third stage, there is a rapid increase in heading loads. That is because a large amount of billet surface is in contact with the upper die surface simultaneously at that instant.

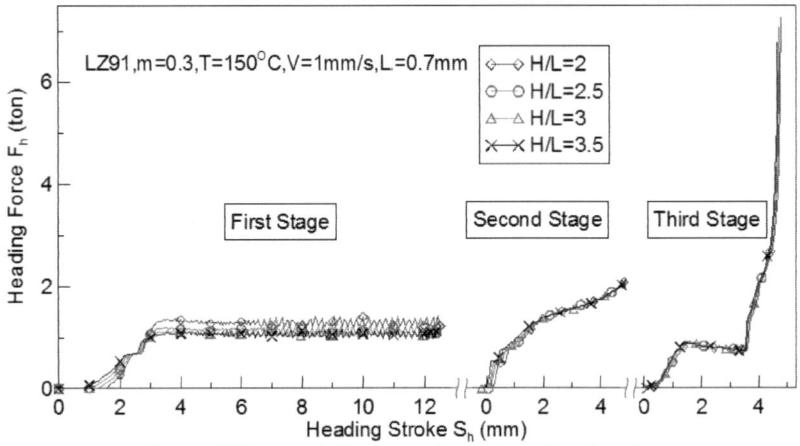

Fig. 8. Effects of die arc ratio on heading loads.

The effects of die speed on loads of the upper die for the three stages of heading processes are shown in Fig. 9. The compression speeds of the upper die, V, are 1, 5, and 10 mm/s. From Fig. 9, it is known that the heading load increases with the increase of the die speed. The maximum load at the first stage at a die speed of 10 mm/s is about 1.2 tons, whereas the maximum load at the second and third stages of heading process are about 3 and 7 tons, respectively. For all the stages of heading process, the heading load increases with the die speed.

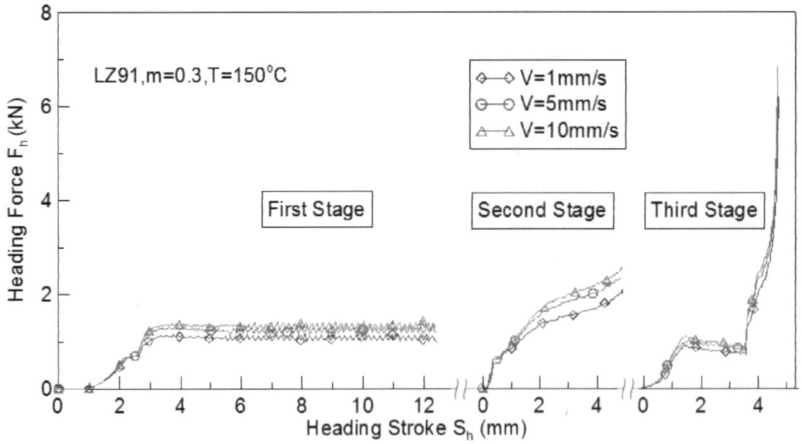

Fig. 9. Effects of die speed on heading loads.

The effects of forming temperature on loads of the upper die for the three stages of the heading processes are shown in Fig. 10. The forming temperature of the billet, T, are 150, 200, and 250°C. From Fig. 10, it is known that the heading load decreases with the increase of the forming temperature.

The maximum load at the first stage at forming temperature of 150°C is about 1.2 tons, whereas the maximum load at forming temperature of 250°C is about 0.4 ton, one third of that at forming temperature of 150°C.

The effects of friction factor on loads of the upper die for the three stages of the heading processes are shown in Fig. 11. The friction factor at the interface of the billet and dies, m, are 0.1, 0.3, and 0.5. From Fig. 11, it is known that the heading load at the first stage increases with the increase of the friction factor, whereas the loads are not influenced greatly by the friction factor at the second and third stages.

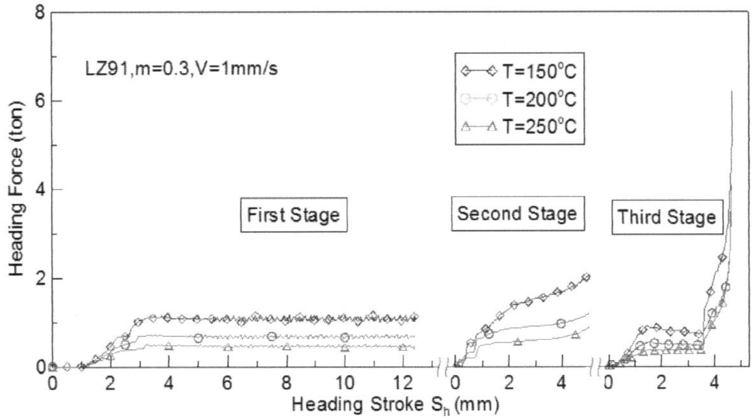

Fig. 10. Effects of billet temperature on heading loads.

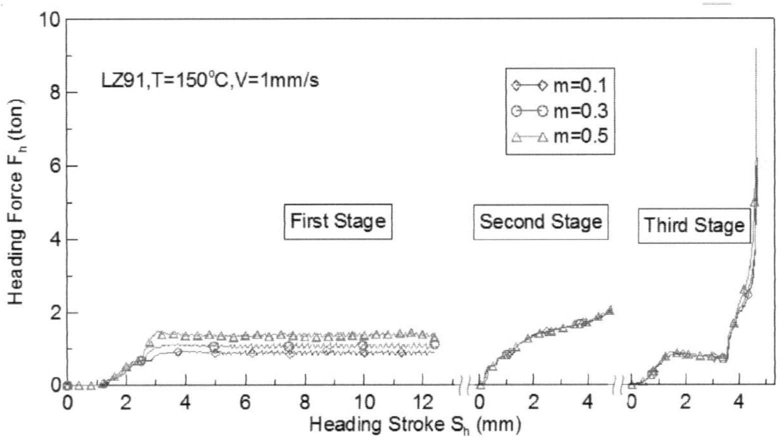

Fig. 11. Effects of friction factor on heading loads.

4. Experiments of warm heading processes

The schematic of a self-developed warm heading apparatus is shown in Fig.12, where the upper die holder is coupled with a universal testing machine, which provides the power to push the upper die downwards. The upper and lower die set can be replaced with other die sets, so the experiments of shaft extrusion, head forging and hole forging stages, as shown in Figs. 2(a), 2(c) and 2(d), respectively, can be conducted using this apparatus. The forming conditions for warm heading experiments are the same as those in Table 1. MoS_2 was used as the lubricant.

The analytical and experimental loads for the three stages are shown in Fig.13. From the figures, it is known that the simulated heading loads considering non-isothermal are closer to the experimental results. The errors of the final loads at each stage between simulation and experimental values are smaller than 10%. The appearance of the formed products after each stage in the experiments of heading processes is shown in Fig.14. The expected dimensions of the final product are shown in Fig.15. The comparisons of the final product dimensions between analytical and experimental results are shown in Table 3. The maximum error is about 4% occurring at the hexagonal hole length A.

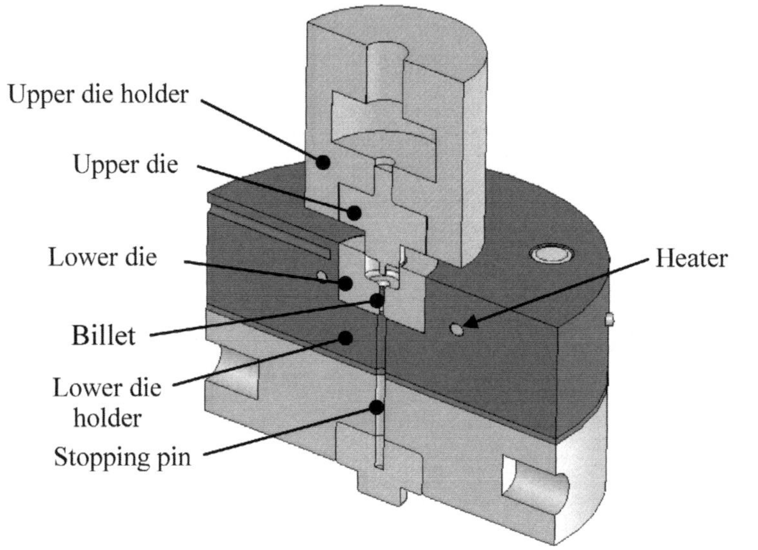

Fig.12. Schematic of self-developed warm heading apparatus.

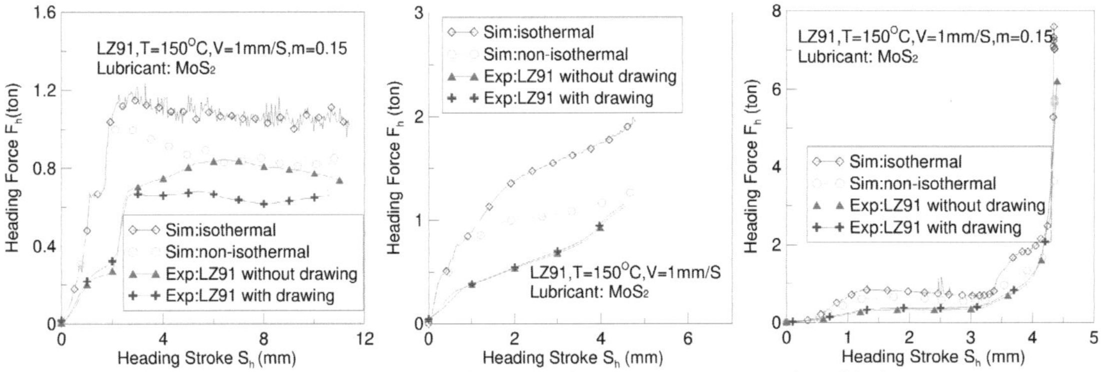

(a) First stage (b) Second stage (c) Third stage

Fig.13. Comparisons of heading loads between analytical and experimental results.

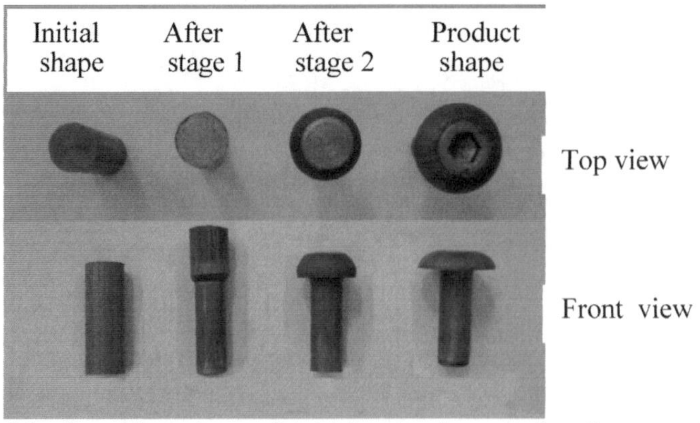

Fig.14. Appearance of the formed products after each stage in warm heading experiments.

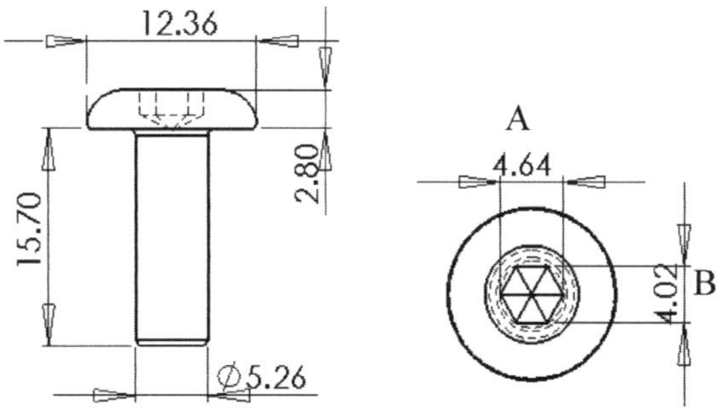

Fig.15. Expected dimensions of the final product. (Unit: mm)

Table 3. Comparisons of the final product dimensions.

	Analysis [mm] X	Experiment [mm] Y	Error [%] (Y-X)/X
Head diameter	12.36	12.7	2.75
Head length	2.8	2.86	2.14
Shaft diameter	5.26	5.22	-0.76
Shaft length	15.7	15.91	1.34
Hole dimension A	4.64	4.47	-3.66
Hole dimension B	4.02	4	-0.5

5. Conclusions

In this paper, simulations of warm heading processes of a LZ91 magnesium alloy screw were conducted firstly. The effects of the friction factor, die speed and forming temperature on the heading load were discussed. An experimental apparatus for heading processes was developed. Finally, warm heading experiments of LZ91 magnesium alloy screws were conducted with a lubricant of MoS_2. The experimental values were compared with the simulation results. Sound products were obtained with forming temperatures of 150°C in the heading process. The maximum error of the final product dimension was smaller than 4% occurring at the hexagonal hole length.

Acknowledgement

The authors would like to extend their thanks to the Ministry of Science and Technology of the Republic of China under Grant no. MOST 100-2221-E-110-033-MY3. The advice and financial support of MOST are greatly acknowledged.

References

[1] Gontarz A, Pater Z and Weroñski W 2004 Head forging aspects of new forming process of screw spike, *J. Mater. Proc. Tech.* **153-154** 736-740

[2] Pater Z, Gontarz A and Weronski W 2004 New method of thread rolling, *J. Mater. Proc. Tech.* **153-154** 722-728

[3] Lin Y F, Kao Y C, Cheng H Y and Wang K J 2007 The study of computer aided thread rolling for micro screw, *Proc. 24th National conf. Chinese Soc. Mech. Eng.*, pp.4286-4290, Taoyuan, Taiwan.

[4] Yoshida K 2009 Cold drawing of magnesium alloy wire and fabrication of micro screw, *J. Jap. Soc. Tech. Plast.* **50** 919-923

Synthesis High Sensivity ZnO Cholesterol Biosensor with One Dimensional Nanostructures

Y M Lu[1*], P G Jhuang[2],J F Tang[3] , S Y Chu[4] and CW Hsu[5]

[1*,2,5]Department of Electrical Engineering, National University of Tainan, Taiwan
[3,4]Department of Electrical Engineering, National Cheng Kung University, Tainan, Taiwan

[1*]Department of Electrical Engineering, National University of Tainan, Taiwan
ymlu@mail.nutn.edu.tw

Abstract. Chelosterol ZnO biosensors were synthesized by hydrothermal method with pre-deposited ALD-ZnO seed layers. The ZnO nanostructures were examined by SEM, XRD, AFM, respectively. The XRD is used to analyze the crystal structures of ZnO seed layers that were grown by ALD process, then using SEM and AFM to analysis the surface morphologies of them. Growing of one dimensional ZnO nanostructures on seed layer of 10nm thickness, reveals high sensitivity of 3.15 uA $(mg/dl)^{-1}cm^{-2}$(121.96 $uAmM^{-1}cm^{-2}$) and low K_m value of 0.29mM in 25-100mg/dl cholesterol concentration range. The response time could be as low as 10seconds. In the study, it is found that increasing the aspect ratio of ZnO one dimensional nanostructure also increases the sensitivity and expends the linear measureable range of detecting cholesterol in solution. The thickness of seed layer has a significant influence on sensitivity for one dimensional nano ZnO cholesterol biosensor.

1. Introduction

The biologic artifacts are chemoreceptors with good receptiveness to chemical substances. To detect external stimuli and signals, humans rely on not only taste and smell but also receptor proteins, cholesterol, enzymes, antibodies, antigens and many others. Neurohumours of the neurotransmitter systems are also involved in conveying tactile signals [1]. The development of biosensors started in 1962, and it was based on the enzyme electrode concept proposed by Clark and Lyons [2]. This enzyme electrode concept is about fixing glucose oxidase (GOD) on a dissolved oxygen electrode, once the GOD contacts with glucose, the glucose will be oxidized. Because oxygen in the solution is involved in the oxidation, its concentration will decrease, and this concentration change is detected by the dissolved oxygen electrode. In this case, the concentration of glucose in the solution can be estimated by measuring the amount of oxygen consumed. In recent years, nanostructures are often used as carriers; for example, the high surface area of zinc oxide (ZnO) nanostructures has been applied to achieve effective cholesterol oxidase absorption and to detect the concentration of cholesterol in order to enhance the efficiency of cholesterol biosensors and to prolong the lifespan of the sensors [3]. The characteristics of ZnO, such as good absorption and excellent biocatalytic efficiency, are suitable for biosensors. In addition, ZnO also possesses other excellent qualities, such as a higher isoelectric point (9.5), which gives ZnO a better bio-compatibility, better chemical stability, excellent electronic transmission ability and no toxicity. Therefore, the application and development of ZnO materials as intermedium nanomaterials for sensing cholesterol are promising [4-7]. In this study, the atomic layer deposition (ALD) process and the hydrothermal method were used to synthesize the ZnO seed layers and ZnO nanostructures respectively. The ZnO nanostructures were then assembled as a cholesterol biosensor. It is suggested in the literature [8] that reducing the concentrations of zinc nitrate hexahydrate ($Zn(NO_3)_2 \cdot 6H_2O$) and hexamethylenetertramine ($C_6H_{12}N_4$, HMTA) can increase the aspect ratio of ZnO nanorods during hydrothermal growth. Therefore, increasing the surface area of the nanostructure, would benefit for enhancing biosensor sensitivity.

2. Experimental Procedures

Firstly, the ITO glass was washed by acetone, isopropyl alcohol and deionized water sequentially. The precursor,$(C_2H_5)_2Zn$ (DEZn, Fullesty Corporation) and deionized water were set in a 1:1 pressure vapor ratio (40 mtorr:40 mtorr or 5.33Pa:5.33Pa) alternatively feeding into the vacuum reaction chamber for depositing 100nm, 30nm, and 10nm thickness of ZnO seed layers by atomic layer deposition (ALD). The process pressure was kept about 1 torr(133.32Pa). The process temperature was set at 150℃, and the feeding order of the precursor into the reaction chamber in one cycle was DEZn (1second) followed by Ar (purge, 5second) and then DI water (1second), and Ar (purge, 5second) and repeated again depending on how many cycles were set (see Figure 1).

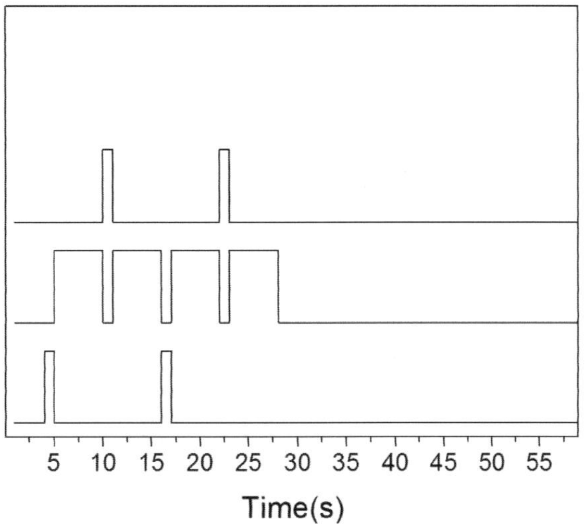

Fig.1 One cycle sequence of ALD-ZnO process.

The crystal structure of obtained ALD-ZnO seed layers had checked by XRD. Subsequently, hexamethylenetertramine ($C_6H_{12}N_4$, HMTA) and zinc nitrate hexahydrate ($Zn(NO_3)_2 \cdot 6H_2O$) were used in a 1:1 molarity ratio (0.025M:0.025M) for growing ZnO nanostructures on the ALD-deposited seed layers hydrothermally. The growing temperature was maintained at 90℃ for hydrothermal growth. For ZnO nanostructures, the SEM was used to observe the surface morphologies and the cross-section. As for cholesterol biosensors, the PBS solution with disodium hydrogen phosphate dehydrate (Na_2HPO_4), sodium dihydrogen phosphate (NaH_2PO_4), sodium chloride (NaCl) and DI water was firstly prepared. Next, potassium hexacyanoferrate (III) ($K_3[Fe(CN)_6)$ and potassium hexacyanoferrate (II) trihydrate ($K_4[Fe(CN)_6)$ were added into PBS solution to be the electron transfer medium. The specimen of the cholesterol biosensor was prepared by pipetting 10uL of cholesterol oxidase (ChOx) onto the ZnO nanostructures for physical adsorption, and then it was placed in an environment of 25℃ to dry naturally. The 15cc of well prepared PBS solution was taken, when detecting the cholesterol in a solution. The tri-electrode electrochemical method was used for the cholesterol biosensor measurement. The schematically experimental process sequences for manufacturing ZnO cholesterol biosensor can be seen in Figure 2. Stage A: clean the substrates by acetone and de-ion water. Stage B: Using the ALD process to grow ZnO seed layers. Stage C: Hydrothermal grows ZnO seed layers structures from the mixture of HMT($C_6H_{12}N_4$) and$Zn(NO_3)_2$ solution at 90°C. Stage D: adsorption cholesterol molecules onto the surface of ZnO nano-structure s.

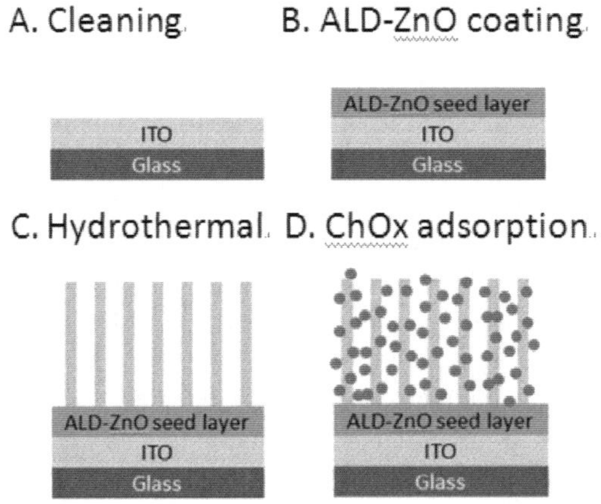

Figure 2 Experimental process sequences for manufacturing ZnO cholesterol biosensor

The overall experimental sequences can be shown by the flow chart shown below:

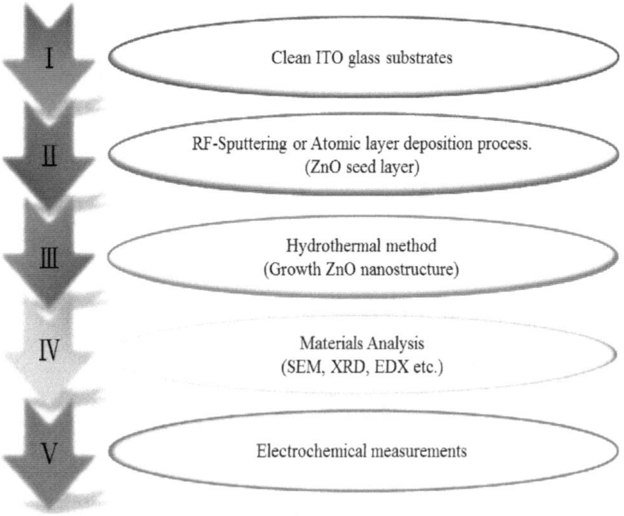

Flow chart

3. Results and Discussion

In the study, X-ray diffraction (XRD) was used to analyze the ZnO seed layers prepared by ALD and the preferred orientation of hydrothermally grown ZnO nanostructures. In figure 3, the diffraction peaks can be easily found by crystal planes having orientations of (100), (002), and (101) of ZnO seed layers, and the preferred orientation was not profound. After hydrothermally growing ZnO nanostructures, a strong diffraction peak can be found in the orientation of (002). A possible reason is that when preparing the ZnO nanostructures from solution, the surface energy would reduce due to thermaldynamic rearrangement. The thermaldynamic rearrangement has three possible stabilization mechanisms: surface reconstruction, surface electron transferring, and surface nonstoichiometry. The surface reconstruction is the one that is most likely to happen for reducing the surface energy [9-14]. The surface reconstruction method may reduce the surface energy of crystals and thereby increase the uniaxial growth capacity by decreasing the surface area of the polar plane.

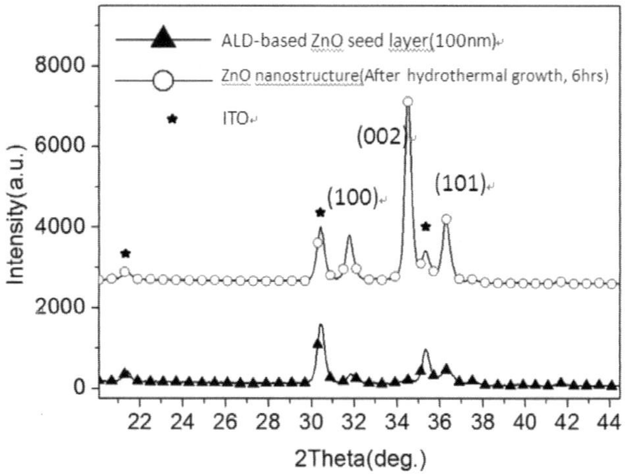

Figure 3 XRD patterns of ALD-ZnO seed layer and nanostructure after hydrothermal growth .

In this study, the SEM was used to observe the hydrothermally grown ZnO nanostructures on seed layers of 100nm, 30nm, and 10nm thickness respectively. In Figure 4, according to the literature[3], increasing the time of hydrothermal growth would increase the aspect ratio of ZnO nanostructures, which in turn would affect the amount of ChOx absorption on the surfaces. Therefore, hydrothermal growth periods of 3hr, 5hr, 9hr and 12hr were performed and the nanostructure formed at 12hr growth period which was chosen to be the best parameter of ChOx absorption. It can be found from the SEM diagram that as the time of hydrothermal growth increased, the length of nanostructures increased as well. Nonetheless, when the thickness of the seed layer was reduced (from 100nm to 30nm), nanostructures also became finer and denser, which might enhance the efficiency of the biosensor due to more surfaces creating. For the 10nm thickness seed layer, the nanostructure displayed a flower-like pattern and about 50 fold larger than previous microstructures. From the top view of SEM image, it can be found the microstructure distributed in a much more dispersed status, the flower-like crystals were about 5um in diameter. Though it is dispersed, the flower-like crystals created more surfaces which may lead to a compromised contribution in the performance of ZnO biosensor.

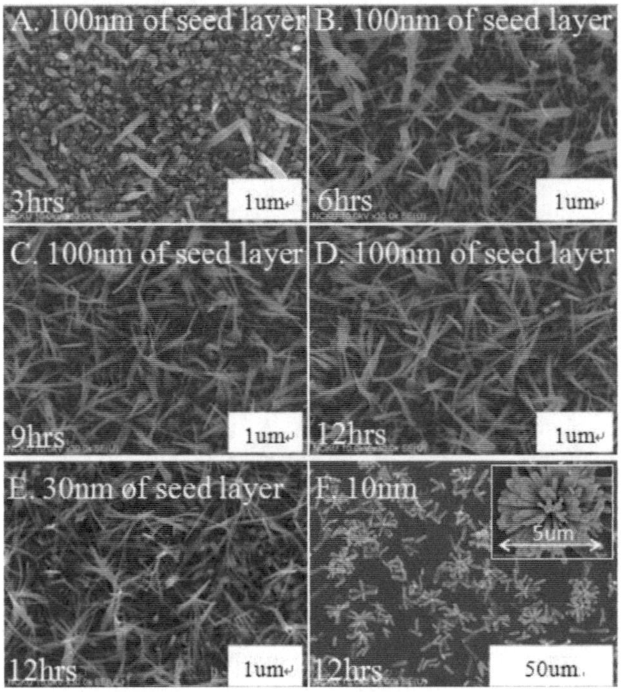

Figure4 HR-SEM images of ZnO nanostructures with different thickness and hydrothermal growth time (A)100nm ZnO seed layer and 3hrs growth time, (B)100nm ZnO seed layer and 6hrs growth time, (C) 100nm ZnO seed layer and 9hrs growth time,(D)100nm ZnOseed layer and 6hrs growth time, (E)30nm ZnO seed layer and 12hrs growth time, (F) 10nm ZnO seed layer and 12hrs growth time,.

For cholesterol biosensor measurement (see Figure 5), we detected cholesterol concentrations ranged from 0mg/dl to 700mg/dl. Figure 5 shows the cyclic voltammetry measurement results of ZnO nanostructure grown on different thickness of ZnO seed layers.

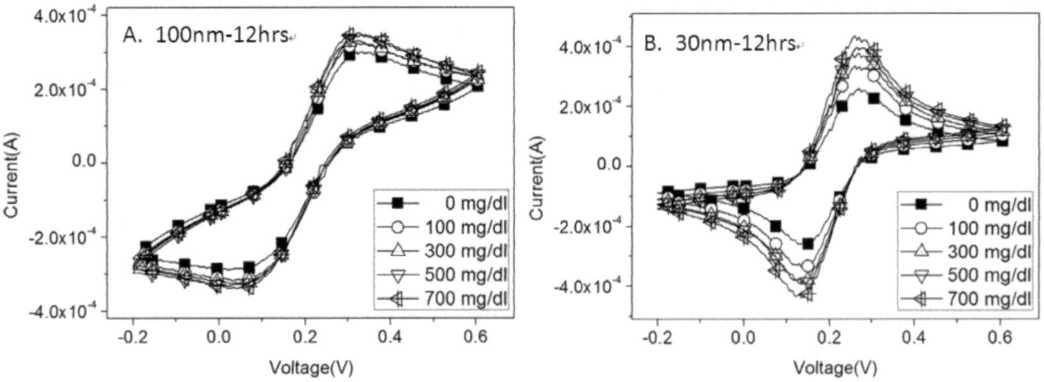

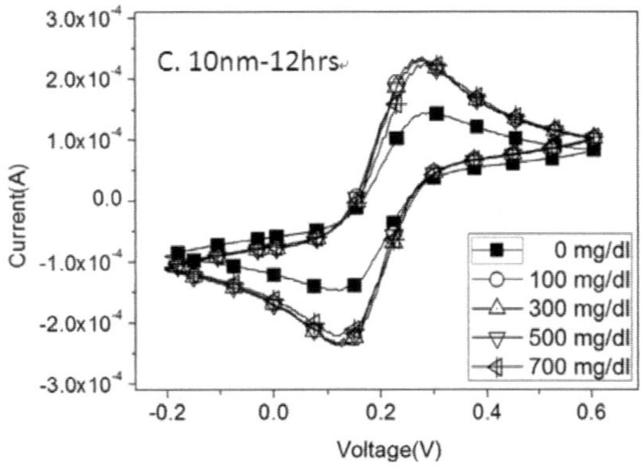

Figure5 Cyclic voltammograms of nano ZnO Cholesterol biosensors at various cholesterol concentrations with different thickness of ZnO seed layer hydrothermally growing ZnO nanostructures at 12hours (A)100nm (B)30nm (C) 10nm

Twelve hours hydrothermal growing time was chosen to grow ZnO nanostructures. According to figure 6 (100nnm and 30nm seedyers), when the concentration of cholesterol increases, the anodic peak current will also increase. It shows that the sensor would be more reliable by using the linear interval detectable range due to the consistent responses of cholesterol concentration. As the cholesterol concentration increases to the point where the ZnO nanostructure surfaces are saturated by cholesterol molecules, the detectable range will not follow this linear response anymore. In other words, this interval detectable range data has its limits for the minimum and maximum cholesterol concentration that can be accurately detected.

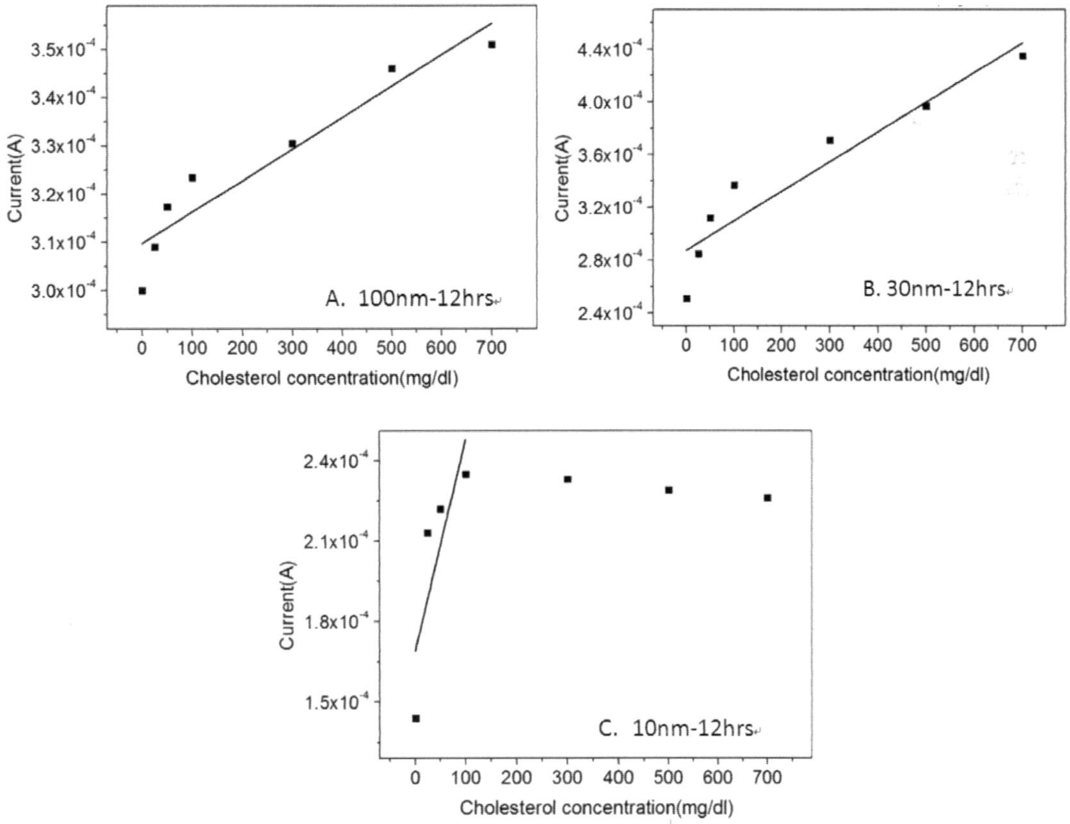

IOP Conf. Series: Materials Science and Engineering **131** (2016) 012004 doi:10.1088/1757-899X/131/1/012004

Figure6 Current v.s Cholesterol concentration of nano ZnO Cholesterol biosensor with different thickness seed layer grown at 12hours

As for the 10nm seed layer, the anodic peak current began to drop when the cholesterol concentration reached 100mg/dl. It is possible that the absorption of ChOx was decreased because the flower-like nanostructure grown from 10nm seed layer distributed more loosely. Consequently, at higher cholesterol concentration condition, excess cholesterol molecules cannot react with ChOx and lead to generate extra electrical resistance, causing the anodic peak current to decrease. The sensitivity can be calculated in terms of the formula: sensitivity (S) = $[\Delta I_{pa}/\Delta C \cdot A]$, where ΔI_{pa} is the current peak variation of oxidization amount, ΔC is the amount of concentration variation, and A is the working area. The cholesterol biosensor using 100nm and 30nm as the base seed layer showed 0.26 $uA(mg/dl)^{-1}cm^{-2}$ ($10.06\ uAmM^{-1}cm^{-2}$) and 0.90 $uA(mg/dl)^{-1}cm^{-2}$ ($34.64\ uAmM^{-1}cm^{-2}$), respectively. It can be found that reducing the thickness of the seed layer was accompanied by increased sensitivity because with a thinner seed layer. The denser ZnO nanostructures formed which is beneficial for absorption ChOx and leads to high sensitivity. As growing ZnO nanostructure from 10nm thickness seed layer, the linear range was only between 0 mg/dl and 100mg/dl, hence this range was used to calculate the sensitivity and a value of 3.15 $uA(mg/dl)^{-1}cm^{-2}$ ($121.96\ uAmM^{-1}cm^{-2}$)obtained. It can be found that the compactness and aspect ratio of ZnO nanostructure do affect the sensitivity of ZnO biosensors and the form of crystal is also another important factor.

The K_m^{app} value was determined by the analysis of the slope and intercept for the plot of the reciprocals of steady-state current vs. cholesterol concentrations, i.e., the Lineweaver–Burk plot of 1/Current vs. 1/Concentration [25]. According to Lineweaver–Burk, the K_m values of ZnO nanostructures grown on 100 nm, 30 nm, and 10nm thickness seed layers were 3.22mM, 2.85mM, and 0.29mM, respectively. The calculated results were demonstrated in Figure 7.

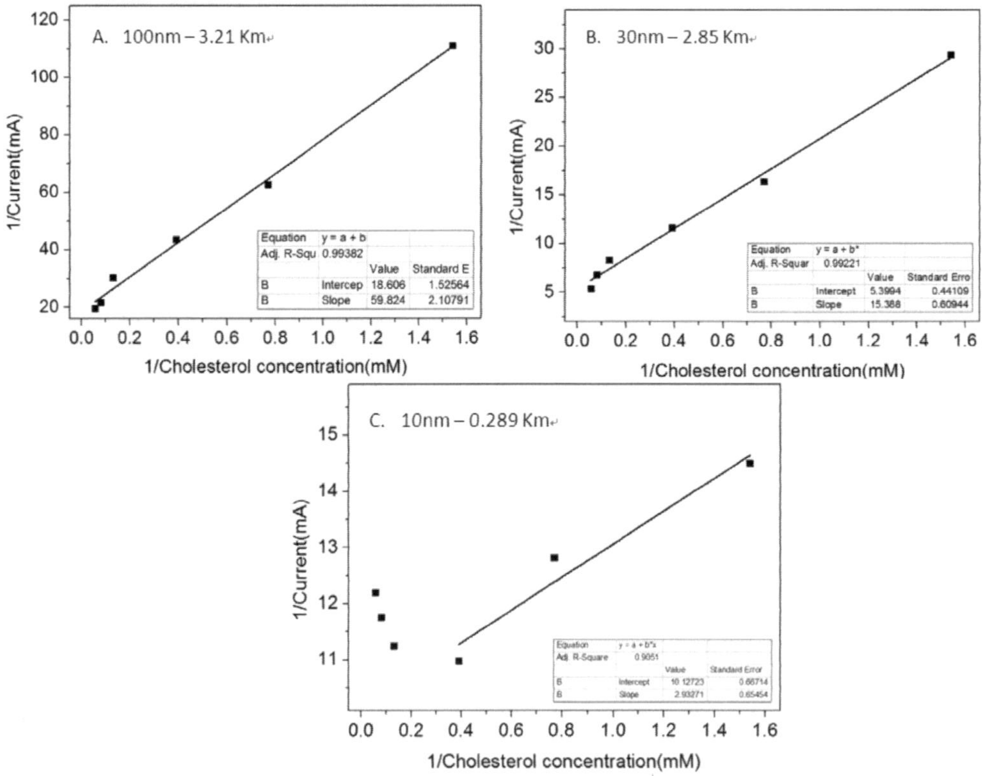

Figure7 Plot of 1/Current vs 1/Cholesterol concentration of nano ZnO Cholesterol biosensor hydrothermally growing for 12hours on different thicknesses of seed layers (A)100nm(B)30nm(C)10nm

The response time of ZnO cholesterol nano-biosensor grown on different seed layer thickness was measured by chronoamperometry. The response time is taken the initial flatting point from the curve of current vs. measuring time plot. As shown in figure 8, for seed layers of 100nm, 30nm and 10nm thickness, their response times were 25seconds, 10seconds, and 10seconds, respectively. It can be found that after reducing the thickness of the seed layer, the response speed increased. Nonetheless, for ZnO grown on 10nm thickness seed layer, no obvious improvement of response time was observed which may be due to looser crystal distribution. The response time of ZnO cholesterol biosensors were dependent on seed layer thickness. For thicker seed layers, electrons have to overcome larger resistance in the transportation according to ohm law. Therefore, the thickness of ZnO seed layers can significantly alter the nano ZnO cholesterol biosensor characteristics, not only the material properties but also the performances of the nano-biosensor, such as the sensitivity, response speed etc.

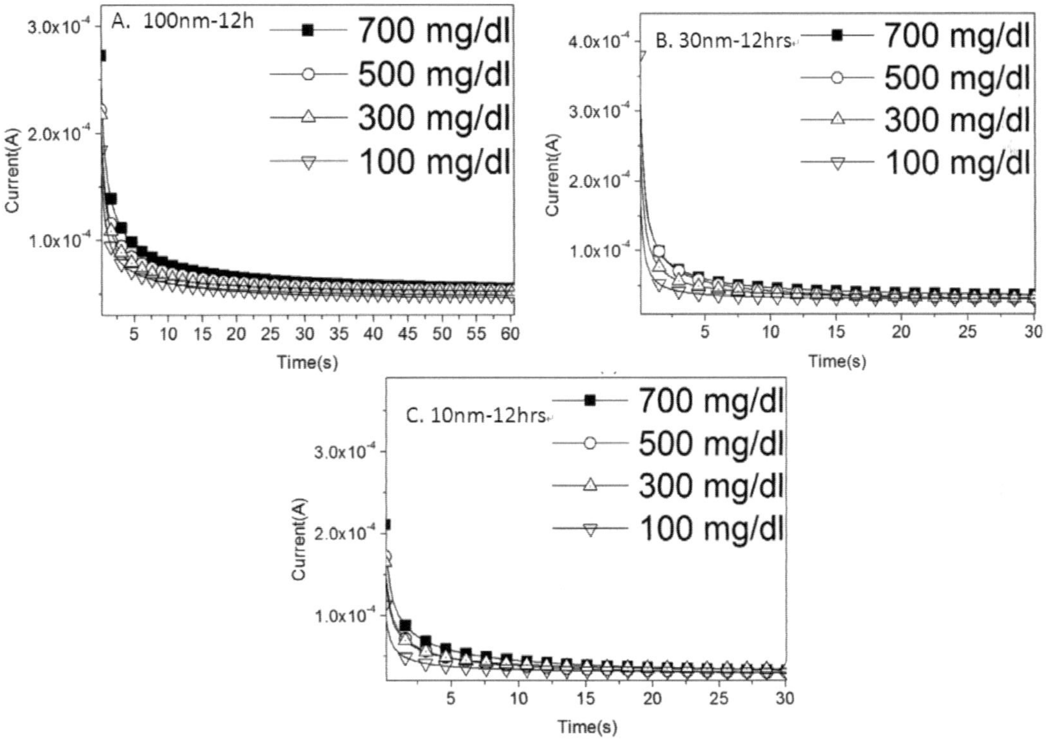

Figure8 Response times of nano ZnO cholesterol biosensor hydrothermally grown on different thicknesses of seed layers for 12 hours (A)25seconds(B)10seconds(C)10seconds

4. Conclusion

In this study, the atomic layer deposition (ADL) was applied to deposit ZnO seed layers of different thicknesses and they were subsequently used to grow ZnO nanostructures for manufacturing cholesterol biosensors with high sensitivity. It is found that using 100nm-thick ZnO seed layer to grow ZnO one dimensional nanostructures which had a sensitivity of 0.26 $uA(mg/dl)^{-1}cm^{-2}$ (10.06 $uAmM^{-1}cm^{-2}$) and a response time of 25seconds. While using 30nm and 10nm thickness ZnO seed layer, the sensitivity was 0.90 $uA(mg/dl)^{-1}cm^{-2}$ (34.64 $uAmM^{-1}cm^{-2}$) and the response time is 10seconds. The thickness of seed layers affected not only on the form of crystal structures but also on the performance of cholesterol biosensors.

References

[1] Arya S K, Datta M,and Malhotra B D 2008 Biosensors and Bioelectronics **23** 1083
[2] Gwenin C D, Kalaji M, Williams P A and Jones R M 2007 Biosensors and Bioelectronics **22** 2869
[3] Ahmad R, Tripathy N and Hahn Y-B 2012 Sensors and Actuators B **169** 382
[4] Israr M Q, Sadaf J R, Nur O, Willander M, Salman S, and Danielsson B 2011 Applied Physics Letters **98** 253705
[5] Solanki P R, Kaushik A, Ansari A A, and Malhotra B D 2009 Applied Physics Letters **94** 143901.
[6] Umar A, Rahman M M, Alhajry A. and Hahn Y-B. 2009 Talanta **78** 284
[7] Israr M Q, Sadaf J R, Asif M H , Nur O , Willander M and Danielsson B 2010 Thin Solid Films **519** 1106
[8] Singh S P, Arya S K., Pandey P, and Malhotra B D 2007 Applied Physics Letters **91** 063901.
[9] Kim D Y, Kang H M, Kim J-M and Kim H G 2011 Applied Surface Science **257** 3776
[10] Gong B, Peng Q, Na J-S and Parsons G N 2011, Applied Catalysis A: General **407** 211
[11] Yuan N Y , Wang SY , Tan C B , Wang X Q, Chen G G and Ding J N 2013 Journal of Crystal Growth **366** 43
[12] Feng Z Z , Jia R , Dou B , Li H F , Jin Z, Liu X Y , Li F , Zhang W and Wu C 2015 Solar Energy **115** 770
[13] Skompska M and Zarebska K 2014 Electrochimica Acta **127** 467
[14] Kahouli M, Barhoumi A , Bouzid A, Al-Hajry A. and Guermazi S 2015 Superlattices and Microstructures **85** 7

Acknowledgments

This work was supported by project grant(103-2622-E-024-003- CC2 ,104-2221-E-024-011) from Department of Natural Sciences, Ministry of Science and Technology, Taiwan.

Carrier dynamics in $Zn_xCd_{1-x}O$ films grown by molecular beam epitaxy

F J Cheng[1], Y C Lee[2,*], S Y Hu[3,], Y C Lin[4], K K Tiong[4], and W C Chou[5]**

[1] Department of Electrical Engineering, Tungnan University, Shenkeng, New Taipei City, Taiwan
[2] Department of Electronic Engineering, Tungnan University, Shenkeng, New Taipei City, Taiwan
[3] Department of Digital Technology Design, Tungfang Design Institute, Kaohsiung, Taiwan
[4] Department of Electrical Engineering, National Taiwan Ocean University, Keelung, Taiwan
[5] Department of Electrophysics, National Chiao Tung University, Hsinchu, Taiwan

Corresponding author's e-mail: * jacklee@mail.tnu.edu.tw
**shenghu2729@gmail.com

Abstract. In this work, the carrier dynamics in $Zn_{1-x}Cd_xO$ thin films with different Cd contents grown by molecular beam epitaxy system have been investigated using photoluminescence and time-resolved photoluminescence measurements. The carrier lifetime can be estimated from the PL decay curve fitted by triple exponential function. The emission energy dependence and temperature dependence of the PL decay time indicate that carrier localization dominate the luminescence mechanism of the ZnCdO alloy semiconductor.

1. Introduction

Recently, ZnO-based semiconductors materials have been rapidly and widely developed for laser diodes (LDs) and light-emitting diodes (LEDs), owing to theirs tunable band gap by alloying with MgO to reach ultraviolet region or by alloying with CdO to achieve the visible region [1-4]. For the application in the region from ultraviolet to green light spectra, the ZnCdO alloy is useful to fabricate ZnCdO-related heterostructures or quantum well structures, which are the key elements in ZnO-based light emitters and detectors [5]. Therefore, it is quite necessary to investigate the optical characteristics in the ZnCdO alloys, such as the properties of band gap engineering, exciton behavior, and luminescence. The luminescence behavior provides not only a fundamental insight into recombination process but also information about the carrier localization.

In this work, we investigate the luminescence characteristics of ZnCdO thin films with different Cd contents grown by plasma assisted molecular beam epitaxy (PA-MBE) system by temperature-dependent photoluminescence (PL) and time-resolved photoluminescence (TRPL) spectra.

2. Experimental

The investigated $Zn_{1-x}Cd_xO$ films were grown on the *c*-plane Al_2O_3 substrates by SVT Associates molecular beam epitaxy system equipped with conventional effusion cells for evaporation of elemental Zn (*6N*) and Cd (*6N*). Oxygen (*5N5*) was supplied via an rf-plasma source after additional gas

purification. In order to reduce the lattice mismatch between $Zn_{1-x}Cd_xO$ and Al_2O_3, the $Zn_{1-x}Cd_xO$ films were firstly grown at 350 °C following a 70 nm thick ZnO buffer layer grown at 650 °C. The Cd concentration of $Zn_{1-x}Cd_xO$ was controlled by adjusting the Cd/Zn beam pressure. The Cd concentrations applied for the four ZnCdO samples are approximately estimated to be 0.06, 0.08, 0.10, and 0.13, respectively by using the energy-dispersive X-ray spectroscopy (EDX) measurement. More details about the growth procedures have been described in the previous work [6].
The PL and TRPL were measured by using a diode laser (PicoQuant), which produces light pulses with 50 ps duration and a repetition rate of 1 MHz. PL measurement was setup using a spectrometer (Zolix omni−λ 500) with a grating of 1200 grooves/mm, and detected using a photomultiplier tube (PMT). TRPL was performed using the technique of time-correlated single-photon counting (TCSPC). The luminescence decay was detected with a high-speed PMT, followed by a computer plug-in TimeHarp counting card, which was triggered with a signal from the diode laser. Janis Research Model CCS-150 and LakeShore Model 321 temperature controller were used to carry out the temperature-dependent PL and TRPL spectra.

3. Results and discussion

Figures 1(a)~(d) show the normalized PL spectra recorded at 12 K of ZnCdO thin films with Cd content of 0.06, 0.08, 0.10, and 0.13, respectively. It is observed that the peak position of PL spectrum shifts to lower energy with increasing Cd content. To further clarify the luminescence mechanisms in ZnCdO thin films the carrier lifetime were measured by time-resolved PL technique. The inset in figure 1(a)~(d) display the luminescence decay profile monitored at the PL peak and the PL decay curve can be fitted by a triple exponential function as [7]

$$I(t) = A_1 * e^{(-\frac{t}{\tau_1})} + A_2 * e^{(-\frac{t}{\tau_2})} + A_3 * e^{(-\frac{t}{\tau 3})} \quad (1)$$

As shown in the insets of figure 1, the red solid curve described from equation (1) can well fit the experimental data and to obtain the value of τ.

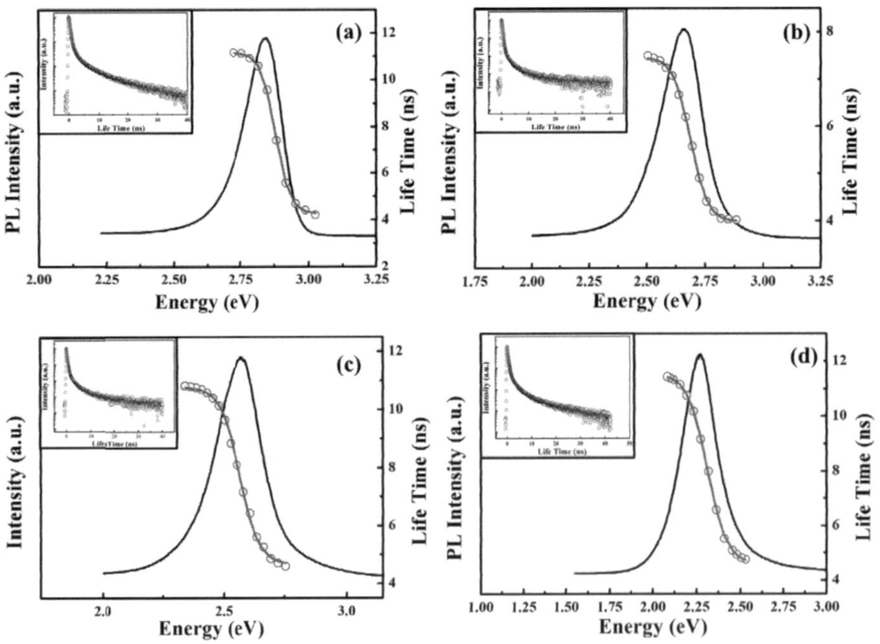

Figure 1. 12 K PL spectrum of ZnCdO thin films with Cd content of (a) 0.06, (b) 0.08, (c) 0.10, and (d) 0.13. The open circles display the emission-energy-dependent lifetime.

The estimated PL lifetime as a function of emission energy is further plotted as open circles in figure 1. The emission energy dependence of carrier lifetime shows that the luminescence decay is faster on the high energy side than the low energy side of PL spectrum. The decrease in the lifetime with increasing emission energy has been known as the characteristic of a localization effect and attributed to the localized carriers transfer process to lower energy states [8,9]. The decay rate of localized carriers is expressed as the radiative recombination rate plus the rate of transfer to lower energy sites, which results in the observed lifetime decreasing as the emission energy increases. The relationship between lifetime and emission energy can be expressed by [8,9]

$$\tau(E) = \frac{\tau}{1 + \exp\left[\left(E - E_{me}\right)/E_0\right]} \tag{2}$$

where τ is the radiative recombination lifetime, and E_0 is the characteristic energy for the depth of the localized states. E_{me} is a definite energy for the decay time that equals the transfer time, which is similar mobility edge. The red solid line in figure 1 indicates the best fit according to equation (2), and the obtained values of E_{me}, and E_0 are listed in table 1. The value of E_{me} decreases from 2.88 eV to 2.15 eV as increasing Cd content. In addition, the obtained E_0 for the $Zn_{0.94}Cd_{0.06}O$ sample is 25.7 meV and that increases to 55.2 meV for the $Zn_{0.87}Cd_{0.13}O$ sample.

Table 1. The values of E_{me}, E_0, and E_a estimated from equation (2) and (3), respectively.

	E_{me} (eV)	E_0 (meV)	E_a (meV)
$Zn_{0.94}Cd_{0.06}O$	2.88	25.7	24.8
$Zn_{0.92}Cd_{0.08}O$	2.69	36.2	35.9
$Zn_{0.90}Cd_{0.10}O$	2.57	40.4	40.1
$Zn_{0.87}Cd_{0.13}O$	2.31	55.2	54.6

The lifetime τ of the PL decay as a function of the reciprocal temperature, for the ZnCdO thin films with Cd content of 0.06, 0.08, 0.10, and 0.13, are further plotted in figure 2(a)~(d), respectively. It decreases slowly at low temperature region, but rapidly at the high temperature region. The temperature-dependent lifetime can be fitted using the function [10]

$$\tau(T) = \frac{\tau_0}{1 + C * \exp\left(-E_a / kT\right)} \tag{3}$$

where τ_0 is the lifetime at low temperature, the coefficient C is a fitting constant, and E_a is the thermal activation energy. It is known that the carriers would be delocalized from the localized states in thermalization process, which results in the reduction of the carrier lifetime. Therefore the activation energy E_a means the localization energy of the localized carriers [10]. The values of E_a are estimated by fitting using equation (3) shown as the solid line in figure 2 and listed in table 1. The obtained values of E_a are in agreement with the values of E_0, which confirm the delocalization of carriers from the localized states in the thermalization process.

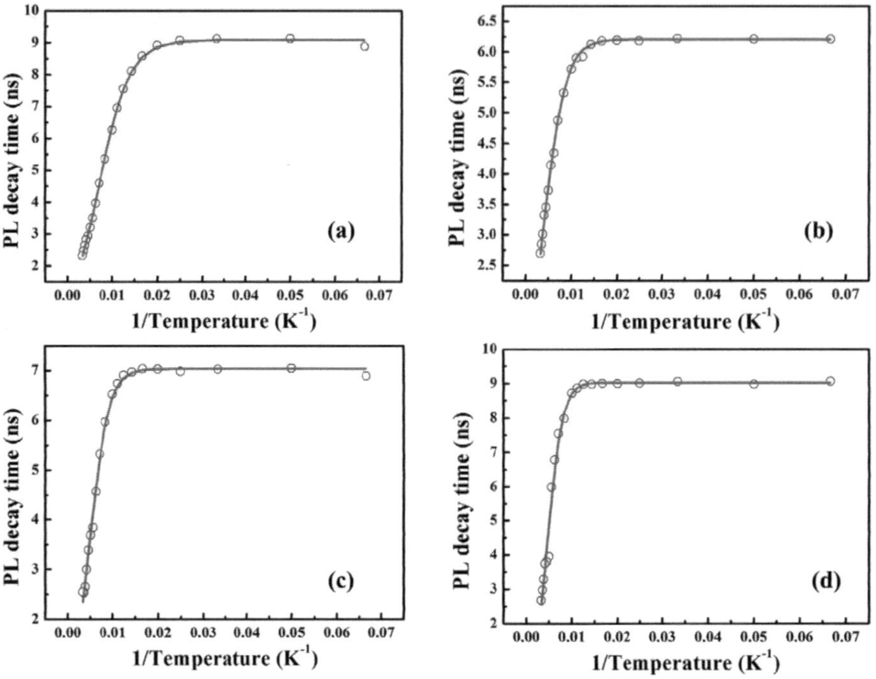

Figure 2. Arrhenius plot of the PL decay time of ZnCdO thin films with Cd content of (a) 0.06, (b) 0.08, (c) 0.10, and (d) 0.13. The solid line is a curve fitted using equation (3).

4. Summary

We have investigated the luminescence mechanism of the $Zn_{1-x}Cd_xO$ films with x=0.06, 0.08, 0.10, and 0.13 by PL and TRPL measurements. The emission-energy-dependent PL decay time shows that the PL decay is faster on the high energy side than on the low energy side of PL spectrum, exhibiting the characteristic of a localization effect. The temperature-dependent PL decay time presents that the carriers would be delocalized from localized states due to thermal activation. Our results indicate that the carrier localization would be increased with increasing Cd content, which dominates the luminesce mechanism in ZnCdO semiconductors

5. References

[1] Özgür Ü, Alivov Ya I, Liu C, Teke A, Reshchikov M A, Doğan S, Avrutin V, Cho S J and Morkoç H 2005 *J. Appl. Phys.* **98** 041301.
[2] Lange M, Dietrich C P, Czekalla C, Zippel J, Benndorf G, Lorenz M, Zúñiga-Pérez J and Grundmann M 2010 *J. Appl. Phys.* **107** 093530.
[3] Wang Y, He H, Zhang Y, Sun L, Hu L, Wu K, Huang J and Ye Z 2012 *Appl. Phys. Lett.* **100** 112103.
[4] Ye Z, Ma D, He J, Huang J, Zhao B, Luo X and Xu Z 2003 *J. Cryst. Growth* **256** 78.
[5] Caglar Y, Caglar M, Ilican S and Ates A 2009 *J. Phys. D: Appl. Phys.* **42** 065421.
[6] Chien K F, Hsu W L, Tzou A J, Lin Y C, Chou W C, Lee L, Chia C H and Yang C S 2013 *J. Cryst. Growth* **378** 208.
[7] Han N S, Shim H S, Seo J H, Kim S Y, Park S M and Song J K 2010 *J. Appl. Phys.* **107** 084306.
[8] Yang M D, Chen K W, Shen J L, Wang J C and Hsu C 2007 *Nanotechnology* **18** 405707.
[9] Murotani H, Yamada Y, Taguchi T, Ishibashi A, Kawaguchi Y and Yokogawa T 2008 *J. Appl. Phys.* **104** 053514.
[10] Shu G W, Lo M H, Yang M D, Hsu C L, Shen J L and Lan S M 2007 *Solid. State. Commun.* **141** 109.

Other Oxides Pre-removed from Bangka Tin Slag to Produce a High Grade Tantalum and Niobium Oxides Concentrate

S Permana[1], J W Soedarsono[1], A Rustandi [1] and A Maksum[2]

[1]Department of Metallurgical and Materials, Faculty of Engineering, University of Indonesia, Depok 16424, Indonesia
[2]Mechanical Engineering Department, State Polytechnic of Jakarta , Depok 16425, Indonesia

E-mail : sulaksana@yahoo.com

Abstract. Indonesia, as the second largest tin producer in the world, has a byproduct from the production of tin. This byproduct is in the forms of tin slag containing tantalum pentoxide (Ta_2O_5) and niobium pentoxide (Nb_2O_5). This study focuses on the recovery of tantalum pentoxide and niobium pentoxide from the tin slag. In the process, one part of the tin slag sample was sieved only (BTS), and the other was roasted at 900°C, water quenched and then sieved (BTS-RQS). Samples BTS and BTS-RQS were characterized by thermo gravimetric analysis (TGA) and X-ray flourence (XRF). One part of BTS-RQS sample was dissolved in hydrofluoric acid (HF) and the other was dissolved in hydrochloric acid (HCl), washed with distilled water, then dissolved into sodium hydroxide (NaOH). Each sample was characterized by using XRF. The BTS sample produced the highest recovery of 0.3807 and 0.6978 % for Ta_2O_5 and Nb_2O_5, respectively, from the particle size of -1.00+0.71 and a fraction of 47.29%, while BTS-RQS produced the highest recovery of 0.3931 and 0.8994 % for Ta_2O_5 and Nb_2O_5, respectively, on the particle size of -0.71+0350 and a fraction of 21%. BTS-RQS, dissolved with 8% hydro fluoride acid, yields tantalum pentoxide and niobium pentoxide with a ratio of 2.01 and 2.09, respectively. For the sample BTS-RQS dissolve first with 6M hydrochloric acid, washed with distilled water, then dissolved with sodium hydroxide 10M, the yield ratios are 1.60 and 1.84 for tantalum pentoxide and niobium pentoxide, respectively. In this study, it is found that the dissolution by using hydrofluoric acid 8% yields the best ratio.

Keywords : dissolution, Niobium pentoxide, Tantalum pentoxide, tin slag.

1. Introduction

Indonesia is the second largest tin producer in the world, with a production area located on the island of Bangka and Belitung. The number of tin smelters operating on Bangka and Belitung are 34 smelters, according to the organization for economic co-operation and development (OECD) in November 2012 [1].

In tin production process, results byproducts such as tin slag-1 and slag-2. Tin slag-1 contains 20-30% Tin, while tin slag-2 contains 2-3% Tin. Tin slag-2 is a remelted tin slag-1 into reverberatory furnace and containing transition metal oxides such as tantalum pentoxide (Ta_2O_5) and niobium pentoxide (Nb_2O_5) [2]. Tantalum pentoxide and niobium pentoxide can be extracted from this slag. Other than from tin slag, tantalum pentoxide (Ta_2O_5) and niobium pentoxide (Nb_2O_5) can be recovered by mining.

Tantalum pentoxide (Ta_2O_5) can be recovered from tantalite ore (oxides) [3], that can be found in the earth's crust of about 2 ppm [4]. While niobium oxide can be economically extracted from the deposit of pyrochlore and columbite, which can generally be divided into two types, namely: primary carbonatite deposits with contain of 0.5-0.7% Nb_2O_5 and enriched pegmatite deposits (carbonite mineral leaching products of primary carbonatite) that contains of 1.0-7.0% Nb_2O_5 [5].

Tantalum and niobium are in the 14 critical materials [6] so the focus of their recovery from tin slag will be an alternative option. For tantalum itself, it is now considered as midterm critical metal where the level of its availability is only until 2020 [3]. Efforts of obtaining tantalum and niobium from tin slags has been carried out by some previous researchers [2,7–13]. Tin slag from Indonesia has been informed to contain $(Ta,Nb)_2O_5$ at about 2.7% [2].

For comparison, the recovery of tantalum pentoxide (Ta_2O_5) and niobium pentoxide (Nb_2O_5) from the mine, we can refer to Anita project owned by Les Mineraux Crevier at the mine site in the Lac-Saint-Jean Quebec, Canada. Anita project has a measured reserves of approximately 25.8 million ton with niobium oxide of 0.196% and tantalum oxide of 234 ppm. Other than that, the mineral inferred resources are 15.42 million ton. These reserves are assumed to be a cut-off grade of 1,000 g/t Nb [14].

Metal extraction can be done by hydrometallurgy, pyrometallurgy and electrometallurgy. On simple hydrometallurgy can be done in liquid-liquid extraction [15–17] or the method of dissolution. This study uses the method of dissolution to reduce the other oxides content.

The challenge of this research is the dominant other oxides content with a total of 99.09% while the precious metal content is of less than 1%. Previous researches, that used tin slag with a total content of Ta_2O_5 and Nb_2O_5 over 2.7%, failed to describe the relationship of process to the decreasing content of other oxides [2,7–13]. The decreasing contents of other oxides on this study is expected to be the leading direction for the next processes in increasing contents of Ta_2O_5 and Nb_2O_5 of tin slags.

We will observe in this study the effect of sieving, roasting at 900^0C-water quenching-sieving to the weight and content fraction of Ta_2O_5 and Nb_2O_5, and the effect of BTS-RQS dissolution with hydrofluoric acid and its relationship with yield ratio of Ta_2O_5 and Nb_2O_5. Furthermore, we will see the effect of dissolution with hydrochloric acid and sodium hydroxide as well as its relationship with yield ratio of Ta_2O_5 and Nb_2O_5.

2. Methodology

2.1. Feasibility study

To calculate the BTS potential reserved of tantalum pentoxide (Ta_2O_5) and niobium pentoxide (Nb_2O_5) we made a comparison between the BTS's of PT Timah Tbk and Anita project. The assumed minimum production of PT Timah Tbk to be of 35,000 mt/year [18], and slag output of about 10% [19] so slag amount was about 3,500 mt/year.

2.2. Materials

Tin slag used in this research was provided by PT Timah Tbk, Indonesia. Initial tin slag characterization was done by XRF (Spectro Xepos Ametek), the results can be seen in Table 1. This study used HF (technical solution), and HCl and NaOH (pa, Merck).

2.3. Procedure of pre-removed of other oxides

First sample, Bangka tin slag sieved with sieve size +1; -1+0.71; -0.71+0350; -0350+0063 and -0.063 mm, then each particle size was characterized by XRF.

Table 1. The initial characterization of Bangka tin slag.

Ta_2O_5 (%)	Nb_2O_5 (%)	SiO_2 (%)	CaO (%)	TiO_2 (%)	Al_2O3 (%)	Fe_2O3 (%)	ZrO_2 (%)	El & Min.OO (%)
0.33	0.64	34.26	15.44	11.92	11.7	8.84	4.78	12.06

*El & Min.OO = element & minor other oxides

Second samples, Bangka tin slag was roasted at a temperature of 900^0C for 2 hours, water quenched and heated in the oven to demoisturized. The samples, then, sieved to the size of +1; -1+0.71; -0.71+0350; -0350+0063 and -0.063 mm and each size particle was characterized by XRF. Both samples, BTS and BTS-RQS were analized with thermo gravimetric, Perkin Elmer STA 6000, to record the sample weight change as a function of temperature.

Dissolving with hydrofluoric acid has done by placing BTS-RQS into 4 pieces of plastic containers each of which weights 20 grams. Then, filled the plastic container with 100 ml of hydrofluoric acid with each of concentration of 4, 8, 16 and 32%. After 2 hours, each sample was rinsed with distilled water and characterized by XRF.

The other dissolutions were to put BTS-RQS into 16 beaker glasses, each of which weights 20 grams, then on every 4 beaker glasses filled with 100 ml of hydrochloric acid with each of concentration of 6, 8, 10 and 12 M, then immersed for 2 hours. After the hydrochloric acid dissolutions completed, the samples were rinsed with distilled water. For every 4 beaker glasses from different concentrations of hydrochloric acid, sodium hydroxide with a concentration of 6, 8, 10, 12M, respectively, was added and immersed for 20 hours. The sample code, type, concentration and solvent dissolution time are showed in Table 2.

Table 2. Sample code, solvent type, concentration (%/M) and dissolution time (hours).

Sample code	Type of solvent, Concentration (% / M) and Dissolution Duration (hours)
F4	Flouride Acid 4 % , 2 h
F8	Flouride Acid 8 % , 2 h
F16	Flouride Acid 16 %, 2 h
F32	Flouride Acid 32 %, 2 h
A6B6	Chrolide Acid 6M, 2 h continue to Sodium Hydroxide 6M, 20 h
A6B8	Chrolide Acid 6M, 2 h continue to Sodium Hydroxide 8M, 20 h
A6B10	Chrolide Acid 6M, 2 h continue to Sodium Hydroxide 10M, 20 h
A6B12	Chrolide Acid 6M, 2 h continue to Sodium Hydroxide 12M, 20 h
A8B6	Chrolide Acid 8M, 2 h continue to Sodium Hydroxide 6M, 20 h
A8B8	Chrolide Acid 8M, 2 h continue to Sodium Hydroxide 8M, 20 h
A8B10	Chrolide Acid 8M, 2 h continue to Sodium Hydroxide 10M, 20 h
A8B12	Chrolide Acid 8M, 2 h continue to Sodium Hydroxide 12M, 20 h
A10B6	Chrolide Acid 10M, 2 h continue to Sodium Hydroxide 6M, 20 h
A10B8	Chrolide Acid 10M, 2 h continue to Sodium Hydroxide 8M, 20 h
A10B10	Chrolide Acid 10M, 2 h continue to Sodium Hydroxide 10M, 20 h
A10B12	Chrolide Acid 10M, 2 h continue to Sodium Hydroxide 12M, 20 h
A12B6	Chrolide Acid 12M, 2 h continue to Sodium Hydroxide 6M, 20 h
A12B8	Chrolide Acid 12M, 2 h continue to Sodium Hydroxide 8M, 20 h
A12B10	Chrolide Acid 12M, 2 h continue to Sodium Hydroxide 10M, 20 h
A12B12	Chrolide Acid 12M, 2 h continue to Sodium Hydroxide 12M, 20 h

3. Results and discussions

3.1. Feasibility study result

With Anita Project's economic parameter basic data, potential reserved calculation of tantalum pentoxide (Ta_2O_5) and niobium pentoxide (Nb_2O_5) from tin slag can be seen at Table 3.

Table 3. Comparative of potential reserves calculation project Anita and Bangka tin slag.

Description	Anita Projek	Bangka Tin Slag
Ta_2O_5 content	234 ppm	0.33%
Nb_2O_5 content	0.196%	0.64%
Measured reserve	25.8 million ton	3,500 ton/year*
Inferred reserve	15.42 million ton	
Mine Life	18 year	

* calculation in minimum production.
Slag Production of PT Timah Tbk 3,500-5,000 ton/year.

3.2. Initial Process of BTS

To simplify observation and discussion, other oxides were classified into two groups, one group was major other oxides (MOO) and another group that consisted of elements and minor oxides (EMO). Table 4 had the details data.

Table 4. Major Others Oxides and Element & Minor Other Oxides.

Major Others Oxides (MOO)	Element & Minor Others Oxides (EMO)
SiO_2, CaO, TiO_2, Al_2O_3, Fe_2O_3 ,ZrO_2	Na_2O, MgO, P_2O_5, S, Cl, K_2O, Sc, V_2O_5, Cr_2O_3, MnO, CoO, NiO, CuO, ZnO, Ga, Ge, As_2O_3, Se, Br, Rb_2O, SrO, Y, MoO, Ru, Rh, Pd, Ag, Cd, In, SnO_2, Sb_2O_5, Te, I, Cs, BaO, La_2O_3, Ce_2O_3, Pr, Nd, Sm, Eu, Gd, Tb, Dy, Ho,Er, Tm, Yb, Lu, Hf, WO_3, Au, Hg, Tl, Pb, Bi, Th and U

Characterization of BTS with XRF can be seen at tabel 5. Particle size of -1.00+0.71 mm had the highest content of Ta_2O_5 and Nb_2O_5, that were 0.3807 and 0.6978%, each with a weight fraction of 47.29%.

The initial purpose of reduction roasting of 900^0C and water quenched of tin slag was to observe the effects of reduction roasting and water quenched on particle size reduction. But the experiment shown another results that there was a decreased content of EMO. Particle size reduction due to roasted 900^0C and water quenched affected only about 2% by weight in 1.00 mm particles size and a significant affected on particles size smaller than 1.00 mm. The percentage decreased content of EMO was from 56.17%-16.44% to 20.84%-11:30%, this can be seen in Table 5 and Table 6.

Table 5. Size and oxides analysis of the initial tin slag.

Particle size (mm)	Weight (gram)	Fraction (%)	Ta_2O_5 (%)	Nb_2O_5 (%)	ZrO_2 (%)	Al_2O_3 (%)	Fe_2O_3 (%)	TiO_2 (%)	CaO (%)	SiO_2 (%)	El & Min.OO (%)
+1,00	479	47.9	0.3239	0.6548	4.732	7.285	7.595	9.761	12.15	22.03	35.4683
-1,00+0,71	472.9	47.29	0.3807	0.6978	4.951	7.77	9.124	11.59	13.59	24.3	27.5965
-0,71+0,350	18.6	1.86	0.3393	0.6484	4.448	7.395	8.796	9.841	11.04	23.4	34.0923
-0,350+0,063	23.1	2.31	0.2758	0.6059	4.212	11.77	8.706	9.872	12.26	35.85	16.4483
-0,063	2.3	0.23	0.2937	0.3715	2.396	4.406	8.112	6.539	5.677	16.03	56.1748
Weight loss	4.1	0.41	-	-	-	-	-	-	-	-	-
Total	1000	100	-	-	-	-	-	-	-	-	-

Table 6. Size and elemental analysis of BTS- RQS.

Particle size (mm)	Weight (gram)	Fraction (%)	Ta_2O_5 (%)	Nb_2O_5 (%)	ZrO_2 (%)	Al_2O_3 (%)	Fe_2O_3 (%)	TiO_2 (%)	CaO (%)	SiO_2 (%)	El & Min.OO (%)
+1,00	441.2	45.8	0.3475	0.7439	5.354	11.25	9.063	12.03	16.4	33.51	11.3016
-1,00+0,71	221	22.9	0.3536	0.7981	5.641	10.55	9.28	11.53	15.48	34.1	12.2673
-0,71+0,350	202.5	21.0	0.3931	0.8994	6.358	9.672	10.53	11.96	15.88	30.12	14.1875
-0,350+0,063	71.2	7.4	0.3502	0.6343	4.412	10.27	9.84	11.51	14.85	32.19	15.9435
-0,063	15.9	1.7	0.3303	0.6867	4.754	9.558	11.63	10.74	13.46	28	20.841
Weight loss	11.8	1.2	-	-	-	-	-	-	-	-	-
Total	963.6	100	-	-	-	-	-	-	-	-	-

3.3. Thermo gravimetric analysis on BTS

To observed thermal decomposition, the initial BTS and BTS-RQS was analyzed using TGA. The linear heating rate of TGA characterization was 10^0C/min, from room temperature to a temperature of 900^0C. The result was showed in Figure 1.

BTS endothermic peak can be seen at about 600^0C, its weight loss started at temperature of 100^0C until the endothermic peak, while the increased of its weight began at a temperature of 700^0C. BTS-RQS weight loss occured from the temperature of 100^0C to about 900^0C, refer to Figure 1(b), Table 5 and Table 6, this weight loss occured due to the oxidation of EMO.

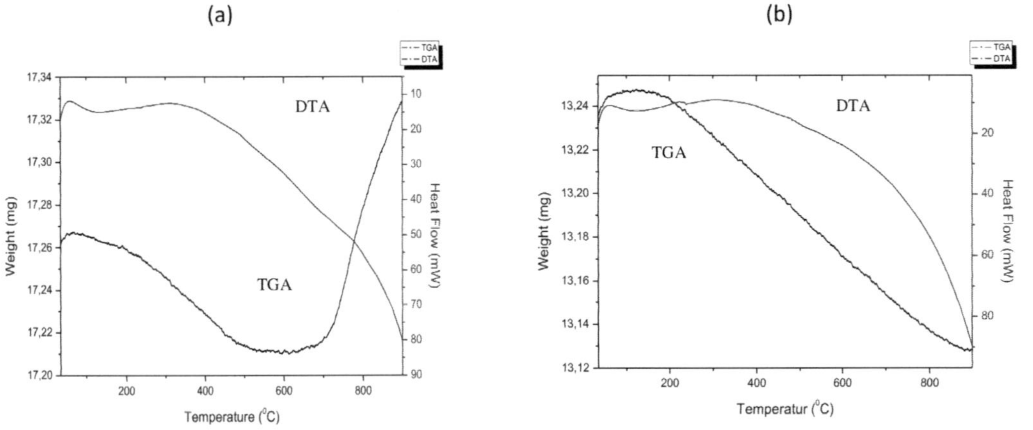

Figure 1. (a) DTA and TGA chart from the initial Bangka tin slag. (b) DTA and TGA chart from the roasted 900^0C and water quenched of Bangka tin slag.

3.4. The effect of HF and HCl+NaOH dissolution

3.4.1. The effect of HF dissolution

BTS-RQS dissolution with hydrochloric acid 8% produceed the highest Ta_2O_5 and Nb_2O_5 content of 0.664% and 1.339%, with Its yield ratio were 2.01 and 2.09, respectively, this can be seen in Figure 2 (a) and Table 7.

This dissolution shown that the higher HF concentration, the higher SiO_2 dissolution rate. This result was in line with SiO_2 dissolution on HF [20]. Dissolution with HF 32% will caused the increased SiO_2 content to 13.06%, this is due to the formation of solid SiO_2. This solid SiO_2 was the reason why weight loss value become 106.5%, which can be seen at Table 7.

Hydrochloric acid dissolved Fe_2O_3 well and TiO_2 slightly. HF 4, 8 and 16% were less successfully dissolving EMO and could not dissolve CaO, Al_2O_3 and ZrO_2. This information can be seen in Figure 2 (b).

(a) (b)

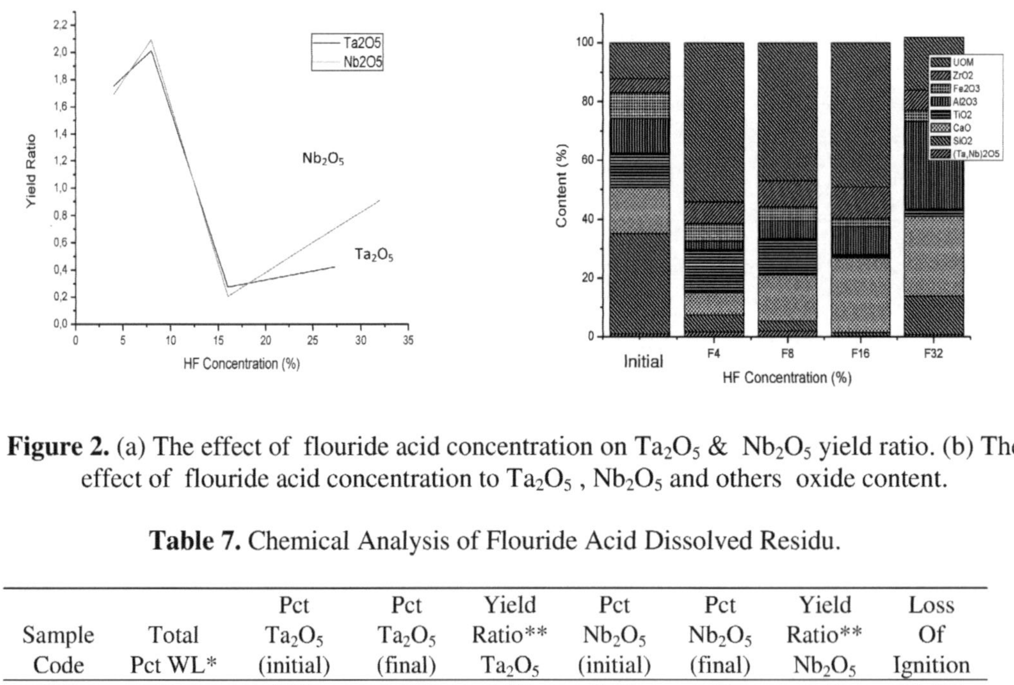

Figure 2. (a) The effect of flouride acid concentration on Ta_2O_5 & Nb_2O_5 yield ratio. (b) The effect of flouride acid concentration to Ta_2O_5 , Nb_2O_5 and others oxide content.

Table 7. Chemical Analysis of Flouride Acid Dissolved Residu.

Sample Code	Total Pct WL*	Pct Ta_2O_5 (initial)	Pct Ta_2O_5 (final)	Yield Ratio** Ta_2O_5	Pct Nb_2O_5 (initial)	Pct Nb_2O_5 (final)	Yield Ratio** Nb_2O_5	Loss Of Ignition
F4	76,0	0,33	0,578	1,75	0,64	1,080	1,69	38,653
F8	56,5	0,33	0,664	2,01	0,64	1,339	2,09	30,589
F16	55,0	0,33	0,091	0,28	0,64	0,133	0,21	34,707
F32	106,5	0,33	0,160	0,48	0,64	0,584	0,91	-1,8748

*WL = Weight Loss. ** Yield Ratio = content before process/content after process.

3.4.2. Effect Dissolution with HCl+NaOH

The dissolution BTS-RQS with HCl 6M and NaOH 10M produced Ta_2O_5 dan Nb_2O_5 content of 0.528 and 1.182%, respectively, with yield ratio for Ta_2O_5 and Nb_2O_5 were 1.60 and 1.85, respectively. HCl and NaOH dissolution graphics can be seen in Figure 3 and Table 8. Contents of Ta_2O_5 and Nb_2O_5

would have about twice the yield ratio if HCl dissolution were followed by NaOH with a particle size smaller than 0.150 mm or the particle size of between 0.180 and 0.150 mm [13].

HCl and NaOH can dissolved EMO well, but was less successful dissolving SiO_2, Fe_2O_3 and TiO_2. HCl and NaOH were not successfully dissolving CaO, Al_2O_3 and ZrO_2, this showed in Figure 4. Leaching of Indonesia tin slag with acid (HCl 2N and HF 0.14 N) and alkaline (NaOH 2N), which used a pyrex reactor with a double wall for hot water circulation, caused a weight loss of 46.3% with Ta_2O_5 and Nb_2O_5 final content of the leaching were 1.9 and 2.9%, respectively [2].

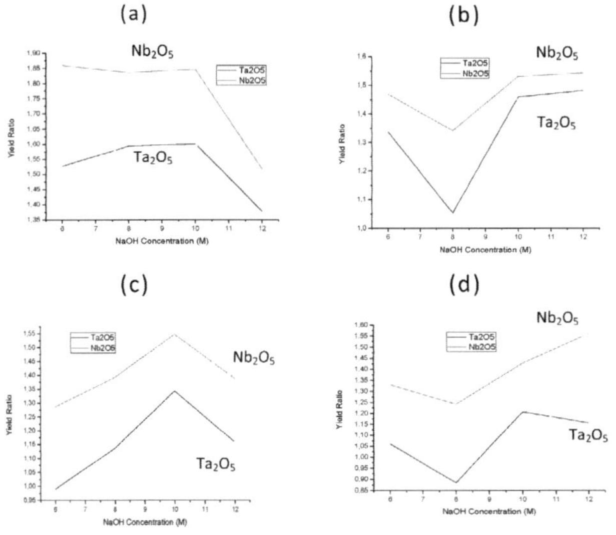

Figure 3. The effect of chloride acid (a) 6M (b) 8M (c) 10 M (d) 12 M on concentrate variation of NaOH and Ta_2O_5 & Nb_2O_5 yield ratio.

Tabel 8. Chemical Analysis of HCl and NaOH Dissolved Residu.

Sample Code	Total Pct WL*	Pct Ta_2O_5 (initial)	Pct Ta_2O_5 (final)	Yield Ratio** Ta_2O_5	Pct Nb_2O_5 (initial)	Pct Nb_2O_5 (final)	Yield Ratio** Nb_2O_5	Loss of Ignition
A6B6	70,5	0,33	0,5037	1,53	0,64	1,19	1,86	-3
A6B8	74,5	0,33	0,5261	1,59	0,64	1,175	1,84	-3
A6B10	71,5	0,33	0,5283	1,60	0,64	1,182	1,85	-3
A6B12	72,5	0,33	0,4547	1,38	0,64	0,9705	1,52	36,304
A8B6	77,0	0,33	0,4412	1,34	0,64	0,94	1,47	30,496
A8B8	74,0	0,33	0,3475	1,05	0,64	0,8586	1,34	44,465
A8B10	76,5	0,33	0,4814	1,46	0,64	0,9795	1,53	26,217
A8B12	79,0	0,33	0,4889	1,48	0,64	0,9876	1,54	20,425
A10B6	80,0	0,33	0,3264	0,99	0,64	0,8234	1,29	42,218
A10B8	82,0	0,33	0,3753	1,14	0,64	0,8916	1,39	35,494
A10B10	83,0	0,33	0,4434	1,34	0,64	0,9901	1,55	22,12
A10B12	83,0	0,33	0,3823	1,16	0,64	0,8885	1,39	33,061
A12B6	91,0	0,33	0,35	1,06	0,64	0,8509	1,33	28,087
A12B8	91,0	0,33	0,2918	0,88	0,64	0,7947	1,24	39,887
A12B10	91,0	0,33	0,3982	1,21	0,64	0,9137	1,43	24,512
A12B12	84,5	0,33	0,3818	1,16	0,64	0,9993	1,56	-3

*WL = Weight Loss. ** Yield Ratio = content before process/content after process.

IOP Conf. Series: Materials Science and Engineering **131** (2016) 012006 doi:10.1088/1757-899X/131/1/012006

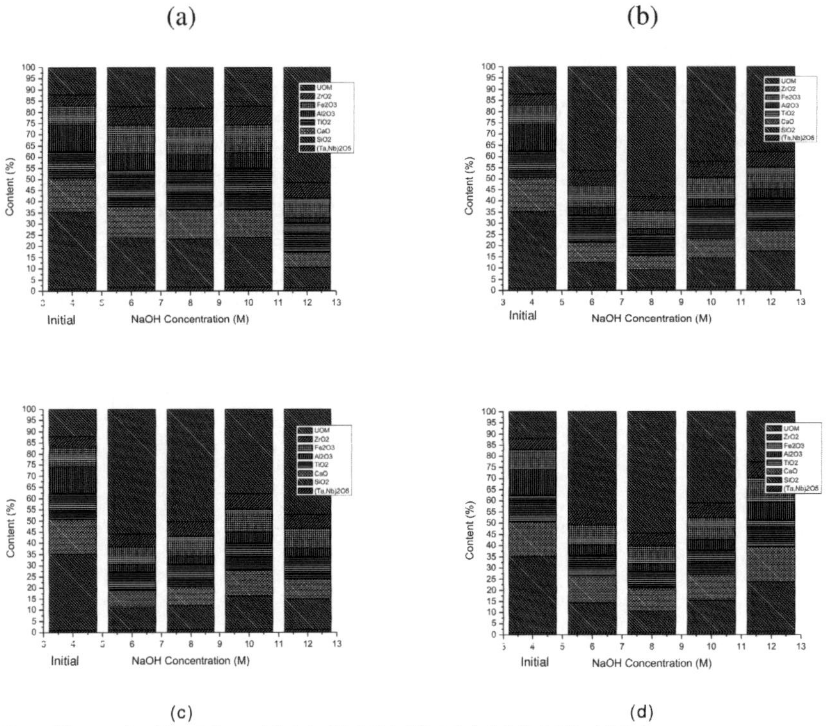

Figure 4. The effect of chloride acid (a) 6M (b) 8M (c) 10 M (d) 12 M on concentrate variation of NaOH to Ta_2O_5, Nb_2O_5 and others oxides content.

4. Conslusions and suggestions

Potential reserved of tantalum pentoxide (Ta_2O_5) and niobium pentoxide (Nb_2O_5) from BTS can be used as an alternative potential revenues for Bangka and Belitung province.

The roasted at 900^0C and water quenched can be reduced EMO content. BTS-RQS produced the highest contents of Ta_2O_5 and Nb_2O_5 with dissolution of HF 8%, also, the dissolution with HCl 6M and NaOH 10M produced the highest contents of Ta_2O_5 and Nb_2O_5.

To dissolved SiO_2 and Fe_2O_3, HF was better than with HCl+NaOH. HF and HCl+NaOH did not dissolve CaO, Al_2O_3 and ZrO_2. BTS-RQS dissolution with HF diluted less EMO, while with HCl 6% and NaOH 6.8 and 10% showed lower contents of SiO_2, Fe_2O_3 and TiO_2.

For further research, researcher will need do roast above 900^0C and water quench to see the effect of it. The next, researcher will need investigate that solid residu from HF 8% dissolution, redissolved in HCl 6M and NaOH 10M to reduce EMO content.

Acknowledgements

This work has been financially supported by the Directorate of Research and Community Engagement, University of Indonesia under 2015 postgraduate grant program. This Bangka tin slag supported by PT Timah Tbk. The authors would like to record their appreciation for this aid.

References :

[1] OEDC 2012 *List of Tin , Tantalum and Tungsten (3Ts) Smelters*
[2] Gaballah I, Allain E, Meyer-Joly M-C and Malau K 1992 A Possible Method for the Characterization of Amorphous Slags Recovery of Refractory Metal Oxides from Tin Slags *Metall. Trans.* **23** 249–59
[3] Laura Talens Peiro, Gara Villalba Mendez R U A 2011 *Rare and Critical Metals as By-products and The Implications for Future Supply*
[4] Bose D K and Gupta C K 2001 Extractive Metallurgy of Tantalum *Miner. Process. Extr. Metall. Rev.* **22** 389–412
[5] Gibson C E, Kelebek S and Aghamirian M 2015 Niobium oxide mineral flotation: A review of relevant literature and the current state of industrial operations *Int. J. Miner. Process.*
[6] Wellmer F and Dalheimer M 2012 The feedback control cycle as regulator of past and future mineral supply *Min. Depos.* **47** 713–29
[7] Cenerazzo J A, Mosheim C E, Zionsville and Corrodo E. Marvisa 1971 US Patent 3.585.024-Up Grading The Tantalum and Columbium Contents of Tin Slag 1–4
[8] Gaballah I and Allain E 1994 Recycling of Strategic Metals from Industrial Slag by a Hydro- and Pyrometallurgical Process *Resour. Conserv. Recycl.* **10** 75–85
[9] Gaballah I, Allain E and Djona M 1997 Extraction of Tantalum and Niobium from Tin Slags by Chlorination and Carbochlorination *Metall. Mater. Trans.* **28**
[10] Subramanian C, Suri A K and Atomic B 1998 Recovery of Niobium and Tantalum from Low Grade Tin Slag - A Hydrometallurgical Approach *Environ. Waste Manag.* 100–7
[11] Mudzanapabwe N T, Chinyamakobvu O S and Simbi D J 2004 In Situ Carbothermic Reduction of a Ferro-Columbite Concentrate in The Recovery of Nb and Ta as Metal Matrix Composite from Tin Smelting Slag Waste Dump *Mater. Des.* **25** 297–302
[12] Brocchi E A and Moura F J 2008 Chlorination Methods Applied to Recover Refractory Metals from Tin Slags *Miner. Eng.* **21** 150–6
[13] Odo J.U., Okafor WC, Ekpe SO and CC. and N 2014 Extraction of Niobium from Tin Slag *Int. J. Sci. Res. Publ.* **4** 1–7
[14] MDN Inc. 2013 *MDN Receives Positive Economic Indicators For The Niobium and Tantalum (ANITA) Project* (Quebec)
[15] Soedarsono J, Burgard M, Asfari Z and Vicens J 1993 Liquid-liquid Extraction of Rare Earth (III) Ions by 25,27-dicarboxy-26,28-dimethoxy-5,11,17,23- tetra- tert-butylcalix[4] arene *XVII International Symposium on Macrocyclic Chemistry*
[16] Soedarsono J, Burgard M, Asfari Z and Vicens J 1995 Liquid-liquid extraction properties of 25,27-dicarboxy-26,28-dimethoxy-5,11,17,23-tetra-tert- butylcalix [4]arene towards the alkali ions *Solvent Extr. Ions Exch.* **13** 755–69
[17] Soedarsono J, Hagege A, Burgard M, Asfari Z and J. Vicens 1996 Liquid-liquid Extraction of Rare Earth Metals Using 25,27-dicarboxy-26,28-dimethoxy- 5,11,17,23-tetra- tert-butylcalix[4]arene *J. Phys. Chem.* **100** 477–81
[18] Pusat Data dan Teknologi Informasi Energi dan Sumber Mineral Kementerian Energi dan Sumber Daya Mineral 2013 *Kajian Supply Demand Mineral*
[19] Sulaiman M Y 1993 Simultaneous determination of thorium and uranium in tin slag *Sci. Total Environ.* **131** 187–95
[20] Judge J S 1971 A Study of the Dissolution of SiO in Acidic Fluoride Solutions *J. Electrochem. Soc.* **118** 1772–5

IOP Conf. Series: Materials Science and Engineering **131** (2016) 012007 doi:10.1088/1757-899X/131/1/012007

Probabilistic Resilience in Hidden Markov Models

Jacopo Panerati and Giovanni Beltrame

Polytechnique Montreal, 2900 boul. Édouard-Montpetit, Montréal H3T 1J4, Québec, Canada

E-mail: {jacopo.panerati, giovanni.beltrame}@polymtl.ca

Nicolas Schwind

Transdisciplinary Research Integration Center, 10-3 Midori-cho, Tachikawa, Tokyo, Japan

E-mail: schwind@nii.ac.jp

Stefan Zeltner

Bloomberg L.P., 731 Lexington Ave, New York, NY 10022, USA

E-mail: zeltner@usc.edu

Katsumi Inoue

National Institute of Informatics, 2-1-2 Hitotsubashi, Chiyoda, Tokyo 101-0003, Japan

E-mail: inoue@nii.ac.jp

Abstract. Originally defined in the context of ecological systems and environmental sciences, resilience has grown to be a property of major interest for the design and analysis of many other complex systems: resilient networks and robotics systems offer the desirable capability of absorbing disruption and transforming in response to external shocks, while still providing the services they were designed for. Starting from an existing formalization of resilience for constraint-based systems, we develop a probabilistic framework based on hidden Markov models. In doing so, we introduce two new important features: stochastic evolution and partial observability. Using our framework, we formalize a methodology for the evaluation of probabilities associated with generic properties, we describe an efficient algorithm for the computation of its essential inference step, and show that its complexity is comparable to other state-of-the-art inference algorithms.

1. Introduction and Related Work

Originally defined in the context of ecological systems, resilience is now a property of major interest for the analysis of many complex systems. Its definition is not always unanimous, but researchers agree that resilience is a characteristic property of those ecosystems that, through their history, are able to absorb extreme spikes and, although transformed, survive. The insect populations of North-eastern American forests [1] are notable examples of such sustainable systems. In the coming years, "resilient by design" networks and robotics system will possess the ability to absorb disruption and to transform in response to external shocks, while still providing their services. Resilience will also be implemented in welfare measures—such as flood prevention plans and healthcare systems—helping making human society a more sustainable system too.

So far, the effort of the artificial intelligence community has been focused on narrowing down the concept of resilience through its formalization, e.g., in constraint-based and non-deterministic dynamic systems [2]. This approach is extremely general and it can be adapted to describe the behaviour of a plethora of real-world systems but it comes at the cost of a limited predictive power. The transition models in non-deterministic dynamic systems resemble those of Markov decision processes but, lacking probability values associated to transitions, they do not tell which future worlds are the most likely.

By exploiting hidden Markov models, we preserve much of the existing research and include a fundamental aspect that was not investigated in earlier works: the unpredictability of the real world. This is essential for any modelling that aspires to be consistent with the idea of a "random world" proposed by Holling [1] and conditional probability distributions can be seen as the stochastic extension of the non-deterministic transition functions. We also enrich the existing discussion about resilience with another important layer of complexity, that is, partial observability. The goal of our work is to be able to efficiently answer queries such as: "what is the probability that Canada's inflation rate will not drop below 1% in the next five years?" or "with 95% confidence, what is the minimum size of a certain wild animal population, only rarely observed?".

In a seminal paper from 1973, Holling introduced the concept of "resilience of ecological systems". Resilient systems are not those systems that simply react to imbalances by quickly returning to equilibria. Instead, when perturbed, they are able to find new sustainable configurations. It is worth noting that Holling defines resilience in the context of what he calls "the random world": an environment that is intrinsically stochastic. Walker *et al.* [3] define resilience as "the capacity of a system to absorb disturbance and reorganize while undergoing change so as to still retain essentially the same function, structure, identity, and feedbacks".

The first attempt to formally define the concept of resilience exploiting the tools of artificial intelligence was provided by Överen, Willsky, and Antsaklis [4] and successively elaborated by Baral *et al.* [5] and Schwind *et al.* [2]. The SR-model, proposed in [2], is the starting point of our work and it is described in more detail in the background section. The most relevant difference between these works and our methodology is the introduction of the ideas of probability theory in the latter. The probabilistic framework on which we develop our analysis is that of hidden Markov models. Their traditional applications are in signal and natural language processing, however, their use to describe more complex dynamic systems has also proven to be fruitful [6].

The two most common types of inference tasks for probabilistic graphical models are marginal and maximum a posteriori estimation. For these, HMMs provide efficient algorithms that have convenient linear time-complexity [7]. In this work, we are interested in the probability of properties lasting over time, therefore, we cannot simply apply the state-of-the-art algorithms for marginals or maximum a posteriori. Nonetheless, we show that the queries we need can be implemented by an algorithm in the same complexity class (i.e. requiring the same amount of computational resources).

2. Background

Before delving into the proposed methodology, we briefly review the theoretical background on the two main aspects of this work: 1) the formal definition of resilience in sustainable systems; 2) and hidden Markov models.

The SR-model is a theoretical framework proposed by Schwind *et al.* [2] combining elements of constraint-based systems and non-deterministic dynamic systems. It provides a formal definition of resilience as a unifying property combining different aspects of three simpler properties: resistance, functionality, and recoverability.

2.0.1. Kinematics and dynamics The SR-model provides two separate formal descriptions for the kinematics and the dynamics of a system: the first consists of unbounded sequences of state-couples called "state trajectories" or SSTs:

$$SST = (CBS_1, \gamma_1), (CBS_2, \gamma_2), \ldots \tag{1}$$

The subscript index skims through discrete time steps. Each CBS_i represents a constraint-based system composed of a set of variables $\mathbf{X_i}$ and a cost function c_i:

$$CBS_i = \langle \mathbf{X_i} = \{X_i^1, X_i^2, \ldots\}, c_i : \mathcal{D}(\mathbf{X_i}) \to \mathbb{R}^+ \rangle. \tag{2}$$

Finally, $\gamma_i \in \mathcal{D}(\mathbf{X_i})$ represents a complete assignment of the variables in $\mathbf{X_i}$:

$$\gamma_i \in \mathbb{R}^{|\mathbf{X_i}|} \tag{3}$$

As a consequence, each SST corresponds unambiguously to a sequence of costs obtained by plugging-in each γ_i into the corresponding cost function c_i:

$$c_1(\gamma_1), c_2(\gamma_2), \ldots \tag{4}$$

The environment dynamics are described through a non-deterministic dynamic system DS:

$$DS = \langle \mathbf{CBS}, \mathbf{A}, t : \mathbf{CBS} \times \mathbf{A} \to \mathcal{P}(\mathbf{CBS}) \rangle \tag{5}$$

where **CBS** represents the set of all possible constraint-based systems CBS_i, **A** is the set of actions available at each time step, and t is a non-deterministic transition function that, given the current constraint-based system and an action, returns the set of possible constraint-based systems at the next time step.

The kinematic description (SSTs and sequences of costs) is essential for understanding the resilient properties and it is preserved in our proposed methodology. In this work, however, the non-deterministic description of the dynamics is discarded for a probabilistic approach.

2.0.2. The resilient properties In this model, resilience is a boolean property of a state trajectories SST. It can be seen as a unifying property, combining different desirable behaviours of a dynamic system and it arises from three simpler properties of state trajectories: resistance, functionality, and recoverability (see Figure 1).

l-resistance The resistance property expresses the fact that a trajectory never incurs in a cost that is larger than a fixed threshold. Therefore, this property is parametrized by this maximum acceptable cost.

Definition 1 *Given a state trajectory* $SST = (CBS_1, \gamma_1), (CBS_2, \gamma_2), \ldots$ *and a positive threshold* $l \in \mathbb{R}^+$, *SST is said to be l-resistant if and only if each cost in its corresponding cost sequence is less than or equal to the threshold* l*:*

$$c_i(\gamma_i) \leq l \quad \forall c_i(\gamma_i) \in (c_1(\gamma_1), c_2(\gamma_2), \ldots) \tag{6}$$

For example, any organization (e.g. the army of a sovereign nation) whose budget is assigned periodically (e.g. yearly) by and external entity (e.g. the Department of Defense) requires to satisfy this propery.

f-functionality The functionality property tells us if the costs of a trajectory are, on average, equal to or below a certain threshold. As in the case of resistance, this threshold parametrizes the property.

Definition 2 *Given a state trajectory $SST = (CBS_1, \gamma_1), (CBS_2, \gamma_2), \ldots$ and a positive threshold $f \in \mathbb{R}^+$, SST is said to be f-functional if and only if the arithmetic average of the costs in its corresponding cost sequence is less than or equal to the threshold f:*

$$\frac{1}{|SST|} \sum_{i=1}^{|SST|} c_i(\gamma_i) \leq f \tag{7}$$

This property is important when we plan operations that span over multiple time steps (e.g a trip lasting several days or weeks) on a fixed budget. From time to time, we might allow ourselves to spend more than the daily allowance: what matters is our total expenditure (or the global average).

$\langle p, q\rangle$-recoverability The recoverability property concerns those systems in which costs over a certain threshold can be accepted, but only as long as the system is able to return within normal conditions before consuming a fixed, restorable, reserve.

Definition 3 *Given a state trajectory $SST = (CBS_1, \gamma_1), (CBS_2, \gamma_2), \ldots$, a positive threshold $p \in \mathbb{R}^+$ and a positive budget $q \in \mathbb{R}^+$, SST is said to be $\langle p, q\rangle$-recoverable if and only if every time the sequence of costs exceeds the threshold, it also returns below (or at) it before that the cumulative offset surpasses the reserve:*

$$\forall k \; s.t. \; c_k(\gamma_k) > p, \exists j > k \; s.t. \; c_j(\gamma_j) \leq p \wedge \sum_{i=k}^{j-1} (c_i(\gamma_i) - p) \leq q \tag{8}$$

An example of a recoverable system is the human muscle tissue: it can perform at maximum intensity for a short time consuming a molecule called *adenosine triphosphate*, or ATP [8], which is available in limited quantities in our body. To recover, the muscle needs to decrease the intensity of the effort, and allow other metabolic pathways to replenish the initial ATP storage.

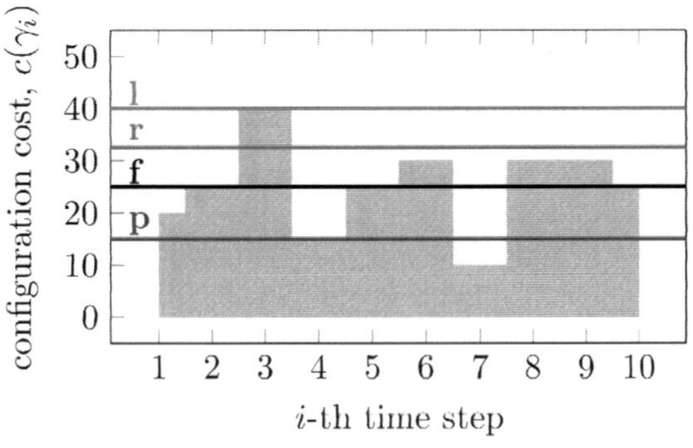

Figure 1. Example of a cost trajectory that is 40-resistant, 25-functional, $\langle 15, 40\rangle$-recoverable

Hidden Markov models (HMMs) can be seen a sub-family of both dynamic Bayesian networks (DBNs) and state-observation models [7]. HMMs have a single discrete state variable S and a single discrete observation variable O. A hidden Markov model is fully specified by the probability distribution of S at time 0, the conditional distribution of O given S at the same time step, and the conditional distribution of S given S at the previous time step [9].

$$HMM = \langle P(S_0), P(O_t \mid S_t), P(S_{t+1} \mid S_t) \rangle \tag{9}$$

HMMs are commonly used for the tasks of signal processing and speech recognition [9] because efficient (i.e. with computational time complexity that is linear with respect to the time horizon of the model) algorithms exist for: 1) the estimation of the probability distribution of S, also called the "hidden" variable, taking only assignments of O as input (filtering and smoothing algorithms); and 2) the identification of the most likely sequence of assignments of S. To formalize resilient properties in a probabilistic context, HMMs offer the probabilistic reasoning of DBNs and the independence assumptions of state-observation models, while minimizing representation size.

3. Methodology

In this section, we identify the shortcomings of the traditional HMM framework and suggest how to improve it. On top of this extended model, we provide the definitions of trajectory of states TS and trajectory of observations TO, and we show how to compute their probability values.

3.1. Framework Extensions

We can imagine the single discrete state variable S of a HMM as a way to represent, by enumeration, the state configuration that was encoded by the set of variables $\mathbf{X}$, and assignment γ, in the constraint-based systems $CBSs$. However, HMMs do not natively provide an important element of the SR-model: a cost function c to state the expensiveness, for the purpose of resilience, of each assignment. In the following, the extension of the HMM framework with a static cost function c, defined over the domain Ω of its (random) state variable S and taking positive values in $\mathbb{R}$, is called c-HHM (see Figure 2) and formally defined as:

$$\langle P(S_0), P(O_t \mid S_t), P(S_{t+1} \mid S_t), c : \Omega(S) \to \mathbb{R}^+ \rangle \tag{10}$$

A static, i.e. non time-varying, cost function means that our preferences do not change over time.

3.2. Trajectories of States, Observations, and Costs

In the SR-model, the definition of the resilient properties is based on the concepts of SSTs and their corresponding sequences of costs. In c-HMMs, we have similar constructs for states and costs, as well as observations. The main aspect of novelty here is that these concepts are now built on top of random variables [10] and, therefore, they also can be associated to probability distribution functions. Given a c-HMM and a finite time horizon T, we define its trajectory of states TS as the sequence of state variables S_i $\forall i \in \{1, \ldots, T\}$. This can be rewritten as:

$$TS := S_1, S_2, \ldots, S_T \tag{11}$$

Because S is a random variable, TS is also a random variable that can take as value any possible sequence of assignments of S of length T, i.e. $ts = s_1, s_2, \ldots, s_T$. Therefore, the number of possible assignments of TS grows exponentially with the time horizon:

$$|\Omega(TS)| = |\Omega(S)|^T \tag{12}$$

Because the mapping provided by the cost function c is entirely deterministic, each assignment of TS is unambiguously associated to a trajectory of costs $tc = c(s_1), c(s_2), \ldots, c(s_T)$ and TC is also a random variable with $|\Omega(TC)| \leq |\Omega(TS)| = |\Omega(S)|^T$. Considerations similar to those we made for trajectories of states are also valid for trajectories of observations ($TO := O_1, O_2, \ldots, O_T$) and their possible assignments ($to := o_1, o_2, \ldots, o_T$). If neither the transition model or the sensor model contain probability values of 0, all possible sequences of states can occur and produce any of the sequences of observations. Therefore, the number of possible configurations of a c-HHM is:

$$(|\Omega(S)| \cdot |\Omega(O)|)^T \tag{13}$$

This represents how difficult inference, in general, could be in our model. We also observe that: 1) the state variable S is called the "hidden" variable because it is never directly observable, meaning that which ts value is actually taken by TS is unknown too; while 2) the values taken by the observations variable O are the pieces of *partial* information that we can always access. Therefore, the kind of probabilistic inference we are interested in is the one dealing with the $|S|^T$ probability values of the conditional probability distribution of TS with respect to a given to:

$$P(TS \mid to) = P(S_1, S_2, \ldots, S_T \mid o_1, o_2, \ldots, o_T) \tag{14}$$

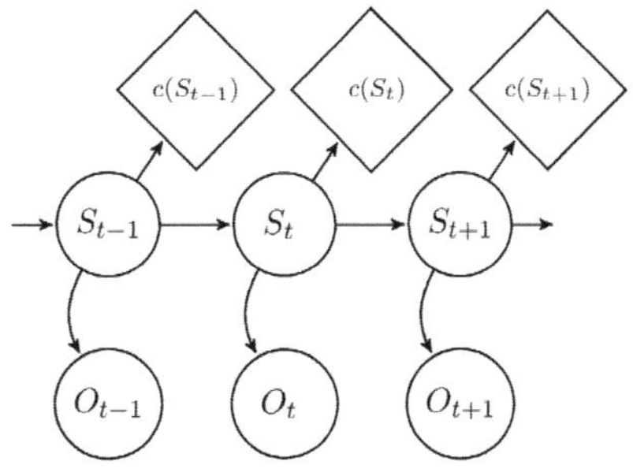

Figure 2. The extended c-HMM framework unrolled over three time steps.

3.3. Probability of Cost Trajectories and Properties

So far, we have seen that the resilient properties, once their parameters are fixed, can be considered as boolean attributes of sequences of costs associated to SSTs. In the context of the c-HMM framework, we say that an assignment of the trajectory of states $ts = s_1, s_2, \ldots$ enforces the property π if and only if its corresponding trajectory of costs $tc = c(s_1), c(s_2), \ldots$ is statisfies the definition of that property, i.e. $\pi(c(s_1), c(s_2), \ldots) = true$. Computing the probability distribution of parametric properties $\pi(k)$, such as resistance and functionality, with respect to their parameters, can provide valuable insights, as shown in Figure 3.

The probability of property $P(\pi)$ is equal to the sum of the probabilities of all the distinct trajectories of costs in which π holds:

$$P(\pi) = \sum_{\forall i \in \{i | \pi(tc_i) = true\}} P(tc_i) \tag{15}$$

In turn, the probability of a fixed assignment of the trajectory of costs tc is equal to the sum of the probabilities of all the distinct trajectories of states that are mapped to it by the cost function c. To simplify the notation, we will also use $C(ts)$ to indicate the $tc = c(s_1), c(s_2), \ldots$ resulting from the application of the cost function c to SST assignment ts.

$$P(tc) = \sum_{\forall i \in \{i | C(ts_i) = tc\}} P(ts_i) \tag{16}$$

By plugging Equation 16 into Equation 15 we can compute the probability of a property π as a function of the probabilities of distinct assignments of the trajectory of states (Equation 17). We observe that all the possible assignments of TS are "distinct" by definition, even if many of them could be mapped by c to identical trajectories of cost.

$$P(\pi) = \sum_{\forall i \in \{i | \pi(tc_i) = true\}} \sum_{\forall j \in \{j | C(ts_j) = tc_i\}} P(ts_j) \tag{17}$$

3.4. Conditioning w.r.t. the Trajectory of Observations

When we are able to observe the system, we become interested in computing the conditional distribution of π with respect to the assignment of the trajectory of observations to. To do so, we need to update Equation 17 by conditioning on both sides using to:

$$P(\pi \mid to) = \sum_{\forall i \in \{i | \pi(tc_i) = true\}} \sum_{\forall j \in \{j | C(ts_j) = tc_i\}} P(ts_j \mid to) \tag{18}$$

Equation 18, shows that computing the probability of a property π, given an assignment of the trajectory of observations to, consists of two different subproblems:

a) identifying which assignments of the trajectory of states TS map to assignments of the trajectory of costs TC such that the property is statisfied;
b) computing the conditional probability of these assignments of TS with respect to the assignment of the trajectory of observations to.

The problem at point a) strictly depends on the nature of the property we are evaluating and it is discussed further later on. Instead, we now approach the second problem which is more general and can be addressed combining different inference methods for HMMs.

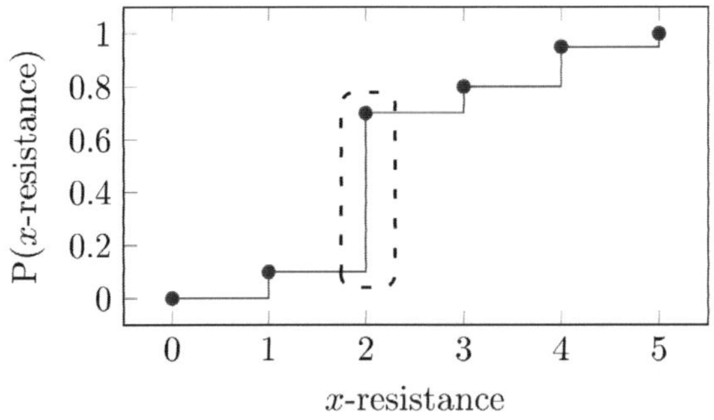

Figure 3. Probability distribution of parametric resistance in an example where $\forall s,\ c(s) \in [1, \ldots, 5]$. The discontinuity around 2 shows the critical threshold for the property.

3.4.1. Conditional probability of a finite assignment of the state trajectory: $P(s_1, \ldots, s_T | o_1, \ldots, o_T)$

A highly inefficient way to find this probability value consists of computing the complete joint probability distribution (JPD) of the c-HMM over the time horizon T—because HMMs are Bayesian networks, their JPD is equal to the chain product of all the conditional probability distributions (CPDs) in their nodes—then, conditioning by the evidence of to, and finally re-normalizing the entire distribution. This is not a good approach because computing the JPD requires time and space complexity of $(|S| \cdot |O|)^T$. A more efficient approach to the computation of the conditional probability of a finite state trajectory assignment $s_1, \ldots, s_T$ with respect to the observation trajectory assignment $o_1, \ldots, o_T$ can be found using the Bayes' theorem:

$$\begin{aligned} P(s_1, \ldots, s_T \mid o_1, \ldots, o_T) = \\ = \frac{P(o_1, \ldots, o_T \mid s_1, \ldots, s_T) \cdot P(s_1, \ldots, s_T)}{P(o_1, \ldots, o_T)} \end{aligned} \tag{19}$$

Equation 19 shows how to decompose the problem into three simpler computations that can be tackled separately.

Probability of: $o_1, \ldots, o_T \mid s_1, \ldots, s_T$ Given an assignment of the state variables, the conditional probability of a sequence of observations can be computed as the product of the appropriate entries of the sensor model.

$$P(o_1, \ldots, o_T \mid s_1, \ldots, s_T) = \prod_{i=1}^{T} P(O_t = o_i \mid S_t = s_i) \tag{20}$$

Probability of: $s_1, \ldots, s_T$ The probability of an assignment of the the trajectory of states TS, notwithstanding the values taken by the observations variables, only depends on the transition model $P(S_{t+1} \mid S_t)$ and the ground belief $P(S_0)$. First, we need to compute the probability of S_1 taking the value of s_1 as $P(S_1 = s_1) = \sum_{\forall s \in \Omega(S)} P(S_{t+1} = s_1 \mid S_t = s) P(S_0 = s)$; then, the probability of the entire trajectory can be computed multiplying the appropriate entries of the transition model:

$$P(s_1, \ldots, s_T) = P(S_1 = s_1) \cdot \prod_{i=2}^{T} P(S_{t+1} = s_i \mid S_t = s_{i-1}) \tag{21}$$

Probability of: $o_1, \ldots, o_T$ The computation of this latter factor is the trickiest: it can be performed efficiently by applying a dynamic programming technique called Forward algorithm [11]. This algorithm iteratively computes the quantity $P(o_1, \ldots, o_T, s_T)$—which is, from now on, re-written as $F(T, s_T)$—using the transition and sensor models. The initialization step of the Forward algorithm is:

$$\forall x \in \Omega(S) \quad F(1, x) = P(S_1 = x) \cdot P(O_1 = o_1 \mid S_1 = x) \tag{22}$$

The distribution of $P(S_1)$ can be computed in the same way we did for Equation 21. Then, the $t+1$ iteration step of the Forward algorithm is:

$$\begin{aligned} \forall x \in \Omega(S) \qquad F(t+1, x) = \\ \sum_{\forall y \in \Omega(S)} F(t, y) \cdot P(S_{t+1} = x \mid S_t = y) \cdot P(O_t = o_1 \mid S_t = x) \end{aligned} \tag{23}$$

Finally, having iterated the algorithm until T, $P(o_1, \ldots, o_T)$ is computed as the sum over all possible values of S_T:

$$P(o_1, \ldots, o_T) = \sum_{\forall x \in \Omega(S)} F(T, x) \tag{24}$$

3.5. Considerations on Time Complexity

Having how to compute $P(s_1, \ldots, s_T \mid o_1, \ldots, o_T)$, we now examine its efficiency. We assume that querying the sensor and the transition models for any of their elements involves a constant and negligible delay. The computation of the $P(o_1, \ldots, o_T)$ factor is the quickest: starting with an initialization value of 1, we need to multiply it T times by the correct entry of the sensor model. Therefore, the time complexity of computing $P(o_1, \ldots, o_T)$ is $O(T)$. The factor $P(s_1, \ldots, s_T)$ is obtained through $|\Omega(S)|$ multiplications and $|\Omega(S)| - 1$ sums, to find $P(S_1 = s_1)$, and $T - 1$ products by entries of the transition model. Its time complexity is equal to $O(|\Omega(S)| + T)$. Finally, the Forward algorithm has a run time of $O(|\Omega(S)|^2 \cdot T)$ [11], plus $|\Omega(S)| - 1$ additions to compute $P(o_1, \ldots, o_T)$. Hence, the computation of $P(o_1, \ldots, o_T)$ using the Forward algorithm is the slowest of the three. Indeed, this is also the overall time complexity of $P(s_1, \ldots, s_T \mid o_1, \ldots, o_T)$:

$$\text{time-complexity: } O(|\Omega(S)|^2 \cdot T) \tag{25}$$

This result arises from the consideration that all the three computations of Bayes' theorem factorization in Equation 19 are independent (from a computational point of view, not with regard to probability) and, given sufficient resources, they can be can be performed in parallel, with the last one strictly dominating the others.

Most importantly, we observe that, if the time horizon is much larger than the number of states (i.e. $T \gg |\Omega(S)|$), this probabilistic inference algorithm has an overall time complexity dominated by $O(T)$. This result places our algorithm in the same time complexity class of other well-known inference algorithms for HMMs: the Forward-backward algorithm for the computation of smoothed marginals distributions, $P(S_i \mid to)$; and the Viterbi algorithm for the computation of the most likely sequence of hidden variables, $\text{argmax}_{TS} P(TS \mid to)$ [7].

3.6. Considerations on Space Complexity

With regard to space complexity, we first have to consider the data structures necessary to represent the c-HMM framework: $P(S_0)$ has size of $O(|S|)$, $P(O_t \mid S_t)$ of $O(|\Omega(S)| \cdot |\Omega(O)|)$, $P(S_{t+1} \mid S_t)$ of $O(|\Omega(S)|^2)$, and c of $O(|\Omega(S)|)$. The input of our inference queries consists of two vectors, $s_1, \ldots, s_T$ and $o_1, \ldots, o_T$, having size of $O(T)$ each. The computation of $P(o_1, \ldots, o_T)$ only requires to iteratively multiply the result of a previous product and finally store this single floating point value, hence, its space complexity is $O(1)$. Similarly, the factor $P(s_1, \ldots, s_T)$ can be computed by repeatedly storing the result of successive additions and and multiplications in the same memory cel and it has space complexity of $O(1)$. Finally, the execution of the Forward algorithm demands a chunk of memory of $O(|\Omega(S)|)$, again dominating the other two factors.

As a results, the performance of the computation of $P(s_1, \ldots, s_T \mid o_1, \ldots, o_T)$ can be optimized by using *ad hoc* memory components for its three parts:

$$\text{space-compl.} \begin{cases} 1) \text{ model: } O(|\Omega(S)| \cdot |\Omega(O)| + |S|^2) \\ 2) \text{ input: } O(T) \\ 3) \text{ algorithm: } O(|\Omega(S)|) \end{cases} \tag{26}$$

This also means that the memory requirements of the inference algorithm itself do not depend on the time horizon T. In most practical cases, in which $T \gg |\Omega(S)|$, the memory bottleneck will be represented by the memories dedicated to the storage of the input sequences $s_1, \ldots, s_T$ and $o_1, \ldots, o_T$. Figure 4 summarizes the time and space complexity relationships of our algorithm and its input.

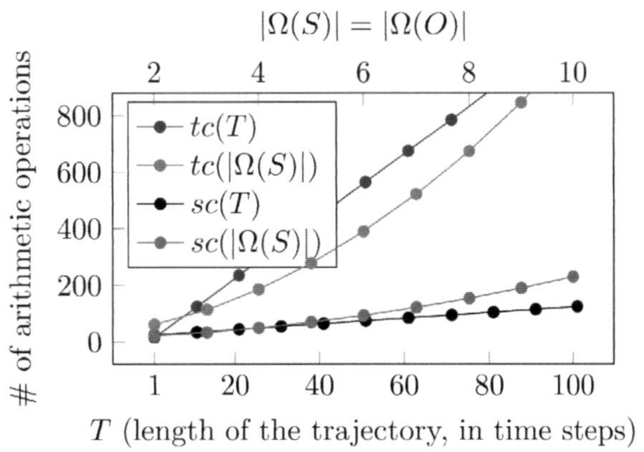

Figure 4. Complexity growth of the inference algorithm with respect to the time horizon T and the size of the state domain $|S|$. In the legend, tc and sc are abbreviations for time-complexity and space-complexity, respectively.

4. Conclusions

In this work, we proposed to formalize the resilient properties of resistance, functionality, and recoverability over the timed probabilistic framework of hidden Markov models. To do so, we extended the HMM definition to include a cost function. Then, we defined an inference algorithm able to answer the queries required for probabilistic property checking over this model. The algorithm has been implemented and tested in the MATLAB-compatible scripting language Octave. Finally, we analyzed the space and time complexity of this inference algorithm, as well as the specific complexity of checking generic properties.

References

[1] Holling C S 1973 *Ann. Rev. of Ecology and Systematics* **4** 1–23

[2] Schwind N, Okimoto T, Inoue K, Chan H, Ribeiro T, Minami K and Maruyama H 2013 Systems resilience: A challenge problem for dynamic constraint-based agent systems *Proceedings of the 2013 International Conference on Autonomous Agents and Multi-agent Systems* AAMAS '13 (Richland, SC: International Foundation for Autonomous Agents and Multiagent Systems) pp 785–788 ISBN 978-1-4503-1993-5

[3] Walker B, Holling C S, Carpenter S R and Kinzig A 2004 *Ecology and Society* **9**(2)

[4] Özveren C M, Willsky A S and Antsaklis P J 1991 *J. ACM* **38** 729–751 ISSN 0004-5411

[5] Baral C, Eiter T, Bjreland M and Nakamura M 2008 *Artificial Intelligence* **172** 1429 – 1469 ISSN 0004-3702

[6] Rabiner L 1989 *Proceedings of the IEEE* **77** 257–286 ISSN 0018-9219

[7] Russell S J and Norvig P 2003 *Artificial Intelligence: A Modern Approach* 2nd ed (Pearson Education) ISBN 0137903952

[8] Gaitanos G C, Williams C, Boobis L H and Brooks S 1993 *Journal of Applied Physiology* **75** 712–719 ISSN 8750-7587

[9] Koller D and Friedman N 2009 *Probabilistic Graphical Models: Principles and Techniques - Adaptive Computation and Machine Learning* (The MIT Press) ISBN 0262013193, 9780262013192

[10] Ross S M 2006 *Introduction to Probability Models, Ninth Edition* (Orlando, FL, USA: Academic Press, Inc.) ISBN 0125980620

[11] Durand D and Hoberman R 2011 Computational genomics and molecular biology, hmm lecture notes Tech. rep. Carnegie Mellon University

A Review of Supply Chain Collaboration Practices for Small and Medium-sized Manufacturers

SY Wee[1], AC Thoo[2], Sulaiman Z[2] and Muharam FM[2]

[1]Doctoral Student, Faculty of Management, Universiti Teknologi Malaysia, Malaysia
[2]Senior Lecturer, Faculty of Management, Universiti Teknologi Malaysia, Malaysia

E-mail: sin88yi@gmail.com

Abstract. For the decades, organizations have endeavored to look for external sources for opportunities to achieve efficient and responsive supply chain with their partners especially for small and medium manufacturers (SMM). In this scenario, supply chain collaboration (SCC) is an interaction between supply chain members with the purpose of utilizes the knowledge and resources of customers and suppliers, and integrates the flows of products and information in order to achieve a common goal and obtain mutual benefit. The essential SCC dimensions for SMMs comprised of information sharing, joint knowledge creation, joint decision making, goal congruence and incentive sharing. The successful implementation of SCC can give SMMs an edge over their competitors. This paper aims to introduce a review of SCC practices for SMM. Overall, the findings provide managerial insights for the SMM in SCC implementation owing to resource scarcity and the need to draw SCC in order to ensure a sustainable competitive advantage.

1. Introduction

Supply chain collaboration (SCC) is defined as "the formation of close, long-term partnerships where supply chain members work together and share information, resources, and risk to accomplish shared objectives" (Cao and Zhang, 2011 pp.166). SCC is one of the most common norms for many companies around the world (Ramanathan and Gunasekaran, 2014). Collaboration encourages all members in the supply chain to participate in planning, information sharing, forecasting, replenishment, incentive sharing and resource sharing (Ramanathan, 2012; Hudnukar et al., 2014). Furthermore, the application of SCC able to cut down the operating costs (Adams et al., 2014), improve in decisions execution (Swink et al. 2007; Kim and Lee 2010), and strengthen the provision of time-and-place utility to the customer (Randall et al. 2010).

In a traditional supply chain, the flow of materials and information are linear and is limited in terms of collaboration and visibility (Qu and Yang, 2015). Each supply chain partner has limited information regarding each other. Therefore, the main focus of SCC is to manage resources and collaborative behavior across partnering organizations in a supply chain effectively (Hudnukar et al., 2014). Nowadays, more and more organizations have begun to coordinate their production and distribution channels and collaborate with supply chain partners instead of managing their internal resources alone (Baraldi et al., 2012).

In the context of small and medium-sized manufacturers (SMMs), cost-effective SCC strategy is critical for their survival and growth. There is a controversy surrounding the benefit which SMMs can derive from SCC. However, studies evidence that SMMs can actually benefit from SCC (Rezaei et al., 2015). For instance, Zeng, et al. (2010) discovered that inter-organization partnership has the most significant positive effect on the firm performance in a study of Chinese manufacturing companies. Furthermore, SCC provides SMMs with access to comprehensive and external expertise that can guide them to resolve business problems and allow them to engage in learning networks. For instance, successful SCC can be shown by market share, market share (Mishra and Shah, 2009) and also the satisfaction of supply chain partners. Therefore, future collaboration projects will mainly based on the success of current collaborative partnership (Ramanathan and Gunasekaran, 2014). In a similar vein, Eyaa et al. (2010) conducted a study in SMMs in Uganda, a developing nation to reveal the perceived benefits of SCC to SMMs. The study shows that information sharing and incentive alignments were found significantly related to supply chain performance. Another study by Katunzi and Zheng (2010) on Tanzania's SMMs showed that collaboration with supply chain partners able to provide local SMMs the chance to make collective agreements in quantity planning, demand and delivery time to customers.

2. Supply Chain Collaboration Practices

Collaboration effort is the main ingredient of SCC that facilitates the flow of communication and cooperative efforts between departments (Flynn et al., 2010a). Hence, a highly collaborative of supply chain involves the collaboration efforts from functional departments, suppliers, customers as well as outbound and inbound logistics to connect and coordinate the flow of raw material supply to end customers. The essential SCC dimensions for SMMs include information sharing, joint decision making, incentive sharing, goal congruence and joint knowledge creation (Simaptung and Sridharan, 2008; Zacharia et al., 2009; Cao and Zhang, 2011; Hudnukar et al., 2014).

2.1. Information Sharing

Information sharing is described as the "nerve center" (Chopra and Meindl, 2007) ,"heart" (Lamming, 1996), "lifeblood" (Stuart and McCutcheon, 1996), , "essential ingredient" (Min et al., 2005), "foundation" (Lee and Whang, 2000) and "key requirement" (Sheu et al., 2006) of SCC. Therefore, information flow is the vital element in any relationship and is the primary flow in a supply chain (Chopra and Meindl, 2007). Information sharing determines the way and magnitude of the material flow such as finished product from suppliers to end user and also the product return and repairs from the end user. The information flow may include of the exchange of transactional data, from product manufacturing, customer feedbacks to products research and development matters.

Information sharing is seen as the "glue" that strengthen the business structure of supply chain partners which allow supply chain to be more agile in dealing with competitive advantages (Hudnukar et al., 2014). Furthermore, since the suppliers are located all over the world, it is essential for the organizations to exchange accurate and timely information between partner organizations in order to achieve mutual goals. Crook et al. (2008) suggested that when independent firms collaborate and share knowledge with others, they can achieve the advantages beyond what could achieve in arm's length exchange. However, organizations often alert of sharing sensitive information within the supply chain. They may concern about the leak of valuable information to their rivals such as demand forecasts, new technology acceptance and new products developments. Therefore, the collaborative supply chain relationship will be able to reduce the chance of information leak. Trust among supply chain partners plays an important role in information sharing through the supply chain (Fawcett et al., 2007).

Haji-Pakir and Alina (2010) discovered that the information sharing process in SMMs is top down which is from the manufacturer to the bottom of the supply chain, but the information is dedicated only to the relevant party. Due to the flat organizational structure of SMMs, the information is able to reach relevant parties more efficient than the highly structured organizations (Grant et al., 2010; Papastathopoulos and Beneki, 2010). In support, Eyaa et al. (2010) found that the SMMs in Uganda easily share information with their supply chain partners by using emails to improve supply chain performance.

2.2. Joint Decision Making

Joint decision making refers to the process of supply chain partners synchronize their decisions making process in supply chain planning and operations with the purpose of supply chain benefit optimization(Simatupang and Sridharan, 2002; 2005). Besides, supply chain decisions include combining plans and information, resolving differences and conflicts and establishing routines, procedures, and rules (Cao and Zhang, 2011; 2013). Some problems occur in decision-making processes when information is widely dispersed or there is unclear authority structure on how the decision should be made with the concern for supply chain member benefits (Hudnukar et al., 2014; Ramanathan and Gunasekaran, 2014). Conventionally, organizations perform business processes individually, but Ding et al. (2011) raised concerns that independent decision-making tends to sub-optimal business performance such as bullwhip effect. Bullwhip effect can happen due to limited knowledge about market demand in each company of a supply chain (Mackelprang et al., 2015). Hence, bullwhip effect causes inaccurate inventory order which leads to unnecessary inventory and total costs (Ding et al., 2011). The activity of joint decision making includes of re-assign the decision rights to sync up supply chain planning and execution that try to fulfil the demand with supply. In addition, an effective decision making is based on supply chain profitability (Corbett et al., 2012) and the rapid response in fulfilling customer demand. The virtual discussion forums and face-to-face meetings are the examples of joint decision making process take part by supply chain partners.

Managers are very often the owners of the SMMs, and most of the decision making is dominated by them (Ghobakhloo et al., 2011). The competence of owner-manager is the most important determinant of an SMM's success (Cocca and Alberti, 2010). However, the studies by Wiengarten et al. (2010) and Eyaa et al. (2010) found that joint decision making may not necessarily lead to improved SMMs performance. Due to the visions, missions, and leadership style of SMMs, sometimes it is not easy to attain the balance of all supply chain members. Therefore, SMMs need align the supervision systems with joint decision making as well as organizational support (Eyaa et al., 2010).

2.3. Incentive Sharing

As highlighted by Simatupang and Sindhran (2005), incentive sharing is a process of risks, costs and benefits sharing among supply chain partners. SCC is a relationship strategy between the supply chain members, development of short-term or long-term planning to achieve lower risk, higher quality, and greater product innovation to enhance performance (Lee and Whang, 2001; Manthou et al., 2004). In the view of successful partnership, the sharing of profits and losses should be equally distributed among the partners. Hence, the supply chain partners need to act in rational ways to achieve the mutually agreed objectives as a collective responsibility (Salmela et al., 2011).

Simatupang and Sridharan (2002) proposed that interdependency between supply chain members can develop in the forms of tasks and tasks, resources and resources, and tasks and resources. From organizational perspective, tasks are viewed as the main activities such as forecasting, planning, purchasing, distributing, replenishing, pricing, paying, and serving. The organizational resources can be referred as inventory, capacity, funds, and capabilities. Despite the fact that the collaboration is based on the collective agreement, but it is still perceived as a self-interest process where supply chain members seek to obtain individual advantages such as eliminating redundant functions, reducing transactional costs, increasing responsiveness, achieving lower inventory, and so forth (Cao et al., 2011).

In order to create an effective incentive sharing scheme, the supply chain partners need to consider three basic questions: how the incentive can be linked with overall performance, how the incentive is to be paid and what level of incentive is to be paid (Scholten, 2015). If the objective of the collaboration is to achieve lower total cost and enhance the profit structure, the partners should select a performance measure based on their objective when determining the incentive (Simatupang and Sridharan, 2005). In fact, the design of incentive sharing scheme should be able to draw attention, encourage, and maintain participating partners in achieving the common outcomes by linking a portion of participating partner's compensation to common outcomes. Jap (2001) discovered that fairness is particularly essential in the allocation of collaboration's outcomes in order to maintain the relationship between supply chain members. The incentive distributed to the supply chain partners should commensurate with the respective investment and risk (Hui et al., 2015).

2.4. Goal Congruence

Nowadays, most of the supply chains are integrated with individual preferences (Prajogo et al., 2012; Wang et al., 2016). Goal congruence refers to the extent in which supply chain members perceive their objectives to be satisfied by the fulfilment of the supply chain objectives (Kohli and Jensen, 2010; Hudnurkar et al., 2014). Goal congruence includes the defining roles and responsibilities of each partner, development of objectives, common goals, performance measures, standardization of information technology, formalization of shared information, and jointly implement the plan (Min et al., 2005).

As highlighted by Poirier and Houser (1993, p.201), "True supplier partnering requires an understanding of each party's needs and capabilities to establish a clear vision for focusing the efforts of people who work for buyer and supplier". A clear strategic goal leads to accomplishment of collaborative agreements. Supply chain members need collectively design the collaboration implementation plans in order to achieve the desire goals (Cao and Zhang, 2011). Additionally, Lambert et al. (1998) raised several concerns about the importance of strategic direction and the business vision of the participating organizations. The study suggested that supply chain partners need to be in line with the SCM vision and key business processes. Supply chain members need to focus on the collaborative relationship to gain the greatest inter-firm improvements and rewards (Krishnapriya and Rupashree, 2014). Hence, without such a roadmap, optimal collaborative results cannot be achieved (Cao and Zhang, 2011). Also, the expectations and needs of the supply chain members need to be taking account into the supply chain strategies and operations to enhance member's profit, cash flow, and return-on-investment (Ramanathan and Gunasekaran, 2012).

2.5. Joint Knowledge Creation

Joint knowledge creation occurs when supply chain members perceived the need for interrelated skills and assets. Generally, there are two types of knowledge creation which are knowledge exploration and knowledge exploitation. Knowledge exploration refers to the search and acquires of new and relevant knowledge. Moreover, knowledge exploitation refers to assimilate and apply relevant knowledge (Bhatt and Grover, 2005). Identify, communicate, and assimilation of knowledge in process, technology, or market knowledge between supply chain members can trigger innovation and long-term competitiveness of the supply chain. It is important to realize that supply chain partners need to identify jointly customer needs. Also, supply chain partners need to build a knowledge base together such as disseminating and sharing interpretation that enable organizations to create new values such as responding to customer's needs, building brand image, developing new products and establishing channel relationship (Slater and Naver, 1995; Johnson and Sohi, 2003; Luo et al., 2009).

Managing dependency in a collaborative partnership has become a critical process. Organizations and its supply chain partners need to learn jointly the intentions and capabilities of their competitors (Cao and Zhang, 2011). The learning-based relationship between supply chain partners can maintain supply chain relationships and achieve the firm's goals. Scholten (2015) found that the jointly created knowledge contributes to more supply chain resilience. The importance of shared learning among the supply chain members can increase the search and acquire suitable knowledge, which in turns enhance the visibility of shared information, avoid disruptions in downstream and upstream of a supply chain.

3. Conclusion

An integrated SCC is needed to form in the supply chain network in order to reduce the perennial supply chain challenges such as functional silos, poor transparency of knowledge and information and the inadequate formation of appropriate upstream and downstream relationships. For the most part, SCC plays a key role in stimulating organizational performance (Cao and Zhang, 2011; Ha *et al.*, 2011). SMMs are encouraged to utilize information sharing on defining mutual objectives and associated performance measures and link their performance systems with incentive sharing scheme to compensate the partners fairly. The clear linkage between SMMs will encourage their partners to improve shared supply chain processes that benefit the supply chain. Hopefully this paper will provide insight to researchers and practitioners to have better understanding towards the effectiveness and efficiency of SCC.

References

[1] Adams FG, Richey, RG, Autry, CW, Morgan, TR, and Gabler, CB 2014 Supply chain collaboration, integration, and relational technology: how complex operant resources increase performance outcomes *Journal of Business Logistics* **35** pp 299-317
[2] Baraldi, E, Gressetvold, E, and Harrison, D 2012 Resource interaction in inter-organizational networks: foundations, comparison, and a research agenda *Journal of Business Research* **65** pp 266-276
[3] Bhatt, GD, Grover, V, and Grover, V 2005 Types of information technology capabilities and their role in competitive advantage: an empirical study *Journal of Management Information Systems* **22** pp 253-277
[4] Cao, M, and Zhang, Q 2011 Supply chain collaboration: impact on collaborative advantage and firm performance *Journal of Operations Management* **29** pp 163-180
[5] Chopra, S, and Meindl, P 2007 Supply chain management: strategy *Planning and Operation Upper Saddle River*
[6] Cocca, P, and Alberti, M 2010 A framework to assess performance measurement systems in SMEs *International Journal of Productivity and Performance Management* **59** pp 186-200
[7] Corbett, CJ, Blackburn, JD, and Van Wassenhove, LN 2012 Partnerships to improve supply chains *Sloan Management*
[8] Crook, TR, Giunipero, L, Reus, TH, Handfield, R, and Williams, SK 2008 Antecedents and outcomes of supply chain effectiveness: an exploratory investigation *Journal of Managerial Issues* **20** pp 161-177
[9] Ding, H, Guo, B, and Liu, Z 2011 Information sharing and profit allotment based on supply chain cooperation *International Journal of Production Economics*, **133** pp 70-79
[10] Eyaa, S, Ntayi, JM, and Namagembe, S 2010 Collaborative relationships and SME supply chain performance *World Journal of Entrepreneurship, Management and Sustainable Development* **6** pp 233-245
[11] Fawcett, SE, Osterhaus, P, Magnan, GM, Brau, JC, and McCarter, MW 2007 Information sharing and supply chain performance: the role of connectivity and willingness *Supply Chain Management: An International Journal* **12** pp 358-368.
[12] Flynn, BB, Huo, B and Zhao, X 2010 The impact of supply chain integration on performance: a contingency and configuration approach *Journal of Operations Management* **28** pp 58-71
[13] Ghobakhloo, M, Arias-Aranda, D, and Benitez-Amado, J 2011 Adoption of e-commerce applications in SMEs *Industrial Management & Data Systems* **111** pp 1238-1269
[14] Grant, K, Hackey, R and Edgar, D 2010 *Strategic Information Systems Management* Singapore: Seng Lee Press
[15] Ha, BC, Park, YK, and Cho, S 2011 Suppliers' affective trust and trust in competency in buyers: its effect on collaboration and logistics efficiency *International Journal of Operations & Production Management* **31** pp 56-77
[16] Haji-Pakir, MI, and Alina, S 2010 Level of supply chain collaboration of Malaysian SME

manufacturers In *Management of Innovation and Technology (ICMIT), 2010 IEEE International Conference on* pp 169-174

[17] Hudnurkar, M, Jakhar, S, and Rathod, U 2014 Factors affecting collaboration in supply chain: a literature review *Procedia-Social and Behavioral Sciences* **133** pp 189-202

[18] Hui, Z, He-Cheng, W, and Min-Fei, Z 2015 Partnership management, supply chain collaboration, and firm innovation performance: an empirical examination *International Journal of Innovation Science* **7** pp 127-138

[19] Jap, SD 2001 Perspectives on joint competitive advantages in buyer–supplier relationships *International Journal of Research in Marketing* **18** pp 19-35

[20] Johnson, JL, and Sohi, RS 2003 The development of interfirm partnering competence: Platforms for learning, learning activities, and consequences of learning *Journal of Business Research* **56** pp 757-766

[21] Katunzi, TM, and Zheng, Q 2010 Tanzanian SMEs' perceptions towards adoption of supply chain management (SCM) strategy *International Journal of Business and Management* **5** p 42

[22] Kim, D, and Lee, RP 2010 Systems collaboration and strategic collaboration: their impacts on supply chain responsiveness and market performance *Decision Sciences* **41** pp 955-981

[23] Kohli, AS, and Jensen, JB 2010 Assessing effectiveness of supply chain collaboration: an empirical study In *Supply Chain Forum: An International Journal* **11** pp 2-16

[24] Krishnapriya, V, and Baral, R 2014 Supply chain integration-a competency based perspective *International Journal of Managing Value and Supply Chains* **5** pp 45

[25] Lambert, DM, Cooper, MC, and Pagh, JD 1998 Supply chain management: implementation issues and research opportunities *The international journal of logistics Management* **9** pp 1-20

[26] Lamming, R 1996 Squaring lean supply with supply chain management *International Journal of Operations & Production Management* **16** pp 183-196

[27] Lee, HL, and Whang, S 2000 Information sharing in a supply chain *International Journal of Manufacturing Technology and Management* **1** pp 79-93

[28] Lejeune, MA, and Yakova, N 2005 On characterizing the 4 C's in supply chain management *Journal of Operations Management* **23** pp 81-100

[29] Luo, X, Wu, C, Rosenberg, D, and Barnes, D 2009 Supplier selection in agile supply chains: An information-processing model and an illustration *Journal of Purchasing and Supply Management* **15** pp 249-262

[30] Mackelprang, AW, and Malhotra, MK 2015 The impact of bullwhip on supply chains: Performance pathways, control mechanisms, and managerial levers *Journal of Operations Management* **36** pp 15-32

[31] Manthou, V, Vlachopoulou, M, and Folinas, D 2004 Virtual e-Chain (VeC) model for supply chain collaboration *International Journal of Production Economics* **87** pp 241-250

[32] Min, S, Roath, AS, Daugherty, PJ, Genchev, SE, Chen, H, Arndt, AD, and Glenn Richey, R 2005 Supply chain collaboration: what's happening? *The international journal of logistics management* **16** pp 237-256

[33] Mishra, AA, and Shah, R 2009 In union lies strength: collaborative competence in new product development and its performance effects *Journal of Operations Management* **27** pp 324-338

[34] Papastathopoulos, A, and Beneki, C 2010 Organizational forms based on information & communication technologies (ICTs) adoption *Research in Business and Economics Journal* **2** pp 1-18

[35] Poirier, CC, and Houser, WF 1993 Business partnering for continuous improvement: How to forge enduring alliances among employees, suppliers & customers Berrett-Koehler Publishers

[36] Prajogo, D, and Olhager, J 2012 Supply chain integration and performance: The effects of long-term relationships, information technology and sharing, and logistics

integration *International Journal of Production Economics* **135** pp 514-522
[37] Qu, WG, and Yang, Z 2015 The effect of uncertainty avoidance and social trust on supply chain collaboration *Journal of Business Research* **68** pp 911-918
[38] Ramanathan, U 2012 Supply chain collaboration for improved forecast accuracy of promotional sales *International Journal of Operations & Production Management* **32** pp 676-695
[39] Ramanathan, U, and Gunasekaran, A 2014 Supply chain collaboration: Impact of success in long-term partnerships *International Journal of Production Economics* **147** pp 252-259
[40] Randall, WS, Pohlen, TL, and Hanna, JB 2010 Evolving a theory of performance-based logistics using insights from service dominant logic *Journal of Business Logistics* **31** pp 35-61
[41] Rezaei, J, Ortt, R, and Trott, P 2015 How SMEs can benefit from supply chain partnerships *International Journal of Production Research* **53** pp 1527-1543
[42] Salmela, E, Happonen, A, and Huiskonen, J 2011 Best collaboration practices in supply chain of technical wholesale items *International Journal of Collaborative Enterprise* **2** pp 16-38
[43] Scholten, K, and Schilder, S 2015 The role of collaboration in supply chain resilience *Supply Chain Management: An International Journal* **20**
[44] Sheu, C, Rebecca Yen, H, and Chae, B 2006 Determinants of supplier-retailer collaboration: evidence from an international study *International Journal of Operations & Production Management* **26** pp 24-49
[45] Simatupang TM and Sridharan ,R 2008 Design for supply chain Collaboration *Business Process Management Journal* **14** pp 401-418
[46] Simatupang, TM, and Sridharan, R 2002 The collaborative supply chain *The International Journal of Logistics Management* **13** pp 15-30
[47] Simatupang, TM, and Sridharan, R 2005 The collaboration index: a measure for supply chain collaboration *International Journal of Physical Distribution & Logistics Management*, **35** pp 44-62
[48] Slater, SF and Narver, JC 1995 Market orientation and the learning organization *Journal of Marketing* **59** pp 63-74
[49] Stuart, FI, and McCutcheon, D 1996 Sustaining strategic supplier alliances: profiling the dynamic requirements for continued development *International Journal of Operations & Production Management* **16** pp 5-22
[50] Swink, M, Narasimhan, R, and Wang, C 2007 Managing beyond the factory walls: effects of four types of strategic integration on manufacturing plant performance *Journal of Operations Management* **25** pp 148-164
[51] Wang, Q, Zhao, X, and Voss, C 2016 Customer orientation and innovation: A comparative study of manufacturing and service firms *International Journal of Production Economics* **171** pp 221-230
[52] Wiengarten, F, Humphreys, P, Cao, G, Fynes, B, and McKittrick, A 2010 Collaborative supply chain practices and performance: exploring the key role of information quality *Supply Chain Management: An International Journal* **15** pp 463-473
[53] Zacharia, ZG, Nix, NW, and Lusch, RF 2009 An analysis of supply chain collaborations and their effect on performance outcomes *Journal of Business Logistics* **30** pp 101–123
[54] Zeng, WJ, and Ma, SH, 2010 The Impact of Supply Chain Relationship Dynamics on Collaboration *Journal of Industrial Engineering and Management* **2**

Evaluation of C/C-SiC Composites as Potential Candidate Materials for High Performance Braking Systems

Rahmat Saptono Duryat

Universitas Indonesia, Kampus UI Depok, Indonesia 16424

E-mail: rahmat.saptono@ui.ac.id

Abstract. This paper is aimed at evaluating the characteristic and performance of C/C-SiC composites as potential candidate materials for high performance braking system. A set of material specifications had been derived from specific engineering design requirements. Analysis was performed by formulating the function(s), constraint(s), and objective(s) of design and materials selection. Function of a friction material is chiefly to provide friction, absorb and dissipate energy. It is done while withstanding load and maintaining the structural adequacy and characteristic of tribology at high temperature. Objective of the material selection and design is to maximize the absorption and dissipation of energy and to minimize weight and cost. Candidate materials were evaluated based on their friction and wear, thermal capacity and conductivity, structural properties, manufacturing properties, and densities. The present paper provides a state of the art example on how materials – function – geometry – design, are all interrelated.

1. Introduction

Braking system is one of the most important parts in mechanical design. Challenging engineering design requirements along with such other critical issues as safety, liability, health, environment, and energy, has driven the development of new friction materials. On the other hand, the discovery of friction materials, which is also pushed by science and technology, has promoted the development friction system design. The development of friction materials is strongly related to the design and applications.

One of the major applications of friction materials is in transportation vehicles. Development of advanced friction materials was essentially 'pulled' by the need of aerospace and military industries that are considered as the frontiers in advanced technology. Carbon/Carbon composites, which were attributed to Brennan Forcht of Chance Vought Aircraft, were initially developed in 1958 to respond the need of US space and defense industries[1]. Carbon/Carbon composites, which were particularly developed to meet a set of material specifications dictated by specific design objectives, then had become one of the important friction materials for lightweight and high performance applications.

Use of Carbon/Carbon composite materials in commercial passenger aircraft braking systems dates back to 1973 when the materials, which were provided by Dunlop, went to the first trial on VC10 aircraft before becoming a standard attachment on supersonic passenger aircraft Concorde in 1974[1,2]. The technology, which is formerly directed to advanced applications, has been recently adopted and developed for a wider range of engineering applications. By the end of the 21th century, about 60% volume of Carbon/Carbon composites had been used in Aircraft Braking System in more than 50 types of aircrafts[1]. Recently, the materials have become the prime choice for aircrafts and other high performance braking systems involving military aircrafts, commercial aircrafts, and racing cars[1-4].

Carbon/Carbon composites are basically a carbonaceous matrix composite reinforced by carbon fiber. Prior to further discussion it is important to specify the carbon materials related to the braking system. According to Blanco[4] carbon materials is best described as the defective form graphite that, along with diamond, are the dominant crystalline form of carbon. Blanco[4] classified carbon materials into several groups based upon its structure. Graphitic carbon materials are carbon materials that have a three dimensional graphite lattice. Non-graphitic carbon materials, which do not have a three dimensional graphite lattice, can be divided into graphitizable and non-graphitizable carbon materials. The graphitizable carbon materials transform into three dimensional graphite lattices when it is heat treated up to 3000°C. Non graphitizable carbon materials are isotropic whereas the graphitizable ones are non-isotropic. Referring to Blanco[4] the key different between the two categories is in the mode of relative arrangement of the constituent graphite layers. In the isotropic, non-graphitizable, carbon materials, the graphene layers are smaller and arranged relative to each other and bonded to each other in all directions. In graphitizable, anisotropic, carbon materials, the graphene layers are larger and arranged parallel to each other and bonded to each other.

Graphitic carbon materials are very attractive for aerospace and other rigorous applications due to such a unique combination of properties that are not demonstrated by other materials. Windhorst[1] revealed such important properties associated with carbon materials as low density, high thermal conductivity and shock resistance, low thermal expansion and high modulus. Blanco[4] specified more detailed characteristics of graphitic carbon materials related to braking systems. Graphitic carbon materials have a high melting temperature, high thermal conductivity, high thermal capacity, and stable coefficient of friction. The materials provide interfacial 'debris' that creates frictional surface and serve simultaneously intrinsically as frictional and structural materials. It was also revealed that the materials do not 'fatigue' as understood in metal.

Properties of Carbon Materials and hence Carbon/Carbon Composites materials are highly dependent on manufacturing routes, raw materials, processes variables, as well as reinforcements, treatments, and other modifications. Windhorst[1] has revealed some factors that affect the quality of the materials involving the quality of polymer matrix composite from which Carbon/Carbon composites are made, the choice of pitch that affects the carbon yield, the choice of carbon fiber, the microstructure of meso-phase (pitch-based), the wave pattern of carbon fabric, the matrix bond strength, carbonization method/medium, and surface treatment of carbon fibers.

Structural integrity and properties at higher temperature is one of the most important attributes for the designed applications. Windhorst[1] revealed that the materials capable of retaining structural integrity at above 1000°C and withstanding temperature in excess of 2000°C without major deformation. It was also revealed that the mechanical properties are much superior compared to the conventional graphite as it can be tailored in 3-D to withstand damage and exhibit minimum delamination crack growth under interlaminar shearing. Referring to Windhorst[1] heat treatment sequences that involve carbonizing at 1000°C followed by graphitization at 2700°C, result in a 54% increase in flexural strength, 40% decrease in the interlaminar strength, and a 93% increase in the flexural modulus. It was also revealed that the choice of graphitization temperature will affect the toughness of composites. In general, compressive properties of composites are dictated by the carbon matrix whereas the tensile properties are related to the reinforcement. The other important characteristics revealed by Windhorst[1] are the increase of tensile strength at 1200°C and the non-catastrophic, gradual, graceful failure.

Despite of its engineering excellence carbon materials exhibit some important limitations. Of the major drawbacks are that the material suffers oxidation at temperature above 320°C[1] and burning away at 500°C[5]. This situation will have a serious consequence to the fiber matrix interface quality and the strength of bundles[1]. Extensive discussion about the carbon gasification reactions, structure and surface oxygen formation, and properties of surface oxygen complexes was presented by Blanco[4] aimed at giving clear understanding about such conditions affecting brake performance as the presence of air (oxygen), carbondioxide, water vapor, atmospheric and operating temperatures, and intensity of aircraft use.

Insufficient friction stability of Carbon/Carbon composite caused by humidity and temperature along with low wear resistance has also important become issues. Chen[6] studied the effects of humidity on the tribology behavior of Carbon/Carbon composites made from different types of fiber and revealed the close relationship among humidity, surface morphology, and tribology. Gomes[7] studied the effects of temperature on the tribology behavior of Carbon/Carbon composite and denoted the probable change of this behavior at high temperature caused by fiber debonding and fracture. Krenkel[3] revealed that these conditions prevent the practical use carbon materials in passenger cars and trains or emergency breaks of cranes or lifts.

Efforts have been done to improve the properties and applications of Carbon/Carbon composites. In 1980s the development was mainly focused on increasing the specific strength and strain to failure, increasing static friction coefficient, and reducing the manufacturing cost[2].

Oxidation has become critical issues in the development of Carbon/Carbon composites materials as the materials are designed to experience oxidizing atmosphere at elevated temperature either for short term (high temperature) or long term (low temperature). Many efforts have been performed to deal with this critical issue. Bacos[8] provided a comprehensive lecture on the protection of Carbon-based Composites. Referring to Bacos[8], protection mechanisms, which are highly dependent on the range of temperature, can be divided into 2 approaches. At low temperature, microscopic approach is taken by using poisoning agents which adsorb on the active sites responsible for the carbon gasification. At high temperature, macroscopic approach is selected by the use of a barrier that prevents oxygen diffusion. Further challenge came to the second approach as the mismatch between the substrate and the layer that behave differently at high temperature result in cracks. Bacos[8] reviewed further approaches to deal with the oxidation and cracks on coating including intermediate layer, inhibited matrix, and fiber protection. It was identified the need of oxidation protection systems that are moisture insensitive and able to work above 1800°C for long term applications. Bacos[8] revealed the possibility of graded coatings and mixed layers to fulfill the need. Functionally graded coating design is one of the solutions that have been developed until recently to protect the C/C composites from the oxidation[8].

A major breakthrough came to pass in early 1990s when C/C-SiC composites were investigated as potential braking materials[3]. In transportation, the potential applications of these ceramic composites involve automotive, aircrafts, high speed trains, as well as emergency brake for lifts and cranes[10].

This paper is aimed at evaluating the characteristic and performance of C/C-SiC composites as the potential candidates for high performance braking system. This paper deals with the general characteristic and performance of C/C-SiC composites, which are particularly related to the engineering design requirements of high performance braking systems.

2. Design Requirements of A Braking System

In general, braking system in transportation consists of **a rotating part (rotor)** clamped by **static parts (stators)**. Braking action takes place when the stators, driven by hydraulic pressure, press the rotor. Specifications of advanced braking system vary from applications to applications. Krenkel[3] has revealed the characteristic of different transportation systems dictating the specification of correspondence breaking systems. In general, their characteristics can be compared by Energy per Brake Disk (MJ) as depicted in Figure 1.

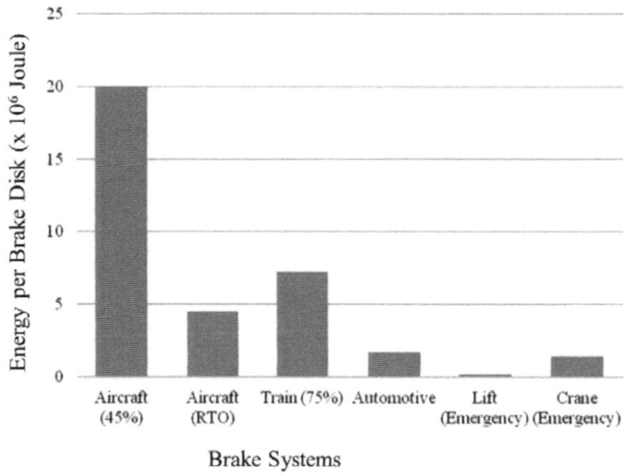

Figure 1. Brake System Characteristics of Various Transportations represented by Energy per Brake Disk based on data available from Krenkel[3].

Selection and development of materials for a particular engineering application must be started with the understanding of design. Ashby[11] has suggested a selection strategy to match materials and processes attributes with design requirements. **The most critical step of material selection and development is to translate design requirements into a set of material specifications**. Referring to Ashby[11], design requirements can be systematically formulated by identifying function(s), constraint(s), and objective(s) of design and material selection. The task is effectively done at the component or sub component level where function(s) and constraint(s) can be defined clearly. The results of analysis were summarized in Figure 2.

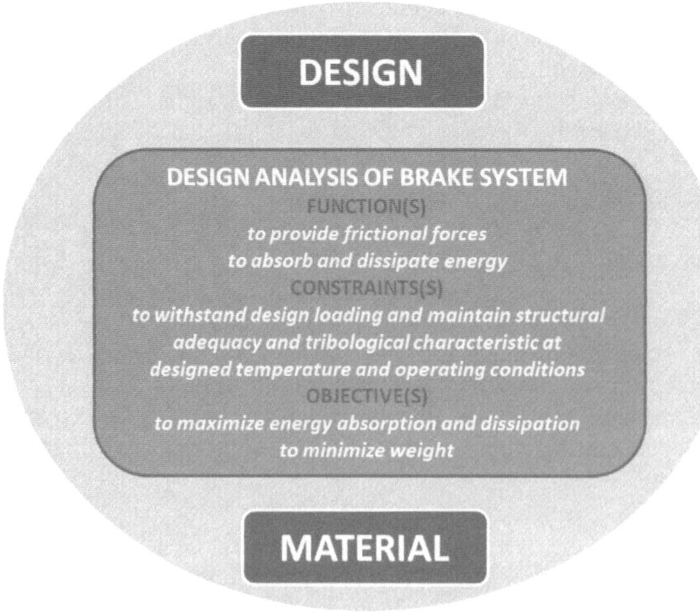

Figure 2. Design Analysis of a Brake System for Materials Specifications

Multifaceted design requirements of high performance braking systems can be learned from an aircraft braking system. Stimson[2] provided a useful background about the design of Concorde aircraft braking system involving working principles and loading conditions, which dictates the selection and development of Carbon/Carbon composites. General configuration and basic components of an aircraft braking system, can be learned from the section of Concorde Wheel and Carbon Brake provided by Stimson[2] (Figure 3 and Figure 4).

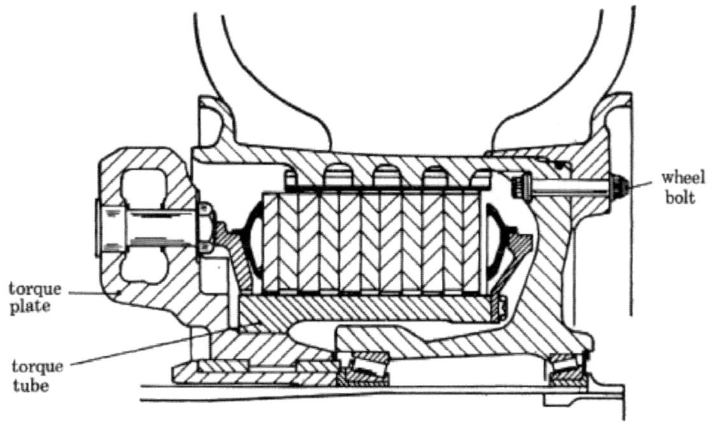

Figure 3. Concorde Wheel (After Stimson[2])

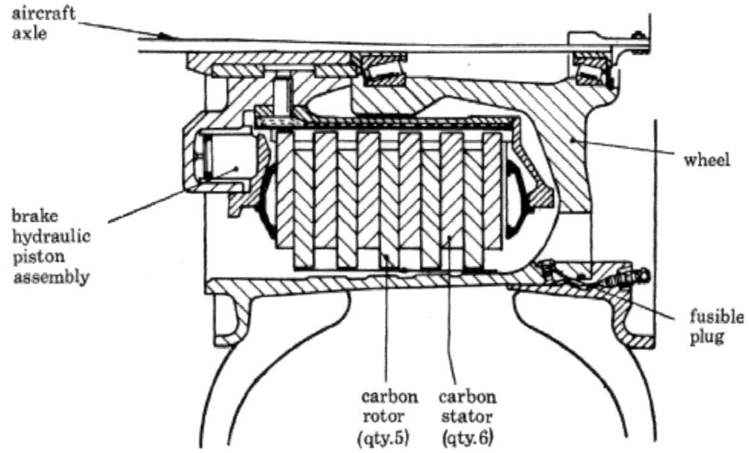

Figure 4. Concorde Carbon Brake (After Stimson[2])

Function of braking system is basically to stop the moving parts of a mechanical system. The primary function of braking materials is (1) to provide frictional forces as well as (2) to absorb and dissipate energy that is converted from the kinetic energy. This so called friction materials must be able to carry the functions while withstanding design loading and maintaining its structural adequacy and the interface characteristics of interacting surfaces in motion at high temperature. The objective of the material selection is to maximize the absorption and dissipation of energy. The selection is also aimed at minimizing weight and cost.

Constraints of the design vary depending on the operating conditions. As an illustration, for Concorde brake design, the disk temperature can be 500°C in normal landing or up to 1300°C in cancelled take off condition[2]. The absorbed energy per disc also varies, depending on the characteristic of the transportation system. With respect to the structural adequacy and integrity, mechanical, thermal, and tribology characteristics must be all together analyzed to identify both bulk and surface stresses material must withstand. A simple qualitative analysis of mechanical loading can be performed based on Stimson[2]. Applied and reactive forces that work on rotor and stator are illustrated in Figure 5. The vertical loading as illustrated in Figure 5 produces deflection on the wheel that must be accommodated in the design. However, the thermal movements of the disk are considerably small. One of the most challenging problems in the design of aircraft braking system is to calculate temperature and thermal stress. Analytical solution of the transient heat conduction problem of friction during the braking for the pad sliding over the infinite disc has been obtained by Yevtushenko[12]. Temperature field and thermal stresses for the friction couple metal-ceramic pad and cast-iron disc were studied. Effects of boundary conditions on the pad at upper surface on the distribution of temperature and thermal stresses were investigated. It was revealed that the obtained solution can be used to model the thermal cracking of the frictional element during braking.

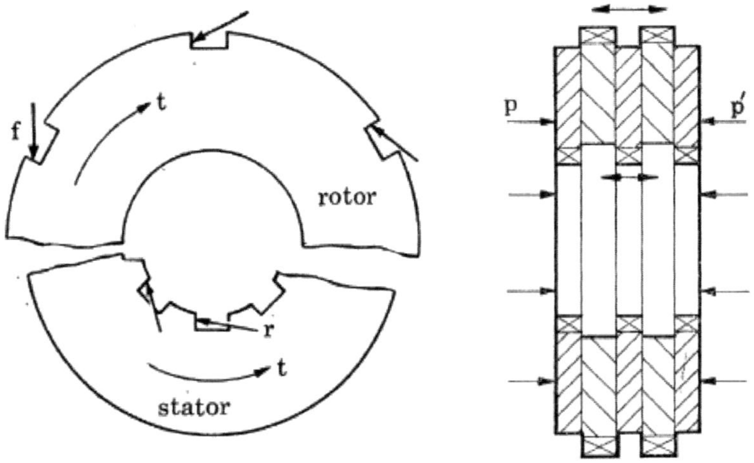

Figure 5. Concorde Carbon Brake (After Stimson[2])

Stimson[2] has specified a set of properties for disk materials in an aircraft braking system. They are (1) high thermal capacity, (2) good strength, impact resistance, and strain to failure, (3) adequate and consistent friction characteristics, and (4) high thermal conductivity. Krenkel[3] has also revealed the important characteristics and performance of materials for advanced braking system in automotive breaking system involving: (1) stable dynamic and static coefficients of friction, (2) high wear stability for lifetime brake systems, (3) low weight to reduce the *unsprung* mass of the transportation system, (4) high degree of freedom in the structural design (e.g. for internal cooling ducts, attachments), and (5) low life cycle costs.

Thermal properties are very critical for high performance braking application. Thermal capacity and thermal conductivity has been identified as the major design criteria for advanced braking systems. Thermal capacity, the quantity of heat necessary to produce a unit change of temperature in a unit mass of a substance, as well as thermal conductivity, the measure of the ability of material to conduct heat are very important. With respect to thermal conductivity, Krenkel[3] revealed that the ability of material to transfer the heat in transverse direction from the outer region to the centre is critical to prevent the friction surface from overheating. Moreover, thermal shock resistance is also important for the braking materials.

Based on the discussion a set of material specifications for braking can be defined as follow: (1) coefficient of friction, (2) wear resistance, (3) thermal capacity, (4) thermal conductivity, (5) thermal shock resistance, (6) resistance to environment (oxidation and degradation) (7) modulus of elasticity, (8) strength, (9) strain to failure, (10) density, (11) manufacturability, (12) cost.

3. Characteristics and Performance of C/C-SiC Composites

Carbon/Carbon-SiC composites are basically a ceramic composite of SiC matrix with Carbon reinforcements. The properties of these materials vary with the variations of their constituents and processing methods. Carbon fiber decreases the brittleness of SiC matrix and hence improving the damage tolerance of the materials. On the other hand, SiC increases thermal stability and resistance to severe environment compared to pure C/C composites materials. The new combinations, therefore, offer the possibility of improving usual C/C composites.

Several routes can be employed to manufacture these composites materials. One of the long established methods is the infiltration of molten Silica into porous Carbon/Carbon composites. The method is known as Liquid Siliconization Infiltration(LSI) or Liquid Phase Infiltration (LPI)[3,5,10].

Krenkel[10] described the LSI process for Aerospace applications. Firstly, preforms of Bidirectionally CFRP of 40-50 MPa are pyrolized at 900°C to 1650°C under inert N_2 atmosphere. This is aimed at converting the polymer matrix precursor into amorphous carbon. At 500°C fiber/matrix debonding start to occur and with the increase of temperature the matrix will crack to relax the stress caused by shrinkage. A translaminar microcrack pattern of dense C/C segments of about 300-500 individual fibers is formed in the materials that are subsequently infiltrated with molten Silicon at about 1600°C. The dense carbon matrix inside the bundles acts as a shielding from the highly reactive Silicon. SiC is formed around the segments with only a small fraction of fiber is converted or damage.

Another common method is Chemical Vapor Infiltration (CVI). Specific techniques are developed to achieve a particular geometry and properties required by design. A brief review on CVI process was revealed by Liu[5] involving the isothermal and the forced flow thermal gradient process. In CVI, the starting materials are porous fiber preforms. Solid are deposited within the network of pores as the substrate is heated and a heterogeneous chemical reaction between gaseous species takes place. Conventional process is limited by the thickness of materials as the reactants will normally take long time to diffuse.

General comparison of mechanical and thermal properties of experimentally and commercially available brake materials are represented in Table 1 based on data available from Stimson[2] and Krenkel[3]. It should be noted that the properties of composites are highly dependent on the composition of its constituents, their configurations, processing, pre-treatments, and post treatments. Demanding sample preparation and tough testing procedures as well as the lack of generally accepted standards will result in a broad variations of measured values and hence difficulties in the evaluation.

Table 1 Thermal and Mechanical Properties of Existing Brake Materials*

Properties	C/C-SiC	C/C Composites	Steel	Copper
Specific Heat (J/(gK)	0.6-2.2	1.42	0.59	0.42
Thermal Conductivity (J/(m.s.K)	7-40	10-150	59	346
Coefficient of Linear Expansion (10^6/K)	0.5-6.5	0-8	14	18
Tensile Strength (MPa)	65-370	66	410	240
Impact Resistance (J)	-	0.7	110	55
Strain to Failure (%)	0.25-0.3	0.55	33	40

*based on Data available from Stimson[2] and Krenkel[3]

3.1. Tribology Properties

Tribology at the interface is very important in determining the performance of friction materials in a high performance braking system. It is generally known that the modification of composite structures will results in the change of the materials tribology at the interface. The phenomena at the interface have been studied extensively but not many results on full-scale test are published openly. Also, the fundamental knowledge about the relationship between materials and their tribology characteristics are not established[11].

Study on the tribology of C/C-SiC composed of the different types of fibers was recently published by Fan and his co-workers[14]. The relationship between structures and tribology properties of aircraft braking materials made from non-woven fiber cloth, short fiber web, and needle fibers are studied from the full scale dynamometer test simulating operating conditions of aircraft braking system. Coefficients of friction at normal landing, overload landing, and rejected take off were reported as 0.27 ± 0.02, 0.26 ± 0.01, and 0.24, respectively. The interactions and mechanism at the interface were studied at micro and sub micro levels. It was revealed that SiC and Si were mostly distributed in the short fiber web layers. Fan[14] also reported that Nano-SiC helps the formation of friction film whereas Micro-SiC enhances the ploughing action of the debris.

Study on wear mechanism of C/C-SiC at a laboratory scale test has also been reported. The most recent works are published by Fan and his co-workers[15]. It was reported that the main wear mechanism were grain-abrasion, oxidation-abrasion, fatigue wear, and adhesive wear, which simultaneously taken place and mutually enhanced the interacting mechanisms. The interaction between C/C-SiC and other braking materials has also been studied. Stadler[16,17] has reported the frictional and wear behavior of sintered metal matrix composites pads and linings in contact with C/C-SiC composites brake discs.

3.2. Thermal Properties

It has been discussion previously that the thermal capacity and conductivity are very important characteristics of C/C-SiC friction materials. It was also revealed that the increase of temperature due to the absorption and dissipation of energy converted from the kinetic energy will influence the temperature distribution and thermal stress at the surface and hence the stability of tribology performance and structural adequacy of friction materials. Moreover, insufficient conductivity together with low thermal expansion coefficient and low stiffness will cause a thermal shock. Therefore, one of the aims of material design is to absorb and dissipate the energy as many as possible away from the matched surface.

Krenkel[3] has revealed the importance of transverse thermal conductivity in the design of materials and geometry for braking system. It was generally assumed that transverse conductivity will reduce the areal braking performance locally resulting in more stable coefficient of friction. One of the advantages of composite material is the ability to tailor the properties in a particular direction. C/C-SiC composites that are generally made from bi-directionally woven fabrics stacked together provide possibilities to be designed for improved thermal conductivity. Krenkel[3] revealed three different possibilities to increase transverse conductivity involving: (1) using fiber with higher thermal conductivity, (2) increasing the angle between fiber and friction surface, and (3) reducing the ceramic content in the C/C-SiC materials.

3.3. Structural Properties

Structural adequacy is very critical in designing materials for braking system and becoming more important as it is the intrinsic properties that cannot be modified locally as what can be done with the surface characteristics. Basically, ceramics is not suitable for structural applications due to its low fracture toughness. Poor tensile behavior, low strain to failure and low tolerance to damage contributes to this behavior. Carbon reinforcement fiber has opened the possibility of using ceramics for structural materials in braking system. It had been discussed that the mechanical properties of

C/C-SiC composites is dependent on the raw materials, manufacturing route, and processing parameters.

Despite the mechanical properties are better than that of monolithic ceramics, Despres[18] revealed the unusual tensile stress and strain behavior of these promising materials as represented in Figure 6. Referring to Despres[18], material exhibits elastic behavior from 0-A that is dependent on the fraction volume of fiber/matrix. At point A, failure occurs in the matrix due to its lower stress bearing capacity compared to the fiber strength. As the failure disseminates, the matrix is extended at high strain value without an increase of stress resulting in a plateau behavior A-B. At B, multiple fractures are maximized and the slope of the curve is mainly dictated by the modulus of fiber assuming that fiber/matrix bond is weak. Then, the slope of the curve increases along B-C.

When the strain to rupture value of the fibers is reached, they start to collapse one after another from C to D. According to Despres[18], the fact that fibers response successively instead of simultaneously are not expected. To improve the quality of the composites, Despres[18] revealed the necessity of adding the third component i.e. the interphase of C/C-SiC composites.

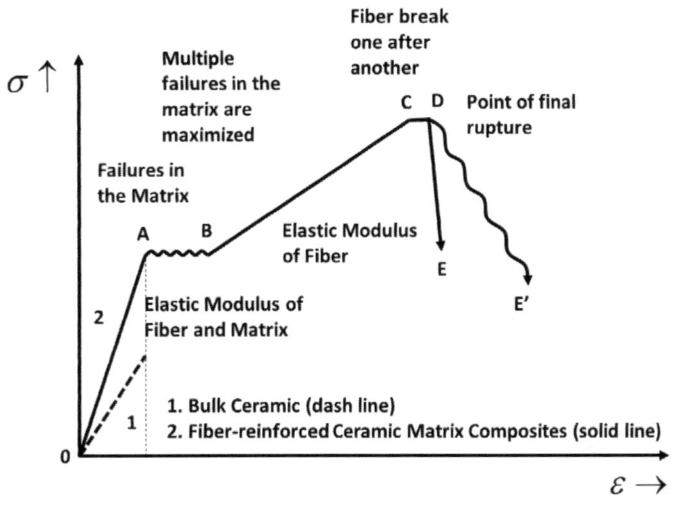

Figure 6. Sketch of the tensile stress/strain behaviour of a bulk ceramic (curve 1) and a fiber-reinforced ceramic matrix composite (curve 2) (After Despres[18])

Camus[19] has studied the damage mechanisms and associated mechanical behavior of a 2D C/C-SiC composite made from CVI process when it is subjected to tension and compression at room temperature. It was reported that the material exhibits an extended non-linear stress/strain related to a multistep development of damage comprising transverse matrix micro-cracking, bundle/matrix and inter-bundle debonding as well as thermal residual stress release. In compression, after initially non-linear behavior involving thermal micro-cracks, the composite material behaves linearly until failure.

With respect to the damage mechanism, Camus[19] proposed three or four development stages involving: (1) an elastic response coupled with partial re-opening of thermal micro-crack, (2) multiple matrix micro-cracking perpendicular to the applied loading, (3) crack opening and related fiber/matrix and mostly bundle/matrix and inter-bundle debonding with (4) progressive transfer of load to the fibers that the failure accordingly until composite failure.

In general, it is learned from the previous works that the structural performance of the C/C-SiC composite materials are generally sensitive to the nature of fibers and interfaces as well as the type and direction of loadings.

4. Conclusion

Selection of friction materials and their development are dictated by the evolving requirements of braking system design. C/C-SiC composites have been developed to respond the limits of established friction materials. A set of material specifications had been derived from design requirements and employed to evaluate candidates. Analysis was accomplished by formulating the function(s), constraint(s), and objective(s) of design and materials selection. Primary function of a friction material is to provide friction, absorb and dissipate energy, while withstanding design loading and maintaining its structural adequacy and tribology characteristic at high temperature. The objective of the material selection is to maximize the absorption and dissipation of energy. In addition, it is aimed at minimizing weight and cost.

The candidate materials were evaluated based on their friction and wear characteristics, thermal capacity and conductivity, structural properties, manufacturing properties, and densities. Cost along with such other local and global issues as safety, liability, health, environment, and energy should also be considered in the final selection of materials but beyond the scope of this paper. Tribology, thermal, and load bearing performance of materials for a particular braking system can be further enhanced through the improvement of material/composite design and processes as well as final geometry modification.

Development of Carbon/Carbon Composites Materials for High Performance Braking System provided a state of the art example on how materials – function – geometry – design, are all interrelated.

5. References

[1] T. Windhorst and G. Blount 1997 *Mat. & Des.* **18**[1] 11-15
[2] I.L. Stimson and R. Fisher 1980 *Trans. R. Soc. Lond. A* **294** 583-590
[3] W. Krenkel and F. Berndt 2005 *Mat. Sc. Eng. A* **412** 177–181
[4] C. Blanco, J. Bermejo, H. Marsh, R. Menendez 1997 *Wear* **213 I** 1-12
[5] W. Liu, Y. Wei, and J. Yideng 1995 *Carbon* **33**[4] 441-447
[6] J.D. Chen, J.H. Chern Lin, C.P. Ju 1996 *Wear* **193** 38-47
[7] J.R. Gomes, O.M. Silva, C.M. Silva, L.C. Pardini, R.F. Silva 2001 *Wear* **249** 240–245
[8] M.F. Bacos 1993 *Journal de Physique IV, Supplement au Journal de Physique* **III**[3] *Colloque* **C9** 819-829.
[9] J.I. Kim, W.-J. Kim, D.J. Choi, J.Y. Park, W.-S. Ryu 2005 *Carbon* **43** 1749–1757
[10] W. Krenkel, N. Heidenreich and R. Renz 2002 *Adv. Eng. Mat, Rev.* **4**[7] 427-436
[11] M.F. Ashby, Y.J.M. Bre´chet, D. Cebon, L. Salvo 2004 *Mat. and Des.* **25** 51–67
[12] A. Yevtushenko, M. Kuciej 2010 *App. Therm. Eng.* **30** 354–359
[13] W. Österle, H. Kloß, L. Urban, A.I. Dmitriev 2007 *Wear* **263** 1189–1201
[14] S. Fan, L. Zhang, Y. Xu, L. Cheng, G. Tian, S. Ke, F. Xu, H. Liu 2008 *Comp. Sc. Tech.* **68** 3002–3009
[15] S. Fan, L. Zhang, L. Cheng, J. Zhang, S. Yang, H. Liu 2011 *Trib. Int.* **44** 25–2
[16] Z. Stadler, K. Krnel, T. Kosmac 2007 *J. Eur. Cer. Soc.* **27** 1411–1417.
[17] Z. Stadler, K. Krnel, T. Kosmac 2008 *Wear* **265** 278–285.
[18] J. Despres and M. Monthioux 1995 *J. Eur. Cer. Soc.* **15** 209-224.
[19] G. Camus, L. Guillaumat and S. Basteb 1996 *Comp. Sc. Tech.* **56** 1363-1372

Corrosion Performance of Inconel 625 in High Sulphate Content

Azzura Ismail

Department of Materials Engineering and Design
Faculty of Mechanical and Manufacturing Engineering
UniversitiTun Hussein Onn Malaysia
BatuPahat, 86400 Johor, Malaysia

*azzura@uthm.edu.my

Abstract. Inconel 625 (UNS N06625) is a type of nickel-chromium-molybdenum alloy with excellent corrosion resistance in a wide range of corrosive media, being especially resistant to pitting and crevice corrosion. However, in aggressive environment, Inconel 625 will suffer corrosion attack like other metals. This research compared the corrosion performance of Inconel 625 when exposed to higher sulphate content compared to real seawater. The results reveal that Inconel 625 is excellent in resist the corrosion attack in seawater. However, at increasing temperature, the corrosion resistance of this metal decrease. The performance is same in seawater with high sulphate content at increasing temperature. It can be concluded that sulphate promote perforation on Inconel 625 and become aggressive agents that accelerate the corrosion attack.

Keywords: Pitting Corrosion, Seawater, Passivity, Breakdown Potentials, Current Density, Passive Film

1. Introduction

Corrosion is the physicochemical interaction between a metal and its environment that results in changes in the properties of the metal, and which may lead to significant impairment of the function of the metal, the environment, or the technical system, of which these form a part. It is known that every substance interacts differently with the environment in which it is used. Furthermore, each ionic species present in the environment behaves uniquely. While no general theory has been established, so far, that accounts with certainty for the corrosion interaction between metals and the environment, it is realized that there is more than one factor that influences the corrosion of metals in aqueous solutions.

It is generally recognized that chromium oxide plays an important role as corrosion protection on passive alloys. For Inconel 625, despite of having nonmagnetic properties, is also has excellent fatigue strength and stress corrosion cracking (SCC) resistance to chloride ions. Most typical applications of Inconel 625 are in chemical processing, aerospace and marine engineering, pollution-control equipment and nuclear reactors. For passivation to occur and remain stable, minimum chromium content should be about 11% by weight, above which passivity can occur and below which it is impossible [1].

Wu *et al.*, studied the passive films formed on Alloy 690 in different pH solutions at high temperatures using potentiodynamic polarization, Auger electron spectroscopy, thermodynamic diagrams and the Mott–Schottkyrelation. They found that the chemical compositions and electronic structures of the passive films were to bestrongly pH-dependent and the passive films were a mixture of Cr_2O_3 and $FeCr_2O_4$ [2]. According to Kawashima *et al.*, the surface films formed on nickel-based alloy at lower potentials in 1.5 M H_2SO_4 solution contain S^{2-}ions other than SO_4^{2-} ions, whereas S^{2-} ions were not incorporated in the passive film.

This makes it necessary to study the effects of the ions in the solution on the corrosion phenomenon, with due consideration to such factors as the type of the metal, temperature, pH, oxygen and ions.

In contrast to general corrosion, attack by pitting sometimes occurs in some particular environmental conditions. These problems result from the presence of specific aggressive species such as chloride, sulphate or nitrate[3]. Each of these species alone can produce pitting corrosion in alloy but with different levels of aggressiveness. For some corrosive resistance alloy such as copper, sulphate ions appear more aggressive than chloride ions and nitrate ions are more aggressive than sulphate ions[4].

El-Naggar[5], who studied the aggressive anions Cl^-, NO_3^- and SO_4^{2-} on the corrosion and passivation behaviour in 0.50 M sodium bicarbonate solutions found that the presence of these aggressive anions stimulates the anodic dissolution rate in both the active and pre-passive potential regions.Pitting corrosion was only observed in the presence of Cl^- anions, while the presence of NO_3^- and SO_4^{2-}anions was only facilitated by the oxygen in the water without themselves participating in the cathodic process. The effect of SO_4^{2-} anion exerts an indirect effect on increasing the cathodic reaction and influence the anodic reaction [5]. Increasing the concentration of SO_4^{2-} ions in the range from 0.0 to 0.2 M in deaerated 0.5 M $NaHCO_3$ solutions shows no pitting and he elucidated that the role of SO_4^{2-} ions is to facilitate repassivation depending on the potentials and passivation currents on the nature of SO_4^{2-} ions [5]. From his study obviously shows that the rate of anodic dissolution increase continuously in the passive region until the concentrations of SO_4^{2-} ions reach 0.20 M, the corrosion rate increase and it decreases with further increase in potential. Therefore, 0.20 M of SO_4^{2-} ions is the limiting concentration concentrations and this can be consider as 0.02 M to 0.18 M SO_4^{2-} ions is low concentration ranges whereas 0.2 M SO_4^{2-} ions is high concentration range as indicative to the changes in the mechanisms of the passivation process [5]. With the evidence of anions effect in increasing and retarding the corrosion rate, therefore, some of these anions in seawater were studied in this research. The effects were studied by increasing the sulphate content (SO_4^{2-}) in seawater in terms of corrosion rate and breakdown potential.

2. Experimental

Inconel 625 is nickel alloy and the main reason for nitrogen alloying in Inconel 625 is to increase the mechanical strength of the steel and to replace some of the expensive nickelalloying element. In addition, increased nitrogen content can also increase the resistance to localised corrosion and retard the precipitation of the carbide and intermetallic phases [6]. Table 1 presents the composition and mechanical properties of Inconel 625.

Table 1. Composition and mechanical properties of Inconel 625

Composition	*Cr*	*Ni*	*Mo*	*C*	*S*	*P*	*Mn*	*Si*	*Others*
	22.0	58.0	9.0	0.05	0.003	0.01	0.03	0.25	*Fe=4.0,Nb=3.15, Ti=0.3,Al=0.3*

Mechanical Properties	Density, ρ (kg/m^3) (X1000)	Elastic Modulus (GPa)	Tensile Strength (MPa)	Yield Strength (MPa)	Hardness
	8.44	205.8	940	430	88(HC)

The corrosion mechanism was characterized according to break downpotential (E_b) of Inconel 625, which is the potential once it reaches a sufficiently positive value, also known as the pitting potential. E_b is also known as the surface potential at which the passive film on the surface breaks down. This leads to an active electrochemical reaction, which causes corrosion or repassivation of the surface in certain conditions. This is the point of localized corrosion for evaluating and considering the potential, which could be an appropriate point according to any given combination of material/ambient/testing methods. The E_b values were identified at 4°C, 20°C, 50°C and 80°C, and each anodic polarisation scan was reversed once the current reached a set current (i_{rev}) of 500μA/cm^2. The scan rate was set at 0.25mV/sec in static condition. The degree of increase in the current beyond i_{rev} gives an indication of the propensity for the propagation of corrosion [7]. An indication of the extent of propagation is therefore obtained by consideration of i_{max}, which represents the maximum current attained should the current not begin to fall immediately after scan reversal.

3. Results and Discussion

The specific electrochemical parameters, are labelled on the graph, which include Open Circuit Potential (OCP), passivation potential (E_{PP}), breakdown potential (E_b), protection potential (E_P), critical current density (i_{cc}) and passive current density (i_p) (Fig. 1). Inconel 625 also exhibited a passive region right from the corrosion potential, E_{corr}. Fig. 2 gives the potentiodynamic of Inconel 625 in real seawater whereas the results in Fig. 3 using artificial seawater which was synthesis in lab with same composition as real seawater. The results reveal that at 20°C, the electrochemical parameters for both conditions are almost a same. Therefore, the experimental was continued using artificial seawater to avoid microbiology activities.

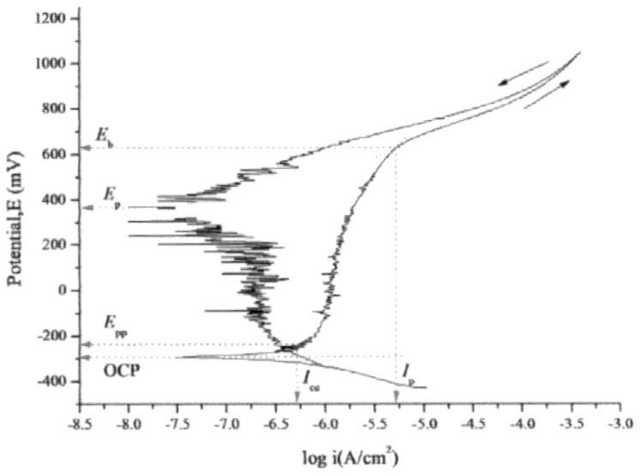

Fig. 1. The typical potentiodynamic polarisation curve for Inconel 625

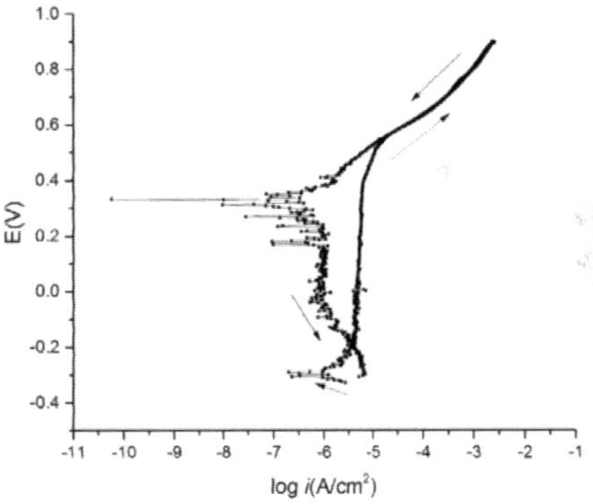

Fig. 2. Inconel 625 in normal seawater at 20°C. The arrow identify forward and reverse scan

3.1. *Inconel 625 in seawater and high sulphate*

For passive alloys, corrosion resistance is provided by a very thin surface film, known as passive film that is an invisible film of oxide, formed by the metal reacting with the ambient environment. Normally these films are free of pores, but their stability may be weakened locally in the environment contained aggressive anions. Fig.3 presents the potentidynamic polarization for Inconel 625 in artificial seawater and Fig. 4 in high sulphate content at 4°C, 20°C, 50°C and 80°C respectively. Despite of having small hysteresis (the forward and reverse scan) value which indicate the small disruption of surface passivity [8], Inconel 625 also present to have a metastable pitting before the passive film breaks at 80°C as shown in red marked.

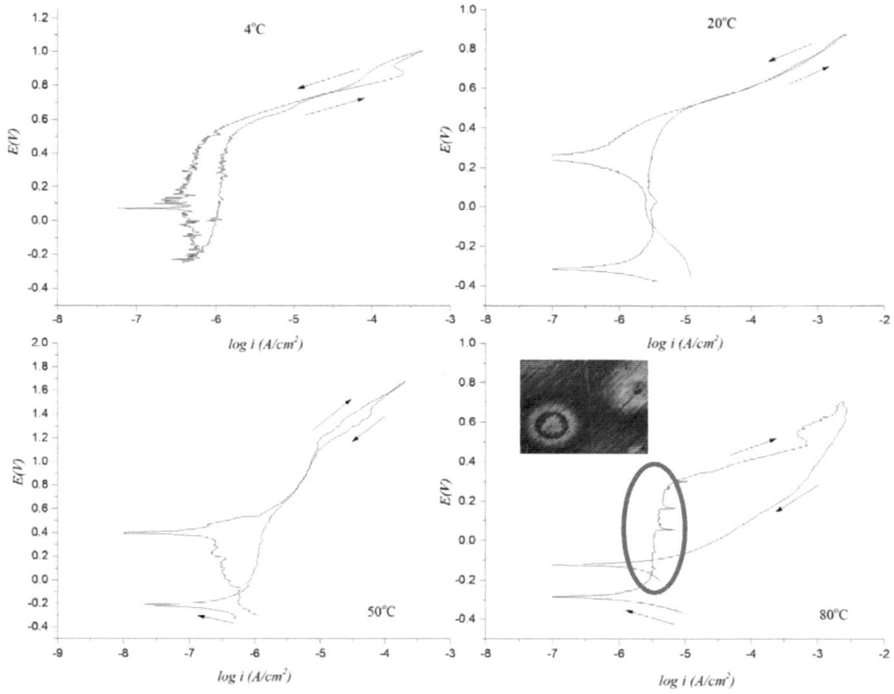

Fig. 3. Inconel 625 in normal seawater at 4°C, 20°C, 50°C and 80°C (metastable pitting image from optical microscopy).

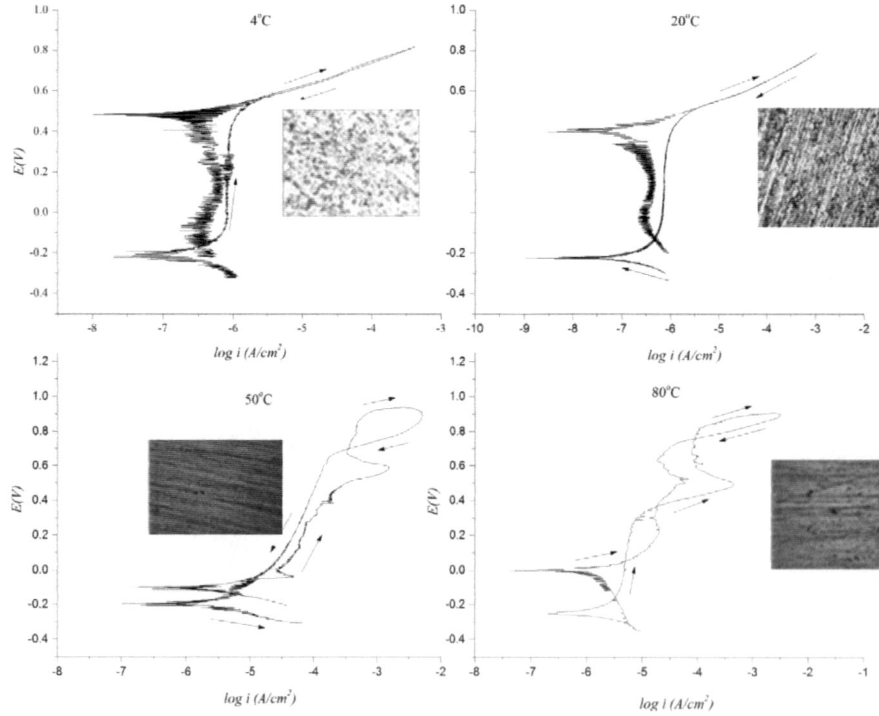

Fig.4. Inconel 625 in high sulphate content at 4°C, 20°C, 50°C, 80°C respectively and its microscopic image.

The corrosion properties were defined from breakdown potentials E_b of Inconel 625 in seawater and seawater with high sulphate content. The highest value represents a more stable of the metal in corrosion media. A set of electrochemistry results are presented in Table 2. The value of OCP increased as the temperature increased reveals that the metal surfaces were attack by pitting corrosion. The E_b value shows that passive film built by Inconel 625 was diminished as the sulphate ion increased. Therefore it can be concluded that sulphate is an aggressive anion that penetrates passive film on metal surface and causing localized attack [9]. It causes the distribution of available pit sites to be shifted to a higher potential, which causes pit propagation in both metastable and stable states. At high temperature, Inconel 625 presents to have secondary E_b and as predicted, the E_b values of Inconel 625 decreases as the temperature increase in both solutions. However, these values are not drastically reduced indicate that Inconel 625 is stable at increasing temperature.

Table 2 The electrochemical data of Inconel 625

Solution	Temp (°C)	OCP (mV)	E_{pp}(mV)	E_b(mV)	i_p(µA/cm^2)	i_{cc}(µA/cm^2)	I_{max}(µA)
Seawater	4	-323	-260	1028	-4.69	--5.57	1608
	20	-296	-233	699	-4.54	-5.19	1660
	50	-192	-120	527	-4.63	-5.48	1470
	80	-182	-122	391	-4.94	-6.12	1470
High sulphate	4	-239	-167	686	-4.62	-5.83	1516
	20	-209	-149	674	-4.89	-6.35	1506
	50	-300	-249	614	-4.52	-5.67	1574
	80	-229	-190	512	-4.94	-6.14	2250

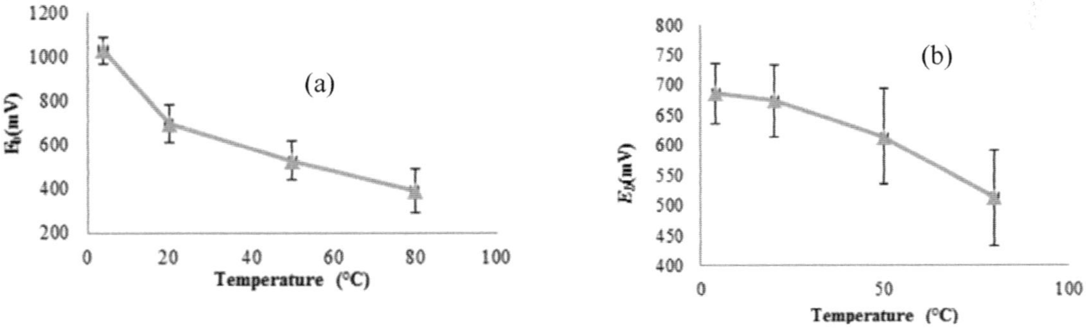

Fig.5. The breakdown potential of Inconel 625 in (a) seawater and (b) high sulphate content as a function of temperature.

4. Conclusion

It can be concluded that, the corrosion attact increased as the temperature increased by shifting the E_b to lower value in both medium. Localised corrosion may attack the metal surface in static condition. Higher sulphate content mitigate the corrosion attact and therefore, sulphate is one of the aggressive ions that accelerate corrosion attack in seawater.

Acknowledgement

The author would like to thank the Office for Research, Innovation, Commercialization and Consultancy Management (ORICC), Universiti Tun Hussein Onn Malaysia and Ministry of Higher Education Malaysia for their funded under Fundamental Research Grant 1460.

References

[1] Pierre-Jean Cunat, *Alloying Elements in Stainless Steel and Other Chromium-Containing Alloy,* International Chromium Development Association, 2009, pg 2473-2481.

[2] Wu X. , Huang J. and Han E.H.,,*Influence of pH on electrochemical properties of passive films formed on Alloy 690 in high temperature aqueous environments.* Corrosion Science, 2009 **(51)**:. pp. 2976-2982.

[3] Takasaki S. and Yamada.Y., Effects of temperature and aggressive anions on corrosion of carbon steel in potable water. Corrosion Science, 2007. **49**(1): p. 240-247.

[4] Duthil, J.P., Mankowski, G. and Giusti A.,*The synergetic effect of chloride and sulphate on pitting corrosion of copper.* Corrosion Science, 1996. **38**(10): p. 1839-1849.

[5] El-Naggar M.M., Effects of Cl−, NO3− and SO42− anions on the anodic behavior of carbon steel in deaerated 0.50 M NaHCO3 solutions,*Applied Surface Science*, 2006. **252**(18).

[6] Manganon P.L., The Principles of Materials Selection for Eingineering Design, Prentice Hall, 1999, pp.512-535.

[7] A.Ismail, Corrosion Behavior Of Austenitic Stainless Steel In High Sulphate Content, Advanced Materials Research Vol. 893 (2014) pp 397-401.

[8] Roberge P.R. *Handbook of Corrosion Engineering*, United States, McGraw Hill, 2008.

[9] El-Naggar M.M., Effects of Cl^-, NO_3^- and SO_4^{2-} anions on the anodic behavior of carbon steel in deaerated 0.50 M $NaHCO_3$ solutions,*Applied Surface Science*, 2006. **252**(18).

IOP Conf. Series: Materials Science and Engineering **131** (2016) 012011 doi:10.1088/1757-899X/131/1/012011

Concrete Slump Classification using GLCM Feature Extraction

Relly Andayani[1], Syarifudin Madenda[2]

[1]Civil Engineering Department, Gunadarma University, Indonesia
[2]Information Technology Postgraduate, Gunadarma University, Indonesia

Abstract. Digital image processing technologies have been widely applies in analyzing concrete structure because the accuracy and real time result. The aim of this study is to classify concrete slump by using image processing technique. For this purpose, concrete mix design of 30 MPa compression strength designed with slump of 0-10 mm, 10-30 mm, 30-60 mm, and 60-180 mm were analysed. Image acquired by Nikon Camera D-7000 using high resolution was set up. In the first step RGB converted to greyimage than cropped to 1024 x 1024 pixel. With open-source program, cropped images to be analysed to extract GLCM feature. The result shows for the higher slump contrast getting lower, but higher correlation, energy, and homogeneity.

Introduction

Concrete is a mixture of Portland cement or other hydraulic cement, water, fine aggregate, and coarse aggregate, with or without additional materials (admixture) [SNI 2013]. Air content, workability and compressive strength are important parameters in making concrete mixture. They should meet sustainability requirements, methods of construction, and the strength of the structure. [Design Control of Concrete Mixture PCA]. Compressive strength is the most important parameters in the resistance of the structure to support the working load (Hemraj R. Kumavat, 2014; F. Falade, 2001).

In order to meet the compressive strength, there are some characteristics in the mix design that must be met, including the workability of fresh concrete. Workability is characteristic of fresh concrete that measure the ease of stirring, transported, poured in the mold, and compacted (KhaledMarar, 2011). Workability measured by Slump testing that shows the size of the concrete viscosity expressed in mm (ACI Committee, 2002). The more viscous the mixture, the smaller the value slump which means the lower the level of workability.

Fresh concrete will lose their slump with the passage of time (delay) (SCG). With the reduced slump value concrete mix can be rejected. It is certainly harm to the readymix. The slump loss value can be fixed by adding water. ASTM recommended water addition as long as it does not cause the water / cement ratio to rise above the maximum allowable value. Additional water is given to the entire batch of concrete mixers, made before the time limit of 1.5 hours since the mixture is stirred. (ASTM C 94, 2003).

Additional of water can improve the workability but decrease the compressive strength (F. Falade, 2001). The addition of water causes the water-cement ratio rose. Water cement ratio is the amount of water compare with the amount of cement used in the concrete mix. It is very influential on concrete compressive strength. The higher water cement ratio the lower the compressive strength can be achieved. (OmotolaAlawode, 2011) (H. Ziad, 2014). Improving workability by additional water declined the strength of concrete, it is advisable to avoid this method (J. Sobhani, 2011).

Due to excessive water cement ratio will alleviate the compressive strength of concrete, the slump test alone is not sufficient to control the quality of the concrete mix. The experience to be able to distinguish mixtures with water cement ratio according to plan or mixed with excessive water cement ratio. The experienced supervisors are quite hard to find in the field. Most of the time the project owner doesn't have that supervisor. It need a tool that could replace the supervisor function in the field.

Use of Information Technology (IT) has penetrated all aspects of research and applications in everyday life. Specifically in the field of civil engineering. Digital Image Processing (DIP) has been widely used to analyse the performance of which is to analyse concrete crack width in concrete structures (Su, 2013) (Ayaho Miyamoto, 2007) (J. Valenca, 2012) (Tung-Ching Su, 2013), analyses the defects in concrete structures (Jian-Hua Tong, 2013) (Bin Sun, 2015) (Xuhao Wang, 2015). On the fresh concrete, the effect of viscosity on the flow identified significant slump (Xuhao Wang, 2015), the monitoring mechanism optimizes concrete mix slump test is used as a parameter in determining the compressive strength of concrete (Hernando Castaneda M), analysis the concrete compression strength (Jui0Sheng Chou, 2012). Image processing is the computer aided reproduction of processes, which humans achieves by mean of their vision system. It's include many process such as image provision, image digitalization, segmentation, and image enhancement (CelalettinBasyigit, 2012) (Alexander Barley, 2014)

This study will classify the slump by extracting the texture feature, normal concrete with compressive strength design 30 MPa is designed with different slump value wich are 0-10 mm, 10-30mm, 30-60 mm, 60-180 mm. Mix design were made according to DoE method.

Material and Method

1.1. Materials

Material used in concrete mix were tested with American Standard Testing Material (ASTM) the world wide international standar for material testing. Aggregate taken from Purwakarta, sand from Cimalaka, Sumedang, and cement type 1 from Indocement.Only properties needed at mix design will be test listed at table 1.

Table 1. Standard testing of material properties

Characteristic	Significance	ASTM References	Requirement or item reported
Grading	Workability of fresh concrete	ASTM C 117 ASTM C 136	Minimum and maximum percentage passing standard sieves
Bulk density;	Mix design calculations	ASTM C 29	Loose weight and compact weight
Specific gravity	Mix design calculations	ASTM C 127 fine aggregate ASTM C 128 coarse aggregate	—
Absorption and surface moisture	Concrete control quality (water-cement ratio)	ASTM C 70 ASTM C 127 ASTM C 128 ASTM C 566	—

1.2. Preparation of sample

Mix Design referred to DOE method, and material proportional justified by the result of slump test. When the slump is lower than design value, limited amount of water is added. Additional water is counted in order to not exceed the water cement ratio maximum (0,6). When slump is higher than design value concrete will be stirred with 90 minutes maximum time limit. If it's steel high, the concrete should be throw away.

Table 2. Material proportion result of mix design

Slump (mm)	0-10	10-30	30-60	60-180
Stone 20 mm	11,9	11,7	10,9	10
Sand	17,6	17,3	16,1	14,8
Cement	4,6	4,7	4,9	5,8
Water	1,75	1,91	2,55	3,24
Water/Cement ratio	0,38	0,40	0,52	0,56

1.3. Slump Testing

Slump test done with ASTM C143.

Fig. 1. Slump testing

1.4. Preparation of images acquisition

Images of fresh concrete specimen were taken in the field using Nikon D-7000 digital camera.Concrete is poured into the container and levelled. Set up of image acquisition is shown as figure 2. And raw image at figure 3.

IOP Conf. Series: Materials Science and Engineering **131** (2016) 012011 doi:10.1088/1757-899X/131/1/012011

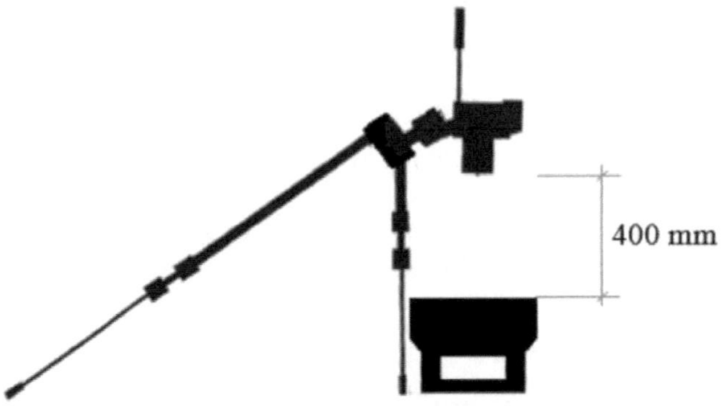

Fig.2. Image acquisition set up

Fig. 3. Raw Image

1.5. Images Processing

The procedures of image processing include converting RGB image to gray. Than cropped by size 1024 x 1024 pixel. Converting and cropping done by mathlab.
Cropped images shown at figure 4.

Fig. 4. Cropped images of fresh concrete with slump of 0-10, 10-30, 30-60, and 60-180 mm

1.6. Result and analysis

Texture can be defined as spatial relationship of pixel value in an image. It can be thought of as a local characteristic pattern of image intensity that identifies the relationship between the pixels. In this study, GLCM method characterized the image on 4 features: contrast, correlation, energy, and homogeneity. The result of GLCM method given in table 3.

Table. 3. GLCM feature

Slump (mm)	Contrast	Correlation	Energy	Homogeneity
0 - 10	102,017400	0,97606	0,00035597	0,259580
10 - 30	76,812300	0,97792	0,000410	0,271020
30-60	47,509100	0,97968	0,00070749	0,320610
60-180	37,398700	0,97977	0,0010239	0,354000

1. Contrast

The function of contrast enhances the contrast of an image. Measure the contrast of intensity between a pixel and its neighbor in the whole image.

$$\sum_{i,j} |i-j|^2 p(i,j)$$

Contrast range are [0 (size (GLCM,1)-1)^2], 0 for a constant image. The result shows that concrete with lower slump has a higher contrast. It mean the lower slump the rough the surface of fresh concrete show by the higher contrast.

2. Correlation

Correlation measures of how correlated a pixel is to its neighbor pixel in the whole image.

$$\sum_{i,j} \frac{(i-\mu i)(j-\mu j)p(i,j)}{\sigma_i \sigma_j}$$

Correlation is 1 or -1 for perfectly positively or negatively correlation image. In this study, higher slump is showed by higher correlation. It mean that the texture is smoother for higher slump.

3. Energy

Is defined as the sum of squared elements in the GLCM and also knows as uniformity or the angular. Ranges of energy are 0-1 and 1 for constant image.

$$\sum_{i,j} p(i,j)^2$$

The result shown that the higher slump the higher energy. Its mean that image with the higher slump has the constant pixel.

4. Homogeneity

Homogeneity returns a value that measures the proximity of the distribution of elements in the GLCM to the GLCM diagonal

$$\sum_{i,j} \frac{p(i,j)}{1+|i-j|}$$

The ranges are 0-1. As the smooth of texture, homogeneity will be higher. Its shown as the result that concrete with higher slump has an higher homogeneity.

1.7. Conclusion and suggestion

The method used in this study GLCM succeeded in extracting concrete image feature that could explain the classification of slump of fresh concrete for 30 MPa concrete compression strength design. The higher the value of the slump (0-10, 10-30, 30-60, 60-180) which is indicated by a low contrast value (102.0174, 76.8123, 47.5091, 37.3987), but the higher correlation (0.97606, 0.97792, 0.97968, 0.97977), energy (0.00035597, 0.00041, 0.00070749, 0.0010239), homogeneity (0.25958, 0.27102, 0.32061, 0.354).

For further research should be investigated the range of contras, correlation, energyand homogeneity which shows the value of a certain slump so that it can be applied as a tool for supervision slump in the field replacing the supervisor function.

Acknowledgement

The author wishes to thank to Prof. Syarifudin Madenda for his idea and support in writing this paper. For making the specimen and acquiring the image the author also acknowledge to Nabil and Aef Syaefullah.

Reference

[1] Hemraj R. Kumavat, Vikram J. Patel, 2014, Factor Influencing the Strength Relationship of Concrete Cube and Standard Cylinder, *International Journal of Innovative Technology and Exploring Engineering (IJITEE)*, **vol. 3**, p 76-79

[2] F. Falade, & G. L. Oyekan, 2001, Slump Test as an Indirect Indicator of the Characteristic Strength of Laterised Concrete, *West Indian Journal of Engineering*, **vol 24**, p 35-41

[3] KhaledMarar, OzgurEren, 2011, Effect of Cement Concten and Water/Cement Ratio on Fresh Concrete Properties without Admixture, *Internatioanal Journal of the Physical Science*, **vol. 6**, p 5752-5765

[4] ASTM, 2003, *Standard Specification for Ready Mixed Concrete,* ASTM

[5] OmotolaAlawode, O.I. Idowu, 2011, Effect of Water-Cement Ratio on the Compresive Strength and Workability of Concrete and Lateric Concrete Mixes, *The Pacific Journal of Science and Technology*, **vol 12**, p 99-105.

[6] J. Sobhani, M. Najimi, A.R. Pourkhorshidi, 2011, Effect of Retempering methods on the Compresive Strength and Water Permeability of Concrete, *ScientiaIranica, Elsevier*, **vol 19**, p 211-217.

[7] Su, Tung-Ching, 2013, Application of Computer Vision to Crack Detection of Concrete Strukture, *IACSIT International Journal of Engineering and Technology*, **vol. 5**, p 457-461.

[8] Ayaho Miyamoto, Masa-Aki Kono, EugenBrughwiller, 2007, Automatic Crack Recognition System for Concrete Structure Using Image Processing Approach, *Asian Journal of Information Technology*, **vol. 6**, p 553-561.

[9] J. Valenca, D. Dias-da-Costa, E.N.B.S. Julio, 2012, Characterisation of Concrete Cracking During Laboratorial Test Using Image Processing,*Construction and Building Material, Elsevier*, **vol. 28**, p 607-615.

[10] Jian-Hua Tong, Chin-Lung Chiu, Shu-Tao Liao,2013, An Elastic Wave-Based Imaging Method with Synthetic Aperture insida Concrete Structure, *Journal of Applied Science and Engineering*, **vol. 16**, p 45-50.

[11] Bin Sun, Xuan Wang, Zhaoxia Li, 2015, Meso-Scale Image based Modeling of Reinforced Concrete and Adaptive Multi-Scale Analysis on Demage Evolution in Concrete Structure, *Computational Material Science, Elsevier*, **vol.110**, p 39-53.

[12] Jui-Sheng Chou, Chih-Fong Sai, 2012, Concrete Compression Strength Analysis using a Combine Classification and Regression Technique, *Automation in Construction, Elsevier*, **vol. 24**, p 52-60.

[13] Alexander Barley, Christopher Town, 2014, Combination of Feature Descriptors for Texture Image Classification, *Journal of Data Analysis and Information Processing*, **vol. 2**, p 67-76.

[14] Tung-Ching Su, 2013, Application of Computer Vision to Crack Detection of Concrete Structure, *IACSIT International Journal of Engineering and Technology*, **vol 5**, p 457-461.

[15] Xuhao-Wang, Kejin Wang, Jianguo Han, Peter Taylor, 2015, Image Analysis Application on Assessing Static Stability and Flowability of Self-Consolidating Concrete, Cement and Concrete Composite, Elsevier, vol 62, p 156-167.

[16] Ziad H., Ali. S. Zregh, 2014, Retempering of Fresh Concrete and its Effect on Concrete Strength, University Bulletin, no 16, vol. 3

E-Service Quality, E-Satisfaction and E-Loyalty of Online Shoppers in Business to Consumer Market; Evidence form Malaysia

Ong Soo Ting[1], Mohd Shoki Md Ariff[1*], Norhayati Zakuan[1], Zuraidah Sulaiman[1] and Muhamad Zameri Mat Saman[2]

[1] Department of Business Administration, Faculty of Management, Universiti Teknologi Malaysia (UTM), Skudai, Johor, Malaysia
[2] Faculty of Mechanical, Universiti Teknologi Malaysia, 81310 UTM Skudai, Johor, Malaysia

E-mail: m-shoki@utm.my

Abstract. The growing usage of internet and online shopping in Malaysia presents a huge prospect in e-commerce market, specifically for B2C segment. As a result, electronic service quality (e-SQ), electronic satisfaction (e-Satisfaction) and electronic loyalty (e-Loyalty) become vital for online retailers to attract and retain online shoppers in this virtual environment. The association between e-SQ, e-Satisfaction and e-Loyalty should be continuously examined to cope with the advancement in information and communication technology, and the changing expectation of online shoppers. However, construct of e-SQ for online retailers in B2C market is still debatable. In this research, E-SERVQUAL was integrated with the other e-SQ scales to measure e-SQ of a prominent online retailer in Malaysia. Specifically, the e-SQ constructs are Efficiency, Privacy and Trust, Fulfilment, Responsiveness, Contact and Website Design. 390 sets of completed and usable questionnaires were gathered using online questionnaire and convenience sampling procedure. The result indicated that the five proposed dimensions of e-SQ constitute e-SQ of online retailer in B2C market. All the dimensions of e-SQ were found to have positive and significant effect on e-Satisfaction of online shoppers. Responsiveness of e-SQ had the strongest impact on e-satisfaction of online shoppers. The shoppers e-Satisfaction was positively and significantly affected their e-Loyalty towards continuous usage of online retailer's website. Managerial and theoretical implications are discussed based on the results of the study.

Keywords: e-SQ, e-Satisfaction, e-Loyalty, Gap Model Online Shopping

1.0 Introduction

Electronic commerce (e-commerce) refers to the process of carrying out business between enterprises and customers using electronic medium such as internet. Companies are extensively using this sales channel to compete with their competitors in the e-commerce market. Business to Consumer (B2C) is one of the natures of transactions in e-commerce. According to Turban, et.al [1], B2C refers to e-commerce model in which businesses sell to individual shoppers. The B2C online market has been growing rapidly over the past several years and consumers are increasingly favoured online shopping. About 70 per cent of Malaysians are active online users, and more than half do their shopping on the web [2]. Malaysian consumers used up to about RM4.5 bil on online shopping, showing a 33% increase from 2010, [3] and it is projected to raise to RM15bil in 2014. In addition, the number of B2C online companies is also increased. Therefore, it is crucial to examine e-SQ, e-Satisfaction and e-Loyalty of online shoppers in B2C market in Malaysia.

Issue of which dimensions of e-SQ serve as better construct of e-SQ for online sellers in B2C market is highly debated. According to Lei Dai et al., [8], e-SQ dimensions, such as efficiency, system availability, fulfillment, privacy, responsiveness, contact and compensation [9] influences the customers' continuous usage of the B2C website. Sitequal [10] of e-SQ instrument features ease of use, design, processing speed and security to assess the quality of electronic services in a B2C website. Web site design quality is crucial for online stores [11] and it is a strong predictor of customer quality judgments, satisfaction, and loyalty for internet retailers [12]. Besides, Brown et al., [13] proposed service quality and trust as additional factors to consider for measuring e-SQ in a B2C e-commerce environment. In addition, building and enhancing online trust could be a better way to address the customers' data privacy concerns in e-commerce. Since privacy and trust are directly related, establishing procedures to enhance user's trust may lessen customers' privacy concern. According to Saleh Bukhari *et al.* [14], although privacy and trust might not always have the same definition, in many cases it presents similar idea in different fields or contexts. Although the privacy dimension used to measure the service quality in B2C website has been recognized in many studies, the emperical validation of trust dimension and also the website design dimension used to measure the e-SQ of B2C website in Malaysia has not been addressed. This brings to the issue of what constitute e-SQ for online retailers in B2C market in Malaysia? Therefore, examining e-SQ construct by combining various e-SQ models to address these issues is important to research, especially in Malaysia's B2C market.

The delivery of the high e-SQ is the key determinants of success or failure of online retailers in B2C e-commerce instead of merely the presence of website and low price offered [4]. E-service quality is overall customer assessment and judgment of e-service delivery in the virtual marketplace [5], and high e-SQ will increase customers' satisfaction and in turn lead to customer repurchases behaviour. Superior e-SQ is the way to enhance e-Satisfaction and increasing e-Loyalty in B2C market [6,7]. Loyal customers will make recommendation to their friends which in turn will help to attract new customers and boost sales. Previous study had indicated that there was strong and significant effect between e-SQ and e-satisfaction in e-commerce [6,7]. However, with the advancement in internet and information technology, online shoppers' expectation towards quality of online retailers' website is changing and increasing, thus examining e-SQ on a continuous basis is needed. Further, most online shoppers now are information technology literate, therefore their judgement of e-SQ of B2C websites could be different, as compared to the previous related studies. Thus, the effect of online retailers on e-satisfaction and e-satisfaction on e-loyalty of online buyers in B2C market should be further researched due to these changes. Hence, the association among e-SQ, e-satisfaction, and e-loyalty should also be continuously examined in order to provide insightful information towards developing a better understanding of e-SQ-e-Satisfaction-e-Loyalty in B2C segment.

This study provides empirical evidence on the association of e-SQ, e-Satisfaction and e-Loyalty of online shoppers in B2C websites. Specifically, this study aims to answer to the following research questions; (i) what constitute e-SQ for online retailers in B2C market? (ii) Does e-SQ positively affect e-Satisfaction of online shoppers in B2C market of Malaysia?; and (iii) Does e-Satisfaction positively affect e-loyalty of the online shoppers? This paper argues on e-SQ issues and dimensions, thus provides additional insight on how online buyers judge website quality of online sellers. It also enriches current literature on the established association of e-SQ, e-Satisfaction and e-Loyalty in the context of online retailers-online shoppers in Malaysia's B2C market. For e-marketers, this study provides useful information on the current issue of determining e-SQ and ways of improving it to fulfil online e-Satisfaction of shoppers and their e-Loyalty towards the online retailers.

2.0 Literature Review

Electronic service quality (e-SQ) is increasingly important in influencing customer evaluations and judgments regarding the quality of e-service delivery in the virtual marketplace. The e-SQ is highly concerned due to it highly related with the success or failure of an internet-based company such as online shopping in B2C. Online shopping in B2C involves online shoppers accessing the online vendors' websites to purchase a product; therefore the extent to which their needs are fulfilled is important. The transactions between online shoppers and online retailers are conducted through websites and the smooth transactions are achieved by high level of e-SQ. Website quality or e-SQ is defined as "the extent to which a website facilitates efficient and effective shopping, purchasing and delivery of products and services" [16] Website's quality plays a key role in determining the web experience of the customers [16]. In other words, a website is seen as one of the online retailers' instruments for delivering quality of electronic services to online shoppers.

A number of scales to measure e-SQ of an online retailer as perceived by its on-going customers were developed. Among the main e-SQ assessment techniques are E-SERVQUAL, WEBQUAL, SITEQUAL and ETAILQ. SITEQUAL which is used to measure the quality of a website was developed by Yoo and Donthu [10]. The dimensions of SITEQUAL are ease of use, aesthetic design, processing speed, and security. The e-SERVQUAL, developed by Zeithaml *et al.*, [9,16] is a method for measuring website e-SQ based on the online shoppers perceptions of how well the website meet their online transaction requirements. The scale contains the core scale (E-S-QUAL) and the recovery scale (E-RecSQUAL). There are four dimensions in E-S-QUAL which are efficiency, fulfilment, reliability and privacy. Responsiveness, contact and compensation are dimensions of e-RecQual. WEBQUAL was developed by Loiacono, Watson and Goodhue [17] in order to evaluate website quality focusing on the website interface. The WEBQUAL was used to predict the re-visit/re-use behaviour of web users based on their perceptions of overall website quality. There are four dimensions in the WEBQUAL, which are usefulness, ease of use, entertainment and complimentary relationship. WEBQUAL 4 developed by Barnes & Vidgen [18] was the mixture and enhancement from the previous WEBQUAL and it composed of 22 items of three dimensions of usability, information and service interaction. The eTailQ, a scale used for assessing and predicting e-tail quality, was developed by Wolfinbarger and Gilly [19]. The authors suggested that four factors - website design, fulfillment/reliability, privacy/security and customer service - are strong predictive of customer judgments of quality and satisfaction, customer loyalty and attitudes toward a website.

Many studies involving B2C websites have been conducted to investigate dimensions of e-SQ that influence online shoppers to engage in B2C websites. Among e-SQ instruments discussed above, e-SERVQUAL is widely used due to e-SERQUAL measures e-SQ throughout the complete customer shopping experience in online purchasing process, taking into account both the pre-website and post-website stages of this process. The works of Yang and Tsai [6] , Rahman and Miazee [20] and Pooja Jain and Dr Narender Kumar [21] utilizing E-SERQUAL in measuring quality of B2C websites indicated that the instrument is highly reliable in predicting how online shoppers' evaluate quality of retailers' websites. However, other e-SQ dimensions should be considered because in the purchasing process of e-commerce, online shoppers demand for innovative and visually appealing websites, well organized user interface before they trust the websites. Thus, web site design quality is important [11] and it is a strong predictor of customer quality judgments, satisfaction, and loyalty of online shoppers [19]. In the research of Lee and Lin [22], trust and personalization dimensions were used in measuring e-commerce website. However, according to Saleh Bukhari *et al.* [15], although privacy and trust might not always have the same definition, in many cases it shows similar idea in different fields or contexts. Therefore, trust should be incorporated in the e-SQ instrument to determine the quality of a website. Thus, efficiency, fulfilment, reliability and privacy of E-SERVQUAL, web design of e-TailQ and trust were proposed to determine e-SQ for online retailers in B2C market.

According to Anderson and Srinivasan [23], e-satisfaction is described as the gratification of the customer that comes from his or her earlier real buying experience with a certain electronic commerce company. Customer satisfaction is crucial for creating long-term relationships with clients and in sustaining profitability of online retailers. E-Satisfaction is a critical component in determining the success of online shopping, especially in B2C market. Schaupp and Bélanger [24] developed E-Satisfaction Scale to measure the level of satisfaction of customer towards products or services that are provide by online retailers. Park and Kim [25] found that customer satisfaction was influenced by the quality of the user interface, because it provides physical proof of the online service provider's capability as well as facilitating easy use of the online service. Website user interface design is strongly associated to customer satisfaction [26].

Chengwen Yao and Shuling Liao [27] found that wise decision, shopping experience, overall satisfaction and satisfaction in online transactions service will determine e-satisfaction of online shoppers. Further, online shoppers' enjoyment will determine their satisfaction in online shopping [28]. According to Mustafa [29], e-satisfaction of online shoppers was determined by the performance of the website and the product provided by the website. Therefore, wise decision, shopping experience, overall satisfaction, satisfaction in online transactions service, enjoyment, performance of the website and the product provided by the website are highly recommended items to be used in measuring e-satisfaction of online shoppers.

The importance of developing e-loyalty of online shoppers has been stressed by most previous researchers to ensure profitable customer relationship. E-Loyalty is important from marketing point of view since the cost of getting new customers was noticeably higher than the cost of retaining the existing ones [30]. In fact, the successful online business ventures from the unsuccessful ones can be discerned by the repeat purchases from loyal customers. Online shoppers' loyalty deals with customer's real repurchase behaviour and it is defined as the favourable attitude of a customer towards an electronic business resulting in repeat purchase behaviour [23].

Yang & Tsai [6] proposed that first choice, positive comments and encourages someone surfing on the website will determine the loyalty of customers in online shopping in B2C market. In addition, like to use the website, recommendation, purchase intention, care about the website and best retail website will determine the e-loyalty of customers [31]. In conclusion, first choice, positive comments, encourages someone surfing on the website, like to use the website, recommendation, purchase intention, care about the website and best retail website are items that should be considered when measuring e-loyalty of online shoppers.

Zeithaml et al. [32] exploratory research indicated the presence of gaps in companies interacting with their customers through the internet. Four gaps were identified in the gap model of e-SQ, which were information gap, design gap, communication gap and fulfillment gap. The company side shows three potential disconnects that can occur in the process of designing, operating, and marketing Web sites, and these disconnects are information, design, and communication gaps. These gaps jointly contribute to the "fulfilment gap" on the customer's side, eliciting a chain of adverse effects on perceived e-SQ, perceived value, and purchase/repurchase behaviour (e-Loyalty). The elimination of e-SQ gaps will lead to e-Satisfaction which results in increased perceived e-SQ, value, purchases and repurchases (e-Loyalty). Therefore the gap model is a good model for online retailers to determine their websites quality (e-SQ) and how the e-SQ effect e-Satisfaction of online shoppers, which in turn effect the e-Loyalty, with the aims of building a long term relationship with their on line shoppers.

Improving the relationship between online shoppers and online sellers and their mutual benefits of satisfaction (buyers) and performance (sellers) through superior e-SQ has been the focus of e-marketers and researchers. Assessment of e-SQ particularly is to improve the online service delivery that can lead to online shoppers' satisfaction. E-SQ is a significant antecedent of online shoppers' assessment of value that will influence their e-Satisfaction. E-SQ dimensions are strongly predictive of the satisfaction of online consumers [33]. E-SQ is strongly associated to online customer satisfaction, and many studies had indicated that there were strong positive relationship between e-SQ and e-Satisfaction of online shoppers in B2C website [6,34]. Thus, it is hypothesized that e-SQ is positively and significantly affect e-Satisfaction of online shoppers in B2C market.

The relationship between satisfaction and loyalty has been found to apply in both offline and online stores [35]. Many studies showed that customer satisfaction is a direct antecedent of consumer loyalty in B2C e-commerce. In accordance with the earlier research [36], it is estimated that a higher level of customer satisfaction will lead to greater loyalty. Numerous studies have revealed that there was evidenced in the context of e-commerce about the positive impact of online satisfaction on loyalty [7,37]. The relationship between e-satisfaction and e-loyalty of customers had been found to be significant in numerous studies [32,] and this will lead to the positive predisposition of long-term loyalty [38]. Therefore, it is hypothesized that e-Satisfaction of online shoppers is positively and significantly affect their e-Loyalty in in B2C market.

3.0 Methodology

The descriptive survey was employed to examine the relationship between e-SQ, e-Satisfaction and e-Loyalty of online shoppers in B2C market. A questionnaire was developed based on the review of Gap Theory of electronic service quality and previous researches related to e-SQ, e-Satisfaction and e-Loyalty in B2C. A total number of 39 items were constructed to measure the independent and dependent variables of the study, as follows:

- Section A of the questionnaire requires respondents to indicate their demographic information.
- In Section B of the questionnaire, 24 items of e-SQ Dimensions (Efficiency, Privacy and Trust, Fulfilment, Responsiveness, Contact and Website Design) were constructed based on E-SERQUAL instrument [9] and previous studies of Wolfinbarger and Gilly [19], Lee and Lin [22], and Saleh Bukhari *et al.* [15].
- In Section C, seven (7) items of e-satisfaction of online shoppers were adapted from previous researches of Lee and Lin [22], Kim and Jackson [38], Yao and Liao [27], Mustafa [29] and Xiaoying Guo1 *et al.*, [39].
- In Section D, eight (8) items of e-loyalty of online shoppers in the questionnaire were adapted based on the works of Yang & Tsai [6], Leung Nga Yin *et al.*, [31] and Lam [34].

The Likert scale ranging from "strongly disagree" (1) to "strongly agree" (5) was used to measure all the independent and dependent variables of this study.

The population of this study is the total number of online shoppers of a prominent local online seller in Malaysia. This online seller is the most popular B2C and it is ranked top (among 11 popular B2C online retailers) in Malaysia and occupied number 96 in Alexa ranking, 2013. The sample size of 390 online shoppers (for 10% margin error) was determined based on the multivariate research procedure, proposed by Hill [40]. Online questionnaire was employed to gather 390 completed and usable questionnaire using convenience sampling procedure. It took six weeks to collect the 390 completed and usable questionnaires. The data collected was kept automatically in Google spreadsheet and exported to Statistical Package for Social Science (SPSS) for further analysis.

4.0 Result and Analysis

Exploratory Factor Analysis (EFA) was performed in order to examine the validity of the instrument for further analysis. As shown in Table 1 (e-SQ) and 2 (e-Satisfaction and e-Loyalty), the KMO for e-SQ, e-Satisfaction and e-Loyalty are 0.886, 0.925 and 0.921 respectively, supported by Bartlett's test of Sphericity of 0.000. The Total Variance Explained for e-SQ is 72.180%. The rotated component matrix of EFA yielded six components, which are the six proposed dimensions of e-SQ. All dimensions and items of e-SQ were retained since the factor loading scores of ≥ 0.5 (Table 1). The rotated component matrix of EFA yielded one component each for e-Satisfaction and e-Loyalty. The

Total Variance Explained for e-Satisfaction and e-Loyalty are 65.169% and 62.892% respectively. All the seven items of e-Satisfaction and eight items of e-Loyalty were retained based on the factor loading of more than 0.5, as shown in Table 2.

Table 1: Rotated Component Matrix[a] for e-SQ

Variable	1	2	3	4	5	6
E1 This website makes it easy to find what I need.	.052	.188	.109	.085	**.775**	-.020
E2 This website is simple to use.	.081	.156	.150	.089	**.855**	.066
E3 This website loads its pages fast.	.310	.114	.123	.112	**.707**	.195
E4 Information provided at this website is well organized	.248	.220	.074	.293	**.659**	.229
P5 This website protects information about my Web-shopping behaviour.	.354	**.509**	.114	.220	.318	-.077
P6 This website protects information about my credit card.	.347	**.721**	.090	.133	.252	.050
P7 This website does not share my personal information with other site.	.077	**.849**	.088	.112	.202	.099
P8 I feel I can trust this website.	.022	**.788**	.258	.015	.115	.224
P9 This website instils confidence in customers.	-.026	**.690**	.378	.126	.075	.185
F10 This website delivers customer electronic orders when promised.	.022	.325	**.767**	.119	-.032	.011
F11 This website has in stock the items the company claims to have.	.061	.154	**.795**	.115	.228	.114
F12 This website is truthful about its offerings.	.158	.138	**.780**	.155	.172	.137
F13 This website a makes items available for delivery within a suitable time frame.	.302	.081	**.774**	.148	.104	-.036
R14 This website provides me with convenient options for returning items.	**.765**	-.007	.182	.199	.158	.157
R15 This website offers a meaningful guarantee.	**.790**	.057	.162	.181	.224	.178
R16 This website takes care of problems promptly.	**.783**	.180	.089	.258	.128	.216
R17 This website tells me what to do if my transaction is not processed.	**.716**	.237	.097	.110	.085	.348
C18 This website provides contacts (telephone and email) to reach the company.	.280	.122	.055	.098	.112	**.765**
C19 This website has customer service representatives available online.	.314	.143	.051	.193	.076	**.802**
C20 To get more information about products of this website, I could turn to the Website's facebook, twitter or others.	.109	.132	.087	.331	.110	**.753**
W21 This website is innovative.	.218	.106	.251	**.650**	.140	.393
W22 This website is visually appealing.	.116	.071	.181	**.831**	.165	.210
W23 The user interface of this website has a well-organized appearance.	.174	.129	.106	**.830**	.137	.045
W24 It is quick and easy to complete a transaction at this website.	.279	.132	.110	**.653**	.087	.193
Total variation explained	72.180%.					
Kaiser-Mayer-Olkin Measure of Sampling Adequacy	0.886					
Bartlett's Test of Sphericity						
Approac. Chi-Square	5802.254	D.F	276	Significant	.000	

Table 2: Rotated Component Matrix[a] for e-SQ, e-Satisfaction and e-Loyalty

Variable	e-Satisfaction	e-Loyalty
ESAT1 My decision to purchase from this website was a wise one.	**.792**	
ESAT2 I am satisfied with the shopping experience of this website.	**.824**	
ESAT3 I have truly enjoyed purchasing from this website.	**.804**	
ESAT4 I am satisfied with the product provided from this website.	**.814**	
ESAT5 I am overall satisfied electronic purchasing with this website.	**.816**	
ESAT6 The performance of this website meets my expectation.	**.822**	
ESAT7 In general terms, I am satisfied with the online transactions service provided by this website.	**.778**	
EL1 I like using this website.		**.797**
EL2 When I need to make online purchase, Lazada is my first choice.		**.794**
EL3 I would recommend Lazada to others.		**.833**
EL4 I intend to purchase electronic with Lazada in the future.		**.801**
EL5 To me, Lazada is the best retail website with which to do business.		**.804**
EL6 I would say positive things or comments about Lazada to other people.		**.784**
EL7 I would encourage someone surfing on Lazada.		**.754**
EL8 I really care about the future of Lazada.		**.775**
Total variation explained	65.169%	62.892%
Kaiser-Mayer-Olkin Measure of Sampling Adequacy	**.925**	**.921**
Bartlett's Test of Sphericity		

Approac. Chi-Square	1545.197	1813.551
D.F	21	28
Significant	.000	.000

In addition, Cronbach's alpha values for e-SQ dimensions - Efficiency, Privacy and Trust, Fulfilment, Responsiveness, Contact and Website Design are 0.837, 0.854, 0.857, 0.885, 0.840 and 0.853 respectively. The overall Cronbach's alpha value for e-SQ is 0.927. For e-Satisfaction and e-Loyalty, the Cronbach's Alpha values are 0.911and 0.915 respectively. Thus, constructs of e-SQ, e-Satisfaction and e-Loyalty used in this study are reliable and valid for further analysis.

The result of Multiple Regression analysis in Table 3 shows that the entire e-SQ dimensions are positively and significantly affect e-Satisfaction of online shoppers. Specifically, the significant effects of e-SQ on e-Satisfaction are: Responsiveness (β 0.213, t 4.308, Sig. 0.000), Privacy and Trust (β 0.190, t 3.994, Sig. 0.000), Efficiency (β 0.189, t 4.078, Sig. 0.000), Contact (β 0.126, t 2.661, Sig. 0.008), Fulfilment (β 0.118, t 2.649, Sig. 0.008) and Website Design (β 0.103, t 2.093, Sig. 0.037). The R square (R^2) for this regression model was 0.483, indicating 48.3% of the variance in e-Satisfaction was explained by the e-SQ dimensions. In addition, the most significant factor of e-SQ affecting e-Satisfaction is responsiveness (β 0.213, t 4.308, Sig. 0.000).

Table 3: Results on the Impact of e-SQ on e-Satisfaction of Online Shoppers

Coefficients[a]							
Model		Unstandardized Coefficients		Standardized Coefficients	t	Sig.	VIF
		B	Std. Error	Beta			
1	(Constant)	.796	.162		4.902	.000	
	Efficiency	.164	.040	.189**	4.078	.000	1.588
	Privacy and Trust	.164	.041	.190**	3.994	.000	1.684
	Fulfilment	.099	.037	.118*	2.649	.008	1.475
	Responsiveness	.171	.040	.213**	4.308	.000	1.818
	Contact	.092	.035	.126*	2.661	.008	1.665
	Website Design	.087	.042	.103*	2.093	.037	1.807
R = 0.695		F – Value = 59.570					
R Square (R^2) =0 .483		Std. Error of the Estimate = .44476					
Adjusted R^2 = .475							
a. Dependent Variable: e-Satisfaction							

The linear regression analysis result in Table 4 shows that e-Satisfaction of online shoppers is positively and significantly affect e-Loyalty of the shoppers (β 0.772, t 23.889, Sig. 0.000). The R square (R^2) for this regression model was 0.595, indicated that 59.5% of the variance in e-Loyalty was explained by e-Satisfaction.

Table 4: Results on the Impact of e-Satisfaction of Online Shoppers on e-Loyalty

Coefficients[a]							
Model		**Unstandardized Coefficients**		**Standardized Coefficients**	**t**	**Sig.**	**VIF**
		B	**Std. Error**	**Beta**			
1	(Constant)	.898	.126		7.125	.000	
	e-Satisfaction	.784	.033	.772**	23.889	.000	1.000
R = .772		F – Value = 570.682					
R Square (R^2) = .595		Std. Error of the Estimate = .39722					
Adjusted R^2 = .594							
a. Dependent Variable: e-Loyalty							

5.0 Discussions

This research proposed efficiency, fulfilment, reliability of E-SERVQUAL, web design of e-TailQ and a combination of privacy (e-SERVQUAL) and trust to determine e-SQ of online retailers in B2C segment. The finding has confirmed the appropriateness of these constructs in predicting e-SQ of the online retailer's website. This is in line with the work of Yang and Tsai [6], Rahman and Miazee [20] and Jain and Kumar [21]. It shows that employing efficiency, fulfilment, reliability of E-SERVQUAL scale [9,32], website design of e-TailQ [19], and trust [15, 22] with privacy [9,32] are predictors of e-SQ in online shopping of B2C. With the total variance explained of 72.180% and all dimensions and items were well loaded in their components matrix of EFA, it can be concluded that these constructs are predictors of e-SQ for the online retailers. Thus, the first research question of this study is answered.

How e-SQ of online sellers' website quality affect online buyers' e-satisfaction is the second research question of this study. All the e-SQ dimensions are positively and significantly affect e-Satisfaction of online shoppers. This finding is consistent with the previous studies [6,33,34], in which e-SQ has a strong positive effect on e-satisfaction in B2C website. Specifically, responsiveness (β 0.213, t 4.308, Sig. 0.000), Privacy and Trust (β 0.190, t 3.994, Sig. 0.000), Efficiency (β 0.189, t 4.078, Sig. 0.000), Contact (β 0.126, t 2.661, Sig. 0.008), Fulfilment (β 0.118, t 2.649, Sig. 0.008) and Website Design (β 0.103, t 2.093, Sig. 0.037) of e-SQ affect e-Satisfaction of online shoppers. For e-marketers, this finding proposes that in order to stay competitive in online shopping of B2C market, they must ensure all these indicators of websites quality are well addressed in their electronic delivery systems.

Responsiveness (β 0.213, t 4.308, Sig. 0.000), exerted the largest impact on e-Satisfaction of online shoppers, which is consistent with the work of Kim and Jackson [38]. Therefore, online retailers are encouraged to concentrate on providing meaningful guarantee, promptly entertain online shoppers' inquiries and problems and convenient options for returning the purchased items.

The third research question of this study is to determine the effect of online buyers' satisfaction on their e-Loyalty towards the online shoppers. It was found that e-Loyalty of online shoppers is affected by their e-satisfaction (β 0.772, t 23.889, Sig. 0.000). The strong association between e-Satisfaction and e-Loyalty of online shoppers was also observed in the previous studies [7,32,37,38,41]. This study provides empirical evidence to support this link in online shopping of B2C market in Malaysia. Hence, online retailers in B2C segment should concentrate on providing e-SQ and quality products to attain high level of e-Satisfaction and e-Loyalty of online shoppers.

6.0 Conclusion

This study extended the formation of e-SQ, e-Satisfaction and e-Loyalty in online shopping of B2C market. It shows that e-SQ significantly influenced e-Satisfaction of online shoppers, which in turn affected their e-Loyalty. This association was observed in other services, for example internet banking, hotel, telecommunication and higher education, to name a few. Further, construct of e-SQ of this study provides usable dimensions to be considered when developing a conceptual model involving e-SQ. It shows that, relying on one e-SQ instrument is insufficient. Hence, combining several dimensions of e-SQ scales would give better understanding on quality of electronic services delivered. In conclusion, this research found that construct of e-SQ requires a combination of dimensions from various e-SQ instruments (Research Question 1), the e-satisfaction of online buyers is affected by e-SQ of online sellers, (Research Question 2) and the buyers' e-loyalty towards the online sellers is influenced by their e-satisfaction (Research Question 3).

Despite meeting the research questions of the study, this study has several limitations. Firstly, e-e-Satisfaction is influenced by e-SQ dimensions positively, and e-Loyalty of online shoppers is affected by the extent to which they are satisfied with the online transaction with the online retailers. The mediating effect of e-Satisfaction between e-SQ and e-Loyalty is suggested for future researches because this association is observed in other services, for example in higher education, internet banking and telecommunication. Further, it is recommended to cover all popular online retailers in Malaysia, for example top 10 B2C retailers to provide reliable and usable model of e-SQ, e-Satisfaction and e-Loyalty of online shoppers in B2C market.

References

[1] Turban E. King D., Lee, J. K. and Viehland, D. (2009) Electronic Commerce 2009: A Managerial Perspective. Prentice Hall.
[2] Igor Pezzilli, Malaysian Flavours (2013). Retrieved from http://www.malaysianflavours.com/2013/04/lazada-turns-one-in-malaysia.html
[3] The Star Online (2012). Online shopping made easy. Retrieved from http://www.thestar.com.my/story.aspx?file=%2f2012%2f7%2f19%2fnorth%2f11688942&sec
[4] Xi Zhang and Yu Tang (2006). Customer Perceived E-service Quality in Online Shopping. Unpublished Master Thesis. LUT.
[5] Santos, J. (2003), "E-service quality: a model of virtual service quality dimensions", Management Service Quality, Vol. 13. No. 3, pp. 233-46.
[6] Yang Hao-erl and Tsai Feng-Shii (2007). General E-S-QUAL Scales Applied To Websites Satisfaction and Loyalty Model. *Communications of the IIMA*. Volume 7 Issue 2.
[7] Grace Lin, T. R., & Chia, C. S. (2009). Factors influencing satisfaction and loyalty in online shopping: an integrated model. *Online Information Review, 33*(3), 458-475. http://dx.doi.org/10.1108/14684520910969907.
[8] Lei Dai , Lihua Huang, Yuandong Yi(2005). How B2C Service Quality Influences Website Continuance?
[9] Parasuraman, A. Zeithaml, V.A., Malhorta, A. (2005). E-S-QUAL, A Multiple-Item Scale for Assessing Electronic and Revolution in Web Site Functionalities. Decision Sciences, 35(3), 423-455.
[10] Yoo, B. & Donthu, N. (2001). "Developing a Scale to Measure the Perceived Quality of an Internet Shopping Site (SITEQUAL),"Quarterly *Journal of Electronic Commerce* Vol. 2 No. 1, 31-46.
[11] Than, C.R., and Grandon, E. (2002) 'An exploratory examination of factors affecting online sales', *Journal of Computer Information Systems*, Vol. 42, No. 3, pp. 87-93.
[12] Wolfinbarger, M. & Gilly, M. C. (2003). "eTailQ: Dimensionalizing, Measuring and Predicting Retail Quality," *Journal of Retailing* Vol. 79,183-98.
[14] Saleh Bukhari, Ahmad Ghoneim and Charles Dennis (2012). Understanding the Factors that Attract Travellers to Buy Airline Tickets Online In Saudi Arabia. *European, Mediterranean & Middle Eastern Conference on Information Systems 2012*
[16] Zeithaml, V.A. (2002). Service excellence in electronic channels. *Managing Service Quality*, 12(3), 135138.
[17] Loiacono, E. T., Watson, R. T. & Goodhue, D. L. (2002). "WEBQUAL™: A Measure of Web Site Quality," in Evans, K. & Scheer, L. (Eds), *Marketing Educators' Conference*: Marketing.
[18] Barnes, S.J. and Vidgen, R.T. (2002). An integrative approach to the assessment of e-commerce quality. *Journal of Electronic Commerce Research* 3(3), 114-127.
[19] Wolfinbarger, M. & Gilly, M. C.(2003). eTailQ: dimensionalizing, measuring and predicting etail quality, Journal of Retailing, Volume 79, Issue 3, Pages 183-198
[20] Rahman, M.M. & Miazee, M.H. (2010). "E-Service Quality and Customer Satisfaction: A Study of Online Customers in Bangladesh".
[21] Jain and Kumar 2011). "Measurement of E-Service Quality: An Empirical Study On Online Railway Ticket Reservation Website Service". *Sri Krishna International Research & Educational Consortium,* Volume 2, Issue 1.
[22] Lee, Gwo-Guang & Lin, Hsiu-Fen. (2005). Customer perceptions of eservice quality in online shopping. *International Journal Retail & Distribution Management*. 33(2), 161-176.
[23] Anderson, R.E., and Srinivasan, S.S. (2003). E-satisfaction and e-loyalty: A contingency framework. *Psychology and Marketing,* 20(2), 123-138.
[24] Schaupp L. Christian, Bélanger France (2005). A Conjoint Analysis of Online Consumer Satisfaction, *Journal of Electronic Commerce Research*, 6(2), pp. 95-111.
[25] Park, C. and Y. Kim. (2003) "Identifying Key Factors Affecting Consumer Purchase Behaviour in an Online Shopping Context" *International Journal of Retail & Distribution Management*, Vol. 31, 1:16–29.

[26] Alam, S. S. and N. M. Yasin, (2009) "An Investigation into the Antecedents of Customer Satisfaction of Online Shopping," *The Australian & New Zealand Marketing Academy (ANZMAC) Conference, Melbourne, Australia*. 30 November – 2 December, Paper number 79, 2009.

[27] Chengwen Yao and Shuling Liao (2011). "Measuring the Antecedent Effects of Service Cognition and Internet Shopping Anxiety On Consumer Satisfaction With E-Tailing Service". *Management & Marketing Challenges for the Knowledge Society,* Vol. 6, No. 1, pp. 59-78.

[28] Hitoshi Okada, Kanokwan Atchariyachanvanich and Noboru Sonehara (2006). "What Keeps Online Customers Repurchasing through the Internet? *ACM SIGecom Exchanges*, Vol. 6, No. 2, December 2006, Pages 47-57.

[29] Mustafa I. Eid (2011). "Determinants of E-Commerce Customer Satisfaction, Trust, And Loyalty In Saudi Arabia". *Journalof Electronic Commerce Research,* VOL 12, NO 1.

[30] Reichheld, F. and P. Schefter, (2000) "E-loyalty: Your Secret Weapon on the Web," Harvard Business Review, Vol. 78, 4:105-115.

[31] Leung Nga Yin, Yanmie and Ma Karman Ka Ching (2012). "Factors Affecting Hong Kong Customers' E-loyalty and Repeated Purchase Intention in Online Chinese Retail Stores and the Moderating Effect of Electronic Word-of-Mouth".

[32] Zeithaml, V.A., Parasuraman, A., and Malhotra, A. "e-service quality (2000): definition, dimensions and conceptual model," Working paper, Marketing Science Institute, Cambridge, MA.

[33] Zha J., Ju F. and Wang L. (2006). Customer Satisfaction in E-commerce: an Exploration of its Antecedents and Consequences, *International Conference on Management of Innovation and Technology*, pp.540-544.

[34] Lam Shuk Ling (2011). "The Study of E-Service Quality, E-Satisfaction And E-Loyalty Of Generation Y In Hong Kong: A Case Of Taobao".

[35] Zhang, X., & Prybutok, V.R. (2005). A consumer perspective of e-service quality. *Journal of Marketing*, 52, 461–477.

[36] Zins, A., "Relative Attitudes and Commitment in Customer Loyalty Models: Some Experiences in the Commercial Airline Industry," *International Journal of Service Industry Management*, Vol. 12, 3:269-294, 2001.

[37] Devaraj, F., Fan, M. & Kohli, R. (2002). Antecedents of B2C Channel Satisfaction and Preference: Validating e-commerce Metrics, *Information System Research*, *13*(3), 316-333.

[38] Kim E. Y., Jackson V. P. (2009). "The Effect Of E-SERVQUAL on E-Loyalty For Apparel Online Shopping". *Journal of Global Academy of Marketing Science,* 19-4 (2009) 57-65.

[39] Xiaoying Guo, Kwek Choon Ling & Min Liu (2012). "Evaluating Factors Influencing Consumer Satisfaction towards Online Shopping in China". *Asian Social Science*; Vol. 8, No. 13.

[40] Hill R.R. (1998). Analytical comparison of optimization problem generation methodologies. *Proceedings of 30th conference on winter simulation*. Pg 609-616.

[41] Gee, R., Coates, G. and Nicholson, M. (2008). Understanding and profitably managing customer loyalty. *Marketing Intelligence and Planning*. Vol. 26 No.4, pp. 359-374.

Attitudinal Factors Affecting Viral Advertising Pass-On Behaviour of Online Consumers in Food Industry

Nurhidayah Mohd Salleh[1], Mohd Shoki Md Ariff*[1], Norhayati Zakuan[1], Zuraidah Sulaiman[1], Muhamad Zameri Mat Saman[2]

1 Department of Business Administration, Faculty of Management,
2 Faculty of Mechanical Engineering,
Universiti Teknologi Malaysia (UTM), 81300 Skudai, Johor, Malaysia

E-mail: m-shoki@utm.my

Abstract. The increase number of active users of social media, especially Facebook, stimulates viral advertising behaviour among them, thus attracting e-marketers to focus on viral advertising in promoting their products. In global market, use of Facebook platform indicated that food services/restaurant of food industry is ranked number 11 with 18.8% users' response rate within the platform. This development calls for e-marketers in Malaysia to use Facebook as their viral advertising channel. Attitudinal factors affecting the viral advertising pass-on behaviour (VAPB) especially among members of social media is of interest to many researchers. The typical attitudinal factors used were attitude toward social media (ATSM), attitude toward advertising in social media (AASM) and attitude toward advertising in general (AAIG). Attitude toward advertised brand (ATAB) is important in fast food industry because users of social media tend to share their experience about tastes and features of the food. However, ATAB is less emphasized in the conceptual model between attitudinal factors and VAPB. These four factors of consumer attitude served as independent variables in the conceptual model of this study and their effect on viral advertising pass-on behaviour among members of Domino's Pizza Malaysia Facebook page was examined. Online survey using a set of questionnaire which was sent to the members of this group via private message was employed. A total of 254 sets of usable questionnaires were collected from the respondents. All the attitudinal factors, except for AASM, were found to have positive and significant effect on VAPB. AAIG exerted the strongest effect on VAPB. Therefore, e-marketers should emphasize on developing a favourable attitude toward advertising in general among members of a social media to get them involve in viral advertising. In addition, instilling a favourable attitude towards advertised brand is also vital as it influences the members to viral the brand among them. Managerial and theoretical implications of the study are discussed based on the results.

Keywords: Viral advertising; Consumer Attitude; TPB, Fast Food Industry, Facebook

1.0 Introduction

Social media involves the group of Internet-based applications that allow the creation and exchange of User-Generated Content [1,2]. Based on the characteristics of its applications, the social media is classified as content communities (e.g., YouTube), social networking sites (e.g., Facebook, MySpace), microblogging (e.g., Twitter), and virtual social worlds (e.g., Second Life) [2]. The social media emerged as a famous platform for users to exchange and share information. It encourages advertisers to give more effort to use an online social networking as a medium to communicate with their consumers [1,3]. Facebook emerged as primary 'hang out' [4] especially among young users to interact, exchange and sharing information with other people. In Malaysia, the active social network users are 13 million users [5]. This provides opportunity for e-marketers to examine attitudinal factors that motivate users of Facebook to pass-on the information in order to facilitate viral advertising campaign. In fact, Facebook platform provides a promising platform to develop viral-driven, multidirectional communication with consumers [4].

It was reported that 70% of restaurants are listed on social media since 2009 [6]. Food services or restaurant is ranked number 11 in the usage of Facebook platform by industries [7]. A study on top 20 restaurants in social media revealed that, on aggregate, 6.7 posts by admins per week, 227 posts by fans per week, 5.2 average response time for fan post and 32% average response rate in the usage of Facebook platform [8]. Fast food operators use the social media to advertise their products and services, as well as to facilitate users of social media to viral the advertised brand. Interestingly, a positive correlation existed between restaurants' revenue and their social media review [6]. In Malaysia online shopping, food and beverage is ranked number three in which 49% online shoppers buy the products online [9]. Pizza Hut international page has 1,428,309 fans and Pizza Hut Malaysia has 1,322,791 fans through Facebook pages [10]. Domino's Pizza page has 9,090,893 fans in the world and 244,668 in Malaysia through the Facebook page [10]. This trend has attracted many researchers and e-marketers to examine viral advertising pass-on behaviour in fast food industry. However, despite the promising of Facebook as a medium to facilitate viral advertising, studies examining how attitude of members of Facebook affects their viral advertising pass-on behavior are less emphasized in Malaysia's fast food industry.

Most previous researches employed attitude toward social media, attitude toward advertising in social media and attitude toward advertising in general in determining how users or members of a social media respond to the viral advertising. In food industry, e-marketers advertised their products and brands in social media, hoping for the users to exchange and spread out the advertised brand among them. This shows that attitude towards advertised brand is vital in determining how a user responses to the viral advertising. However, the use of this type of attitude in fast food industry, particularly in Malaysia, is less emphasized. Therefore, how the combined effect of these four types of consumer attitudes on viral advertising pass-on behaviour in fast food industry should be carried out.

This paper examines the effect of attitudinal factors of members of Domino's Pizza Facebook on viral advertising pass-on behavior. It features attitude toward advertised brand with the three widely used attitudinal factors – attitude toward advertising in social media, attitude toward advertising in general, and attitude toward social media – to determine their effect on viral advertising pass-on behaviour among group members of Domino's Pizza in Malaysia. Thus, this research extends the current literature of consumer attitude and viral advertising behaviour by addressing specific advertised brand to better explain members' viral advertising pass-on behaviour in social media. Further, this study is significant because it examines the effect of attitudinal factors on VAPB in a fast food industry of the Malaysia market. Viral advertising in social media provides many opportunities in advertising landscape, therefore e-marketers should examine how attitude of the members of a social media group influence them to get involve in viral advertising. Thus, this research provides valuable input for designing appropriate viral advertising strategy particularly for fast food operators.

2.0 Literature Review

Marketing and advertising concepts and tools are changing from time to time. Currently online marketing [11] and viral advertising in social media [12,13] are becoming increasingly important and crucial in determining the success or failure of online business. The social media has significantly changed the way consumers' process marketing information by sharing the information with other users in the purchasing decision making. Consumers exchange useful marketing information and provide personal advice that helps them making a purchase decision. Therefore, e-marketers started using viral advertising in the process of communication with customers and provide details of their brand or products in the social media [14]. In fact, interpersonal advices are becoming important trend in electronic environment business and viral advertising has become a significant conversation tool for e-marketers [13]. Compare to the traditional advertising, viral advertising increases e-marketers ability to reach users of social media through an information delivery medium that is a lot more intimate and personalized [14]. The advertisement placed in the social media could produce electronic word of mouth that could sometimes last for years [14], thus increase the likelihood of consumers to purchase their products.

Viral advertising is a type of advertising that infects customers having an advertising message which goes by from one customer to the next [13,15]. As unpaid media content, viral advertising is considered as personal advertising [13] because users voluntarily share and viral the advertising message to inform, explain, or even influence the other users to use or buy a product. In viral advertising campaigns, viral messages concerning a brand spread to probable consumers, who then pass on the information to other potential customers continuously [12,16]. The essential objective of viral advertising is to change message content into viral and make that distributed by reliable sources including family or friends [13].

Factors impacting viral advertising pass-on behavior of consumers are of interest to e-marketers and online business researchers. Previous researches highlighted that user's attitudes toward social media should be relate positively to their use of the information available on the sites, which in turn could enhance their tendency to join a group and engage in a viral message or advertising [12]. This indicates that when a user has a favourable attitude toward a social media, for example Facebook, he/she is willing to participate actively in Facebook group to provide or pass on product related information [12]. According to the new theory of planned behaviour (TPB), six constructs – attitude, behavioural intention, subjective norms, social norm, perceived power and perceived behavioural control - are collectively represent a person's actual control over the behaviour [17]. Attitude is the amount to which a person has a favourable as well as unfavourable evaluation of the behaviour of interest, i.e. the outcomes of carrying out the behaviour. It is assumed to have two components, which are, belief about consequences of the behaviour (behavioural belief) and the corresponding positive or negative judgments about each of these features of the behaviour (outcome evaluation). Attitudes influence an individual behaviour, for example viral advertising pass-on behaviour, by affecting his or her intentions [12]. In the context of social media, individuals are more likely to rely on the website's content as a source of information and facts when they have favourable behaviour toward the site [12, 18]. Therefore, e-marketers are interested in examining attitudinal factors affecting behaviour of users of social media because it helps them strategizing the viral advertising pass-on behaviour.

Review of current researches (2010 – 2014) on attitudinal factors affecting viral advertising pass-on behaviour showed that three types of attitudinal factors were widely used. These attitude factors are attitude toward social media [4,12] attitude towards advertising in social media [4,19] and attitude toward advertising in general [4, 20, 21]. Attitude toward advertised brand has attracted many researchers [19,21] because this factor influences directly attitude of users of social media on their behaviour and interest towards the particular brand. Based on Theory of Reasoned Action [23], these attitudinal factors are proposed as independent variables in the conceptual framework of this study, as depicted in Figure 1. These factors are determined based on the argument that users' intention to perform certain behaviour, in this case viral advertising pass-on behaviour, is influenced by their positive or negative feelings [22] or emotion [23] when they are considering making a purchase decision.

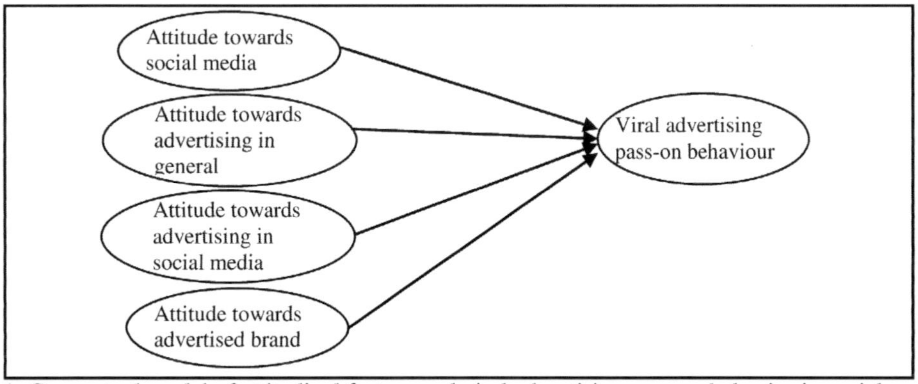

Figure1: Conceptual model of attitudinal factors and viral advertising pass-on behavior in social media

Attitude is "a psychological tendency that is expressed by evaluating a particular entity with some degree of favour or disfavour" [12]. Attitude is defined as consumer's positive or negative feelings when they are about to make the purchasing decision [22] or towards a behaviour, for example viral advertising pass-on behaviour. It is frequently related to emotion and based on the Theory of Reasoned Action [23], attitudes influence individuals' behaviour by affecting their intentions towards something [12]. This relationship is based on the assumption that individuals attempt to make rational decisions based on information available to them, and therefore their behavioural intention to perform or not to perform viral advertising pass-on behaviour is the immediate determinant of their actual behaviour [24]. In the social media setting, individuals are more likely to rely on the media or website's content as a source of information and facts when they have favourable behaviour toward the site [12,18]. In general, relevant and useful viral information affect consumer attitudes in a positive way [13].

Users' favourable or unfavourable attitudes toward social media are positively related to their use of the information available on the websites [12, 25]. When a favourable attitude is formed toward the social media, individuals are willing to participate actively in the media to provide or pass on product related information [12].

Attitude toward advertising is defined as "a learned predisposition to respond in the consistently favorable or unfavorable manner to advertising in general" [26]. In general, attitude towards advertising indicates how a respondent or an audience reacts to any particular advertisement [27]. Early researches measured attitude toward advertising in general based on the "overall favorability and favorability factor" [26]. Two dimensional – the institution dimension and the instrument dimension were employed to determine consumers' attitude towards advertising in general [27].

Based on the view of Dianoux, et.al [27], attitude toward advertising in social media refers to the how users or members of a social media reacts to the advertisements appeared in the media. It is reported that more favorable attitude existed among members of group of social media [12]. Thus, the users who are members of Facebook group tend to have more positive attitude toward advertising in social media.

In social media, advertisements, in whatever form – logos, verbal or pictorial content, graphics and texts, etc. - are usually refer to as advertised brand [19]. Attitude toward advertised brand can be defined as audiences' affective reaction to the advertised brand [21]. It is a tendency to reply in either a favourable and unfavourable manner towards the specific brand after the exposure of advertising on that person [20]. Attitude towards advertised brand in social media, therefore, is individuals' or members' of a social media group affective reaction towards a brand that being advertised in the media. The consumers even enjoy advertisements for brands they like and may actively search for them and expose themselves to them voluntarily [19] and this could lead them to viral the advertised brand.

Viral advertising is unpaid peer to peer communication of attention seeking content originating sponsor online to persuade or effect an audience to pass over the content to others [11,12,14]. This view allows advertisers to use their Facebook group member lists to deliver the viral message with regards to contests or promotions [12]. Viral advertising is unique in that it enables users to select the ads and pass them along to their social connections. For example, via Facebook page, they can share videos, picture or even text, which should help companies to spread word of good messages about their brands quickly [12]. Therefore, viral advertising pass-on behaviour is of important to e-marketers to develop a strong and favourable brand image among users or group members of social networking sites.

Based on the TPB, attitudes of individuals can influence their intention and actual behaviour. It means that in a social networking site setting, an individual behaviour can be influenced by his or her attitude toward the social media, information or anything advertised in the social media. The individual are more likely to rely on the website's content as a source of information and facts when they have favourable behaviour toward the site [12]. For example, individuals' willingness to engage in viral advertising pass-on behaviour can be influenced by how they perceive, see, judge and like the social media or information and advertisement features in the social media. It means that their attitudes toward social media is positively related to their use of the information available on these sites, which in turn could enhance their tendency to join group and engage in viral message [12]. When a user has a favourable attitude toward social media, for example Facebook, he/she should be more willing to participate actively in the Facebook group to provide or pass on product related information [12]. This fact highlights that attitude towards social media, advertising in social media, advertising in general; and attitude towards advertised brand in social media will facilitate individuals to discuss and share the idea and information among them. Group members of a social media, especially teenagers, is reported to have more favourable attitude towards social media and advertising in social media [12] and in turn, facilitate them to viral the advertisement among them [28]. It implies that members of Facebook group tend to have more positive attitude toward social media and advertising in social media allowing them to share the information about advertised brands and form viral advertising pass-on behaviour among them. Previous studies indicated that the attitudinal factors (ATSM, AASM, AAIG and ATAB) positively and significantly influences the viral advertising pass on behaviour [12,19, 20, 26, 27].

$H1_{(a)}$: *Attitude toward social media is positively and significantly affects the viral advertising pass-on behaviour.*

$H1_{(b)}$: *Attitude toward advertising in social media is positively and significantly affects the viral advertising pass-on behaviour.*

$H1_{(c)}$: *Attitude toward advertising in general is positively and significantly affects the viral advertising pass-on behaviour.*

$H1_{(d)}$: *Attitude toward advertised brand is positively and significantly affects the viral advertising pass-on behaviour.*

3.0 Methodology

This study attempted to investigate whether attitude towards social media, advertising in social media, advertising in general and attitude towards advertised brand of group members of Facebook influence their willingness to attend to viral advertising. In this study, four attitudinal factors – ATSM, AASM, AAIG and ATAB serve as independent variables and the viral advertising pass-on behaviour – VAPB, is dependant variable. A questionnaire was used using five points Likert Scale ranging from 1- Strongly disagree to 5 - Strongly agree to measure both independent and dependent variables of this study.

- Six items for Attitude toward social media (ATSM) were developed based on the research of Chu [12].
- Attitude toward advertising in social media (AASM) was measured using six items from the study of Chu [12] and Ketelar [19].
- Attitude toward advertising in general (AAIG) with six items, which were adopted from Chu [12] Dianoux, et.al [27], and Mehta [20].
- Four items of Attitude toward advertised brand (ATAB) were adopted from Keteler [19], and Wei Jia Tan, et.al [20].
- Four items of Viral Advertising Pass-on Behaviour (VAPB) were constructed based on the researches of Chu [12] and Ketelar [19].

It is important to note that utilising six items of ATSM, AASM and AAIG, and four items for ATAB and VAPB showed high validity and reliability of the constructs, based on the previous researches [12, 19, 20, 27]. Further, those studies were carried out in marketing and social media/Facebook setting, particularly on viral advertising pass-on behaviour; therefore they are suitable with the context of this study.

In demographic section, 5 questions, which are, age, gender, time and frequency of using the internet (daily), time spend on Facebook (daily) and a filter question - Have you 'like' Domino's Pizza Malaysia Facebook page.

The population of this study is members of Domino's Pizza Malaysia Facebook page. Since the total population is unknown, 260 sets of questionnaire are needed for 5% margin error [29] since the total numbers of questions in the questionnaire are 26. Convenience sampling was utilized to draw the sample from the population due to time and cost constraints. A web-based survey was employed by creating the questionnaire using Kwisurvey Form, www.kwiksurveys.com and the link of the questionnaire was shared and distributed in Domino's Pizza Facebook page in order to make sure all the respondents are members of the Facebook page. A filter question (a question asking whether the respondents have click button 'LIKE' on Domino's Pizza Malaysia Facebook page was included) before the respondents were allowed to proceed to the survey in order to make sure the questionnaire reached the correct respondents. Furthermore, an advertisement of Domino's Pizza was displayed in the questionnaire so that attitude towards advertised brand can be determined. The data collected was automatically stored in Kwisurvey analysis and finally exported to Statistical Package for Social Science (SPSS) for further analysis.

For instrument assessment, the normality test, linearity and multicollinearity tests were performed to make sure the underlying assumptions of correlation and regression analysis were met. It can be concluded that all the variables are normally distributed based on the skewness and kurtosis values [30]. Linearity test was conducted to ensure the correlations between two variables are linear. Scatterplot was used to demonstrate that fits the data. The scatterplots for ATSM, AASM, AAIG, ATAB and VAPB had shown linear relationship between all the variables. For the multicollinearity test, the results of VIF value were all below 10.00. Hence, there is no multicollinearity problem exists in this study. The results indicate the VIF values for ATSM, AASM, AAIG and ATAB were 1.562, 1.791, 1.308 and 1.179 respectively.

The KMO measures of the sampling adequacy test and Bartlett's test of Sphericity were performed to confirm the suitability of the data for factor analysis. A high KMO value (close unity or > 0.6) and small p-value for the Bartlett's test ($p\text{-value} < 0.05$) were desired. In the first round of Exploratory Factor Analysis (EFA), the KMO of 0.839, Bartlett's test (Chi-Square 3120.960, Sig. 0.000) with the total variances of 61.687% were recorded. The rotated component matrix yielded the four factors of users' attitude, i.e. the proposed independent variables. However, the factor loading for ATSM5 (I thought the viral advertisements on Facebook were very good) and AAIG 6 (On average, brands that are advertised are better in quality than brands that are not advertised) were 0.482 and 0.357 respectively. Therefore, both items were omitted. The KMO of 0.834, Bartlett's test (Chi-

Square 2979.437, Sig. 0.000) with the total variances of 65.531% were achieved in the second round of EFA. Further, the factor loading for all items of attitudinal factors were >0.5, thus they were retained for further analysis. The data used is considered independence and valid for further analysis.

The KMO of 0.692, Bartlett's test (Chi-Square 265.831, Sig. 0.000), and the total variance of 55.857% were recorded for VAPB. Three items of VAPB with factor loading >0.5 were retained and item VAPB 3 (If somebody asks me for advice about an interesting Facebook viral advertising campaign by Domino's Pizza, I will encourage him or her to participate) was deleted since the factor loading was 0.385. The second round of EFA for VATB without VAPB 3 was performed and the results (KMO – 0.701; Bartlett's test (Chi-Square 242.542, Sig. 0.000), total variance explained 71.572%) showed that the VAPB of this study is independence and valid for further analysis.

Table 1: EFA results (second round) for independent and dependent variables

CODE	VARIABLES	FACTORS				
		1	2	3	4	VAPB
ATSM1	Facebook is part of my everyday activity	.178	**.763**	.056	.112	-
ATSM2	Facebook has become part of my daily routine.	.197	**.777**	.201	.136	-
ATSM3	I am proud to tell people I am on Facebook.	.243	**.752**	-.051	.072	-
ATSM4	I would be sorry if Facebook shut down.	.247	**.768**	-.069	-.022	-
ATSM6	I spend my time on Facebook when I should be doing other things.	.509	**.521**	.209	.010	-
AASM1	I liked very much the viral advertisement communication on Facebook.	**.698**	.306	.259	.000	-
AASM2	The viral advertisements on Facebook were appealed very much to me.	**.744**	.354	.127	-.048	-
AASM3	I thought the viral advertisements on Facebook were fitted very well with the brand.	**.761**	.292	.157	.146	-
AASM4	I thought the viral advertisements on Facebook were very pretty.	**.850**	.140	.149	.129	-
AASM5	I thought the viral advertisements on Facebook were very good.	**.874**	.133	-.036	.138	-
AASM6	I thought the viral advertisements on Facebook were very original.	**.725**	.137	.384	.165	-
AAIG1	Advertising helps me keep up-to-date.	.292	-.162	**.726**	.146	-
AAIG2	Too many products do not perform as well as the ads claim.	.127	.046	**.798**	.152	-
AAIG3	Advertising is more manipulative than it is informative.	.103	.068	**.856**	.164	-
AAIG4	Much of advertising is way too annoying.	-.021	.242	**.481**	-.014	-
AAIG5	I like to look advertising.	.225	-.023	**.580**	.075	-
ATAB1	I had positive attitude toward the brands.	.225	.067	.154	**.705**	-
ATAB2	I was interested in the brand.	.068	-.023	.048	**.847**	-
ATAB3	I knew a lot about the brand.	.113	.245	.366	**.686**	-
ATAB4	I knew a lot about the brand compared to other brands	-.027	.057	.047	**.879**	-
VAPB1	I would consider passing along viral advertising of Domino's Pizza to someone I know.	-				**.866**
VAPB2	If I find an interesting Facebook viral advertising campaign by Domino's Pizza, I want to tell my friends about it.	-				**.860**
VAPB4	I would recommend my friends and family to participate in an interesting campaign by Domino's Pizza.	-				**.811**
Total variation explained		65.531%				71.572%
Kaiser-Mayer-Olkin Measure of Sampling Adequacy		0.834				0.701
Bartlett's Test of Sphericity						
Approac. Chi-Square		2979.437				242.542
D.F		190				3
Significant		0.000				0.000

In order to measure the degree of consistency and correlations among the items, reliability test was performed. All alpha values for the ATSM (0.841), AASM (0.910), AAIG (0.770) and ATAB

(0.822) of the independent variables and the VATB (0.800) of the dependent variable were above 0.7; therefore all the dimensions and items in the questionnaire were reliable in this study.

4.0 Result and Analysis

The correlation between all the variables in this study was presented in Table 2. It shows that positive and significant correlations existed between all the variables. ATSM, AASM, AAIG and ATAB of attitudinal factors correlated positively with the viral advertising pass-on behaviour of group members of Facebook. The correlation between the attitude factors and viral advertising pass-on behaviour revealed a strong and positive relationship between AAIG and VAPB (.502, Sig .000) among the group members of Facebook.

Table 2: Pearson's Correlation test between attitudinal factors and viral advertising pass-on behaviour

Correlations (N = 254)						
		ATSM	AASM	AAIG	ATAB	VAPB
ATSM	Pearson Correlation	1	.593**	.206**	.227**	.323**
	Sig. (2-tailed)		.000	.000	.000	.000
AASM	Pearson Correlation	.593**	1	.412**	.274**	.387**
	Sig. (2-tailed)	.000		.000	.000	.000
AAIG	Pearson Correlation	.206**	.412**	1	.350**	.502**
	Sig. (2-tailed)	.000	.000		.000	.000
ATAB	Pearson Correlation	.227**	.274**	.350**	1	.472**
	Sig. (2-tailed)	.000	.000	.000		.000
VAPB	Pearson Correlation	**.323****	**.387****	**.502****	**.472****	1
	Sig. (2-tailed)	.000	.000	.000	.000	

**. Correlation is significant at the 0.01 level (2-tailed).

The effect of attitudinal factor on the viral advertising pass-on behaviour among group members of Facebook was tested using multiple regressions analysis. The result, as presented in Table 3, indicated that ATSM (β 0.134, t 2.153, Sig. 0.032), AASM (β 0.089, t 1.334, Sig. 0.183), AAIG (β 0.333, t 5.870, Sig. 0.000) and ATAB (β 0.301, t 5.577, Sig. 0.000) positively affect the members' viral advertising pass-on behaviour. Therefore, $H1_{(a)}$, $H1_{(c)}$ and $H1_{(d)}$ were supported. However, $H1_{(b)}$ was not supported, indicating no significant effect of AASM on the members' viral advertising pass-on behaviour.

Table 3: Result of the effect of attitudinal factors on viral advertising pass-on behaviour

Coefficients[a]								
Model		Unstandardized Coefficients		Standardized Coefficients	t	Sig.	Collinearity Statistics	
		B	Std. Error	Beta			Tolerance	VIF
1	(Constant)	.448	.229		1.961	.041		
	ATSM	.101	.047	.134*	2.153	.032	.640	1.562
	AASM	.075	.056	.089	1.334	.183	.558	1.791
	AAIG	.344	.059	.333**	5.870	.000	.765	1.308
	ATAB	.295	.053	.301**	5.577	.000	.848	1.179
	F Statistic: 39.246 R Square (R^2): 0.387							
	a. Dependent Variable: VAPB							

5.0 Discussions

This study attempted to determine the effect of attitudinal factors on viral advertising pass-on behaviour among group members of Facebook, a social networking site. The result indicated that

ATSM, AAIG and ATAB of attitudinal factors were positively and significantly affected VAPB. The positive effect of ATSM on VAPB is consistent with the studies of Daugherty, et.al. [25], Mehta [28], and Chu [12]; the effect AAIG on VAPB is sync with the works of Chu [12] and Ketelar [19]; and the positive impact of ATAB on VAPB is supported with the researches of Wei Jia Tan, et.al [20] and Keteler et. al. [19]. This scenario was also observed in Domino's Pizza Facebook Page in the United State. Generally, it shows that when members of Facebook have favourable attitude towards the social media, advertising in general and advertised brand, they tend to get involve in the viral advertising pass-on behaviour, which is sync with the TPB theory of human behaviour [17]. For fast food operators like Domino's Pizza of Malaysia, it is important to associate the company with the Facebook (ATSM) since the members of Facebook treat using Facebook as their daily routine. Information about the products, menu and delivery should be continuously updated since the members are interested in the brand and knew a lot about the brand (ATAB). Further, avoiding manipulative and annoying advertisement in the Facebook is also crucial (AAIG) since they dislike advertisement featuring excessive claim. Such effort would facilitate members of Facebook to viral the advertised brand or information featured in the Facebook.

AAIG (β 0.333, t 5.870, Sig. 0.000) was found to have stronger effect on the VAPB. This finding is inconsistent with the study of Chu [12], who found AASM having stronger effect on the VAPB. It shows that attitude toward advertising in general is crucial for the users of Facebook to commit themselves in viral advertising. It is expected that members of Facebook to have favourable attitude towards an advertisement, before they form a favourable attitude towards the advertised brand. In this study, ATAB (β 0.301, t 5.577, Sig. 0.000) is equally important in facilitating viral advertising pass-on behaviour among members of a networking site. As ATAB is the additional attitude variable included in the study, it shows that the advertised brand of Domino's Pizza has attracted the members to viral it among them.

Generally, all the four attitudinal factors provide 65% of the total variance explained of EFA analysis and 38.7% of (R^2 0.387). It seems that attitudinal factors used in this study are sufficient to establish a conceptual model of viral advertising pass-on behaviour from the consumer attitude perspective. However, other factors, such as attitude towards the advertised firms could be considered to determine construct of consumers' attitude affecting viral advertising pass-on behaviour in social media. Further, some items of attitudinal factors were removed due to low factor loading. Hence, these items should be further examined on their suitability to measure the attitudinal factors of the users of social media.

6.0 Conclusion

This study highlighted that attitudinal factors - attitude toward social media, attitude toward advertising in social media, attitude toward advertising in general and attitude toward advertised brand – affected the viral advertising pass-on behaviour among group members of Facebook. It can be concluded that consumer attitude is important variable in predicting individual behaviour, such as viral advertising pass-on behaviour. When individuals form favourable attitude towards a social networking site, advertising in general and the advertised brand, they tend to get involve in viral advertising. In contrast, the unfavourable attitude formed will influence them not to commit in viral advertising, thus making it difficult for e-marketers to stimulate viral advertising of their brands.

This study only involved members of Domino's Pizza Malaysia Facebook page. In reality, viral advertising pass-on behaviour is also observed among non-members. Therefore, comparing attitude of these two types of Facebook users towards viral advertising is interesting to research. Further, it is important to differentiate attitude of users of a social networking site according to their level of involvement in the usage of the site for better understanding of attitudinal factors affecting viral advertising pass-on behaviour. Future study should also extend this coverage of fast food operators to include major restaurants to explore how members' attitude influences their viral advertising pass-on behaviour.

References

[1] Chi (2011). Interactive Digital Advertising VS Virtual Brand Community: Exploratory Study of User Motivation and Social Media Marketing Responses in Taiwan, Journal of Interactive Adv., Vol 12 No. 1 (Fall 2011), pp. 44-61.
[2] Kaplan & Haenlein (2010), "Users of the world, Unite! The Challenges & Opportunities of Social Media," Business Horizons, 53(1), 59-68.
[3] Anthony Curtis (2013), The Brief History of Social Media, http://www2.uncp.edu/home/acurtis/NewMedia/SocialMedia/SocialMediaHistory.html
[4] Shu-Chuan Chu, 2011). Viral Advertising in Social Media: Participation in Facebook Groups and Responses among College-Aged Users. Journal of Interactive Advertising. Vol. 12, No. 1.
[5] wearesocial.net. Simon Kemp (2013), The State of Social in Asia, News on 26 March 2013, http:// http://wearesocial.net/tag/social-networks/
[6] Joel Windels, 2012. How Social Media is Revolutionising the Restaurant Industry, Brandwatch Blog.
[7] socialmedia.org.nz. Michael Carney (2014), Growth of Facebook Platform by Industries 2013, http://socialmedia.org.nz/facebook-platform/industry
[8] pivotcon.com. 2013.
[9] ecommercemilo.com. Andrian Oh (2013), Categories Online Shopper through industries rank Statistic 2014, http://www.ecommercemilo.com/search/label/socialmedia
[10] Domino's Pizza, Facebook Page Statistic in Malaysia 2014, from http://www.socialbakers.com/facebook-pages/6657899956-domino-s-pizza/in-malaysia
[11] M. Petrescu. (2011), "Viral Advertising: Conceptual and Empirical Examination of Antecedents, Context and Its Influence on Purchase Intentions", Florida Atlantic University Boca Raton, Florida, May 2012, 1-199.
[12] Chu (2011). Viral Advertising in Social Media: Participant in Facebook Groups and Responses Among College-Aged Users, Journal of Interactive Advertising, Vol 12 No. 1 (Fall 2011), pp. 30-43.
[13] Aydin, D.A. Gulerarslan, S. K and Dogan, T (2013). Value of Sharing: Viral Advertisement, International Journal of Social, Human Science and Engineering, Vol 7, No. 5, 2013.
[14] Paqutte (2013). Social Media as a Marketing Tool: A Literature Review, Major Papers by Master of Science Students, Paper 2, University of Rhode Island.
[15] Palka, Pousthchi and Wiedemann (2010), "Mobile word-of-mouth. A grounded theory of mobile viral marketing", Journal of Information Technology, 24, 172-185.
[16] Dobele, 2007, "Why Pass On Viral Message? Because They Connect Emotionally," Business Horizon,50(4),291-304.
[17] Ajzen, I. (2006). The Theory of Planned Behavior, behavioral Change Model. January 6, 2016. Boston University School of Public Health
[18] Poh, Desmond M.H. and Stewart Adam (2002), "An Exploratory Investigation of Attitude toward the Website and The Advertising Hierarchy of Effects," http://ausweb.scu.edu.au/aw02/papers/refereed/poh/paper.html
[19] Ketelar P. (2012), "Why do people pass on viral advertising on social network sites? Investigating the effects of social and attitudinal factors", Journal of Advertising Research, 2012,1-34.
[20] Wei Jia Tan, Choon Ling Kwek & Zhongwei Li (2013), The Antecedents of Effectiveness Interactive Advertising in Social Media, International Business Research, Vol. 6, No.3; 2013.
[21] Mannochehr Najmi, Yashar Atefi & SeyedAlizera Mirbagheri (2012), "Attitude toward Brand: An Integrative Look at Mediators and Moderators", Academy of Marketing Studies Journal, Volume 16, November 1, 2012, 111-130.
[22] Chiu YB, Lin CP, Tang LL (2005). Gender differs: assessing a model of online purchase intentions in e-tail service. Int. J. service Industry Manag. 16(5).
[23] Fishbein, M., & Ajzen, I. (1975). Belief, attitude, intention and behavior: An introduction JIBC, April 2015, Vol. 20, No.1-24-to theory and research: Addison-Wesley Reading, MA.
[24] Ajzen, I. and Fishbein, M., (1980) *Understanding Attitude and Predicting Social Behavior*. Prentice-Hall, Cliffs, NJ.
[25] Daugherty, Terry, Matthew S. Eastin and Laura F. Bright (2008), "Exploring Consumers Motivations for Creating User Generated Content," Journal of Interactive Advertising, 18 (2).
[26] Abhilash Mehta & Scott C. Purvis (1995), "When Attitudes toward Advertising in General Influence Advertising Success", Conferences of the American Academy of Advertising, Norfolk, VA, 1995, 1-17.
[27] Christian Dianoux, Zdenek Linhart and Galjina Ognjanov (2010), A cross-European perspective on attitudes toward advertising in general and attitude toward an ad, 41th Annual Conference of European Marketing Academy, 22-25 Lisbonne.
[28] Mehta, 2010). Mehta, Abhilasha (2000), "Advertising Attitudes and Advertising Effectiveness," Journal of Advertising Research, 40 (3), 67-72.
[29] Hair, J. F., RE Anderson, R.L. Thatham, and W.C Black,- Multivariate Data Analysis, Fifth Edition, NY, Upper Saddler River: Prentice Hall:1998.
[30] Bruin, J. (2006). Newtest: command to compute new test. UCLA: Statistical Consulting Group. Restricted December 1, 2013 from http://www.ats.ucla.edu/stat/stata/ado/analysis

Online Sellers' Website Quality Influencing Online Buyers' Purchase Intention

Tan Shea Lee[1], Mohd Shoki Md Ariff[1*], Norhayati Zakuan[1], Zuraidah Sulaiman[1] and Muhamad Zameri Mat Saman[2]

[1] Department of Business Administration, Faculty of Management, Universiti Teknologi Malaysia (UTM), Skudai, Johor, Malaysia
[2] Faculty of Mechanical, Universiti Teknologi Malaysia, 81310 UTM Skudai, Johor, Malaysia

E-mail: m-shoki@utm.my

Abstract. The increase adoption of Internet among young users in Malaysia provides high prospect for online seller. Young users aged between 18 and 25 years old are important to online sellers because they are actively involved in online purchasing and this group of online buyers is expected to dominate future online market. Therefore, examining online sellers' website quality and online buyers' purchase intention is crucial. Based on the Theory of planned behavior (TPB), a conceptual model of online sellers' website quality and purchase intention of online buyers was developed. E-tailQ instrument was adapted in this study which composed of website design, reliability/fulfillment, security, privacy & trust, and customer service. Using online questionnaire and convenience sampling procedure, primary data were obtained from 240 online buyers aged between 18 to 25 years old. It was discovered that website design, website reliability/fulfillment, website security, privacy & trust, and website customer service positively and significantly influence intention of online buyers to continuously purchase via online channels. This study concludes that online sellers' website quality is important in predicting online buyers' purchase intention. Recommendation and implication of this study were discussed focusing on how online sellers should improve their website quality to stay competitive in online business.

Keywords: Website Quality, Purchase Intention, Online Shopping, eTailQ, TPB

1.0 Introduction

According to Malaysia Crush 2010, Malaysia's online purchasing market size was RM 1.8 billion in the year of 2010, and it is estimated to increase to RM 5 billion in the year of 2014. Furthermore, 45% or RM 825 million out of RM 1.8 billion was spent on local online sellers' websites. Online purchasing is a technology breakthrough as it has just started to assault the retailing sector with online shopping services in Malaysia [1]. Online buyers who use Internet as a shopping approach are increasingly in line with the changing habits of purchase among the young generation [2]. Online shopping is very convenient for people as they can gain access to different types of products from different suppliers by just a click. There are more than 8 million of Malaysians have actually experience the online purchase [3]. In addition, a study by International Data Corporation (IDC) Asia-Pacific indicates that the future forecast for online purchasing in Malaysia is very promising [4]. The major implication of all these data and forecasts to the local online sellers are: (i) they should have a good understanding on the quality of their websites from online buyers' perspective to ensure continuous online purchase; and (ii). This aids online sellers to have better understanding on what their online buyers' expectation in the usage of their website. Online buyers' purchase intention and factors affecting their intention to continuously purchase online is important to research. This should facilitate online sellers to know their positioning and targeting groups well to stay competitive in online market. In fact, the importance of enhancing quality of online websites to ensure sustainability of online shopping has been well addressed in the evolution of e-commerce.

From the internet users' perspective, it can be seen that young consumers are dominating the usage of internet. In Malaysia, most internet users are youths (15-20 years old; 50 percent) and young adults (20 - 29 years old; 39 percent), as reported by Haque and Khatibi [1,5]. These groups of users consist of those between 15 – 29 years of age, and they play an important role in online shopping. They are active online buyers and it is expected that they will dominate online shopping market segment, thus providing huge online market for online retailers [1,5]. Therefore, to ensure online buyers' intention to continuously purchase online, online sellers need to examine how well their websites fulfil requirements of online buyers.

It was observed that most studies of websites quality of online sellers were carried out to examine their effect on purchasing behaviour among young users, for examples in New Zealand [6] and India [7]. However, studies addressing online purchase intention and repeat purchase are less emphasize, particularly among young consumers in Malaysia. Therefore, it is vital to investigate quality of online sellers' website that may affect continuance of online buyers to purchase online in the Malaysian context. Such study will enriches current literature of website quality and purchase intention especially in different countries and group of online buyers. In addition, it will help online sellers to have better understanding of behavioural intention of online buyers, thus creating a long term profitable relationship with the buyers. Therefore, this research is carried out to answer to the research question of how local online sellers' website quality affecting intention of young online buyers to continuously purchase online?

2.0 Literature Review

Internet Technology permits buyers to purchase online and online shopping is increasingly adopted as a convenience way of buying product [8], especially among young consumers. Online shopping is a process of buying products through Internet and it involves online buyers accessing online sellers' websites to search, select, purchase, use, and dispose of goods and services, in satisfying their needs and wants [5]. Companies need to catch-up on e-commerce to avert losing out on the growing online shopping market, and e-commerce is considered as an unstoppable force [9]. Online sellers need to design quality websites in order to increase the companies' visibility online, thus increase the possibility for online buyers to access and buy products online. Quality of Online sellers' websites must be enhanced in order to be truly successful online retailers.

There are various methods or modes for customers to purchase products and services online. The most common mode is buying from the online catalogues at fixed prices. The process starts with online buyers entering to an e-shop, and register as a member. Then, it will bring the buyers into an online catalogue or the buyer's account. Online catalogue is a list of products or services with the description or provide relevant information and price in the online sellers' website. Therefore, most of the online stores have provided search navigation tools for buyers to search and find their product easily and conveniently. At this point, the design of online sellers' websites is important to allow buyers browsing the information required smoothly and easily.

Price is one of the considerations factor when a buyer makes online purchase. Therefore, a buyer will spent time on comparing the prices. In the websites, the online sellers will provide comparisons of prices and products with competing vendors. If the buyer satisfied with the price, then he or she will continue to shop and select the item and place it in a shopping cart. Shopping cart is an electronic commerce tool for buyers to place products inside a "shopping basket" in order to shop at online environment. When the shopping is completed, the buyer goes to a checkout page and selects a shipment and payment options. Online buyers can choose payment options, for example by credit card, PayPal or other online payment methods. Then, there is a requirement for the buyers to check all the details in order to ensure all purchasing data are correct and accurate. After checking all the details, the buyer needs to submit the order. To this point, reliability and security of online sellers' websites are important. In addition, trust of online buyers towards online sellers is also vital because it allows buyers to perform the online transaction with less security and privacy concerned.

There is a post purchase activities in the online shopping. The online seller ships the items that a buyer ordered. Then, the buyer need to wait for the items delivered. In this stage, both the online seller and the buyer might check the shipping status and the seller needs to measure that the item is successfully delivered to buyer. If the buyer satisfied with the item, this transaction is done. However, if the buyer is not satisfied with the item, then the buyer might return back the item to seller and it is money back guarantee. At this stage, website customer service is important to permit online buyers handling the online purchase problems encountered.

Table 1 presents different scales to measure websites quality of online sellers based on the four major e-SQ instruments – e-SERVQUAL, WebQual, e-TailQ and SiteQual [10,11, 12, 13, 14,15]. E-SERVQUAL deals with customer's interactions with a website, the extent to which the website promotes effective shopping, efficient purchasing and delivery [10,11]. The scale contains core scale (E-S-QUAL) which embraced efficiency, fulfilment, reliability and privacy and recovery scale (E-RecSQUAL) which constituted of responsiveness, compensation and contact. WEBQUAL [12,13] composed of 3 dimensions which are information quality, quality of interactivity, and usability of site/usability and design. E-tailQ [14] consists of website design, reliability, security and customer service. SITEQUAL [15] contained four dimensions involved ease of use, aesthetic design, processing speed, and security. All these dimensions are compared and listed as shown Table 1.

Table 1: A review of website quality dimensions based on major e-SQ instruments

Dimensions of website quality	**e-SERVQUAL**	**WebQual**	**SiteQual**	**e-TailQ**
Efficiency	√			
Reliability	√			√
fulfilment	√			
System availability	√			
Privacy/or security	√		√	√
Responsiveness	√			
Compensation	√			
Contact	√			
Usefulness		√		
Ease of use		√	√	
Entertainment		√		
Complimentary relationship		√		
Aesthetic design			√	
Processing speed			√	
Web design				√
Customer service				√

E-tailQ was developed by adopting focus group and large population of respondents. Each of the E-tailQ factors, variables and items indicate high level of inter-item reliability and validity [5,6,14]. Website design [16,17,18], website reliability/fulfilment [19,20,21], website security and privacy [16,21,22] and Website customer service [6,23] were widely used in determining quality of website for online retailing. As mentioned by Durkan [24], customer trust on website is important prior to the development of favourable attitude of online buyers. Customer trust is linked with security and privacy. This is supported by Djahantighi [25], who stressed that online trust is the obstacle to online retailers to succeed in online business.

Theory of Planned Behaviour [26] is used to forecast a person's intention towards online shopping. The intention is to explain factors that influencing a person's behaviour. Intention is assumed to capture the motivational variables that influence an individual behaviour. In TPB, attitude, subjective norms and perceived behavioural control influence the behaviour intentions of consumers. In online shopping setting, the motivational variables that influence an individual behaviour include how a person perceives quality of online sellers' website. Based on the TPB and assessment of the different scales of website quality, a conceptual model of this study is proposed, as depicted in Figure 1.

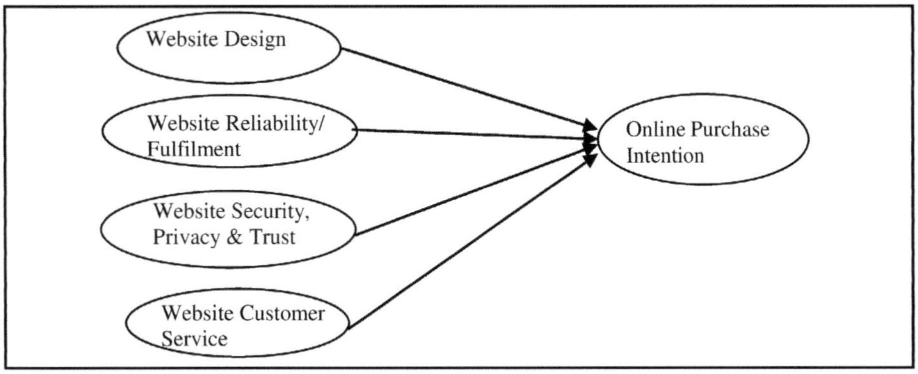

Figure 1: the conceptual model of online sellers' website quality and online buyers' purchase intention

Based on Kim and Lee [20], process and architecture perspectives are important in analysing design of a website. The market transactions are speculated that embraced a number of processes and the system is regarded as a sequence of processes. The architectural comprises the system as a collection of webpage documents. There are four components of design in the architecture perspective. The four components of design encompassed content which represent information shown in website, structure which represent arrangement of information in the webpage, interaction that represent two way communication between sellers and buyers and presentation which represent the way of web page provides details of products to buyers or users. In online purchasing, when a customer is satisfied with the website quality of online sellers, then it will increase his or her involvement in online purchasing [27. This caused customers return to the website more often and create more favourable image of the online sellers. Measure of website design include no time wasting, quick and simple transaction, personalization of the website, understand needs and wants of online buyers and comfortable surfing of the site [5,6,14,34]. Previous studies highlighted that website design has a positive and significant effect on customer online purchase intention [17, 19,28]. Website design is an important factor influencing online buyers' intention to continuously use a website and it is considered to be one of consumers' attitude areas with regards to online buying [18].

Website Reliability/Fulfilment represents the ability of the online sellers' website to fulfil orders correctly, deliver promptly, and keeps personal information secure [19,20]. It is about meeting the promised service in a reliable and dependable manner [6,29]. Therefore, the service provided in the website should fulfil online buyers' requirements. In other words, reliability takes place when the buyer is aware and knowing that there is a high probability the online sellers will deliver exactly what is being promised in their websites. Information content provided on the online sellers' webpage should be correct and reliable as it will lead to high probability for customers to return to the site more often. Website reliability is a crucial factor in the information technology oriented services, such as online shopping. Measure of website reliability/fulfilment include received of the ordered product, timely and correctly received of product and error free transaction, [5,6,14,34]. Previous researches indicated that website reliability and fulfilment positively and significantly influence online buyers' intention to continuously purchase online [5,6].

Website Security, Privacy and Trust are the main concern of the online buyers who want to purchase online. Even though online buyers like to shop online, they want to keep their personal and financial information privacy. According to Malhotra, Kim, and Agarwal [22], privacy concerns is a major issue for many web consumers due to the demands for buyers' information. For example, online transactions typically require providing a billing and shipping address. Perceived security is defined as a hazard that creates an event lead to economic hardship to data or network resources in the form of destruction, disclosures, modification of data, denial of service, and fraud, waste and abuse [31]. Hence, security is an important concern for online buyers and one of the most imperative factors in the success of e-commerce [32]. Further, when a customer trusts the online sellers, he or she believes in the online sellers' capability to protect one's privacy and it is safe to complete an online transaction. Website security is vital for the sense of trust in the online shopping domain, and an online buyer's trust toward a website starts to build up if only the online seller fully meet their needs and expectations [33]. Therefore, there is a link between security, privacy and trust as it plays important role in consumer mind. In online shopping reality, online buyers are concerned about online payment security, reliability, and privacy policy of the online store [29]. Measure of website security, privacy and trust include protection of privacy information, safety usage of credit card, adequate security features, trust, and positive experience of online purchasing [5,6,14,34]. Previous studies indicated that these factors influenced how online buyers rate quality of online sellers, and subsequently effect their online purchase intention [5,6,32,33].

Website customer service deals with prompt responses to consumers' concerns and enquiries [5,6,14]. An online customer's buying experience involves information search and recognition, product evaluation, purchasing decision making, making the transaction, and delivery of the purchased products, therefore, customer service actually plays a vital role in online store. Customer service in online shopping provides assistance and solving customer's inquiry. In addition, online purchasing process involves delivery and verification of the purchased products, thus, website customer service is required to handle complaints or feedbacks from the online buyers. It is important to provide customer service in an online shopping environment as most of the customers significant attributes of shopping online are convenience and accessibility [14,34]. Lack of immediate response, particularly to e-mail enquiries, will create inconvenience situation to online buyers, thus influence their online purchase decision. This means that if the customer service provides immediate response to customers enquires, it will cause buyers to visit and re-visit online sellers' website for online shopping [5,6]. Measure of website customer service include online sellers' willingness and readiness to respond to customer needs, willingness of customer service personnel to help, willingness and sincere in helping online buyers and on time answered of inquiries [5,6,14,34].

3.0 Methodology

A survey questionnaire was employed in this study to respond to the research question discussed earlier. Section A of the questionnaire consists of 7 demographic questions of the online buyers. Section B with 18 questions (Table 2) developed to determine online buyers' perception of web quality of the local online sellers who sell local, imported and franchise products. The website quality is measured by website design, reliability/fulfillment, website security, privacy & trust and website customer service [5,6,14,34]. Section C with 6 questions (Table 2) used to determine online buyers' purchase intention. Two filter questions of "Have you ever made any purchase online" (a list of local online sellers is provided for respondents to choose) and 'are you in the age of 18 – 25 years' are asked to choose the right respondents for this study before they can proceed to fill-up the questionnaire. Thus, the respondents who are fit with these two questions are considered unique and appropriate for this study. All questions in Section B and C used Likert Scale of 1 = very unlikely to 5 = very likely.

Primary data of this research was obtained from online buyers aged between 18 and 25 years old through online questionnaire (google docs), which is placed in the platform. The population of this research is online buyers aged between 18 and 25 years old who have online purchasing experience with any online sellers in Malaysia. According to D'Silva et al., [35], youth constitute half of the Malaysian population, thus researching their online purchase intention is justified. The nonprobability convenience sampling procedure was chosen as the sampling design in this study. Since the population of this study is unknown, multivariate research procedure was employed in this study. The total questions in this questionnaire are 24 questions, thus a minimum of 120 sets of questionnaire are needed for 10% margin error and 240 sets of questionnaire are needed for 5% margin error [36]. 240 sets of completed and usable questionnaire had been successfully collected within 3 weeks. The data collected was kept automatically in Google spreadsheet and exported to Statistical Package for Social Science (SPSS) for further analysis.

4.0 Result and Analysis

Out of 240 respondents, 52.5% or 126 are male and 47.5% or 114 are female. Most of the respondents (61.3% or 147 respondents) are in the age of 22 to 25 and 38.8% or 93 respondents come from those ages between 18 and 21 years. In term of education level, 115 respondents (47.9%) are undergraduate students, 80 respondents (33.3%) hold secondary school level qualification, 25 of respondents (10.4%) hold diploma qualification, 17 of respondents (7.1%) are postgraduate and 3 respondents (1.3%) hold primary school level qualification. For marital status, 228 respondents (95%) are single and 12 respondents (5%) are married.

KMO value for all dimensions of web-based quality is 0.865 and KMO value for online purchase intention is 0.805, which are greater than 0.6 (Table 2). In addition, the results are supported by the value of the Bartlett's test of Sphericity of 0.000. Thus, the adequacy of sample requirement was met to proceed with the Exploratory Factor Analysis (EFA).

Table 2 shows the Rotated component matrix for website quality, the independent variable. All the four dimensions remained as proposed and no items were dropped since all factor loading for the items are >0.5. The total variance explained by these four web-based factors is 67.434%. For dependent variable, only one dimension was extracted; i.e. online buyers' purchase intention with the total variance explained of 51.797% (Table 2). All items of online purchase intention were retained since all factor loading for the items are >0.5.

The Cronbach's Alpha value for website security, privacy and trust is 0.874, 0.841 for website customer service, 0.805 for website design and 0.755 for reliability/fulfilment. The entire reliability coefficient for website quality is 0.919. Thus, the instrument used in this study is valid and reliable for further analysis.

The relationship between online sellers' website quality and online buyers' purchase intention is presented in Table 3. The result shows that all the four independent variables (website design, website reliability/ fulfilment, website security, privacy & trust, and website customer service) are positively correlated to the dependent variable (online buyers' purchase intention).

The research question of this study is how local online sellers' website quality affecting intention of young online buyers to continuously purchase online? To answer to this question, Table 4 shows multiple regression result between online sellers' website quality and online buyers' purchase intention. The Coefficient of correlation of R=0.696 indicated that 69.60% of the overall online purchasing intention could be predicted by the web quality dimensions. In term of R square, there was 48.4% of variation in online purchasing intention (Y) is being explained by website quality dimensions (X). All the four factors of online sellers' website quality positively and significantly influenced online purchase intention of online buyers. Specifically, website design ($\beta = 0.172$, t 2.845, P<.005), website reliability/fulfilment ($\beta = 0.151$, t 2.121, P<.035), website security, privacy & trust (β

= 0.176, t 2.458, P<.015) and website customer service (β =0.473, t – 7.092, P<.000) affect the online buyers' intention to continuously use online as their purchasing channel. The magnitudes of influence induced by website customer service (β =0.473, t – 7.092, P<.000) is the most significant predictor of website quality affecting online buyers' purchase intention.

Table 2: Rotated Component Matrix[a] for Website Quality

CODE	VARIABLES	COMPONENTS				
		1	2	3	4	OPI
WD1	The online sellers' websites do not waste my time.	.288	-.290	.599	.308	-
WD2	It is quick and simple to complete a transaction on the websites	.150	-.020	.614	.354	-
WD3	The level of personalization of the websites is about right, not too much or too little.	.253	.151	.740	.125	-
WD4	The websites understand what I needs and wants.	.074	.306	.733	.020	-
WD5	I feel comfortable when surfing the websites.	.209	.211	.770	.090	-
WR1	I received my product which I ordered from the sites	.091	.090	.191	.835	-
WR2	I received my product by the time that is promised by the online sellers.	.038	.375	.076	.649	-
WR3	I received the correct product that I bought in the websites	.234	.405	.159	.705	-
WR4	I feel that transaction with online purchase is error free	.305	.219	.184	.726	-
WS1	I think the websites can protect my privacy information effectively.	.732	.352	.353	-.112	-
WS2	The websites which I give my credit card number to get my product are safe.	.825	.172	.262	.060	-
WS3	The websites provide adequate security features.	.773	.244	.175	.274	-
WS4	I feel that I can trust these websites.	.702	.163	.198	.416	-
WS5	I have positive experience when purchasing online.	.688	.209	.140	.284	-
WCS1	I feel that the online sellers are willing and ready to respond to customer needs.	.171	.734	.146	.276	-
WCS2	I feel that customer service personnel are always willing to help	.258	.774	.054	.283	-
WCS3	When I faced a problem, the websites show willingness and sincerely in helping me	.196	.758	.181	.166	-
WCS4	When I sent inquiries via the websites, the inquiries are answered on time.	.475	.560	.097	.153	-
OPI 1	I think that the idea of using the websites to continuously purchase a product or service is encouraging	-				**.735**
OPI 2	I think that using the online sellers' websites to continuously buy a product or service would be a good idea.	-				**.754**
OPI 3	I prefer the idea of purchasing a product/service online	-				**.779**
OPI 4	It is likely that I will return to the online seller's websites					**.638**
OPI 5	It is likely that I will consider to repurchase a product /service from the websites in the short term	-				**.635**
OPI 6	Given the chance, I intend to use the online sellers' website to buy a product/service	-				**.764**
Total variation explained		67.434%				51.797%
Kaiser-Mayer-Olkin Measure of Sampling Adequacy		0.865				0.805
Bartlett's Test of Sphericity (Sig: 0.000)						
Approac. Chi-Square		2566.024				439.932
D.F		153				15

Table 3: Pearson's Correlation test between online sellers' website quality and online buyers' purchase Intention

Correlations (N = 240)						
		WD	WR/F	WS, P & T	WCS	OPI
WD	Pearson Correlation	1				
	Sig. (2-tailed)					
WR/F	Pearson Correlation	.529**	1			
	Sig. (2-tailed)	.000				
WS, P & T	Pearson Correlation	.598**	.630**	1		
	Sig. (2-tailed)	.001	.000			
WCS	Pearson Correlation	.394**	.641**	.639**	1	
	Sig. (2-tailed)	.000	.000	.000		
OPI	Pearson Correlation	**.460**	**.498****	**.577****	**.649****	1
	Sig. (2-tailed)	.000	.000	.000	.000	

**. Correlation is significant at the 0.01 level (2-tailed)

Table 4: Result of the effect of online sellers' website quality on online buyers' purchase intention

Coefficients[a]							
Model		Unstandardized Coefficients		Standardized Coefficients	t	Sig.	VIF
		B	Std. Error	Beta			
1	(Constant)	1.213	0.19		6.369	0.000	
	Website design	0.153	0.054	0.172*	2.845	0.005	1.667
	Website reliability/ fulfilment	0.156	0.056	0.151*	2.121	0.035	2.112
	Website security, privacy & trust	0.139	0.057	0.176*	2.458	0.015	2.344
	Website customer service	0.421	0.059	0.473**	7.092	0.000	2.026
	R: .696 R Square (R^2): 0.484 Adjusted R Square (R^2): .475	F Statistic: 55.143 Std. Error of the Estimate = **.35837**					
	a. Dependent Variable: Online buyers' purchase intention						

5.0 Discussions

In brief, the online sellers' websites quality affecting online buyers' purchasing intention are website design, website reliability/ fulfilment, website security, privacy & trust, and website customer service. In general, this finding is consistent with the works of Ruchi, et.al. [37] and Mohd.Shoki [5]. Specifically, website design of online sellers will influence the intention of online buyers to continuously purchase online [5,6,17,18]. Thus it is important for website designers and e-marketers to stress on speed, personalization and comfortable of online buyers when they are surfing the websites. Website reliability/fulfilment is positively and significantly influence online buyers' purchase intention [5,6,17]. It shows that fulfilling orders of online buyers correctly, timely and error free transaction are highly valued by the online buyers. It was found that website security, privacy & trust positively and significantly affect intention of the buyers to continuously purchase online. The finding is consistent with previous researches [5,6,32,33]. This finding is commonly observed worldwide because online buyers expect their privacy information and credit card particulars are protected and safe transaction and trusted services delivered by the online sellers. Website security, privacy and trust imply that it is very vital for online sellers to provide protection of privacy, feeling of safety and adequate security to instil customers' confidence in obtaining the ordered merchandise as well as to safeguard their personal and transactions information. In other words, customers rely heavily on security and privacy control level. If online sellers can assure these features in their websites and during online transaction, the possibility for online buyers to get involve in repeat purchase would be higher. Website customer service is also positively and significantly influence online buyers intention to continuously deal with the online sellers [5,6,32,33]. In fact, it appeared to be the most dominant predictor of online buyers' purchase intention, which is consistent with the work of Syed [21]. This finding calls for e-marketers to focus on responsiveness, willingness to help the buyers and prompt reply to buyers' enquiries for continuous online purchase.

Two focal points are important to highlight in this study. First is on issue of security, privacy and trust. The impact exerted by these website quality features on consumer purchasing behaviour and purchase intention is unarguable. The similar point is addressed among online buyers worldwide, for example among Saudi's women [40] and Indonesia consumers [41] who are concerned with issues of trust, internet trustworthiness and risk in their online shopping [5,6,7,34] and online consumers in New Zealand [6] and India [7]. It can be concluded that, when online buyers rated high on security and privacy of online sellers' websites, they trust the websites as well, then they are willing to continuously purchase online. Secondly, the interrelationship between security, privacy and trust should be further investigated. In previous researches, it was observed that some scholars employed security and privacy of website quality separately or as a combined dimension. Further, most studies employ trust in the conceptual model of website quality when security and privacy are included in the model. Consumer trust is formed when they perceived the websites as having security and privacy features allowing for safe transaction. This study combined trust with the security and privacy to form a website quality dimension and it seems that the total variance explained of 67.434% could be further improved.

6.0 Conclusion

Online shopping involves online buyers accessing online sellers' websites to perform an online transaction. Website quality of online sellers therefore is crucial to ensure repeat purchase or a continuous online purchase performed by online buyers. This effect is proven in this study in which website design, reliability/fulfilment, security, privacy & trust and website customer service are determinant of online buyers' purchase intention. Enhancing these features in online sellers' websites will lead to the purchase of products online on a continuous basis. This is the ultimate aim of any online businesses, thus ensuring and improving website quality is vital.

Some limitations are encountered in this study. Respondents of this study are limit to online purchasing buyers. However, this includes active and non-active buyers. Active buyers are those who are familiar with the online websites and they may perceive that online purchase is ease, convenient and less risk. In contrast, non-active buyers are those who have limited experience with online purchase and they may evaluate online purchase troublesome, inconvenient or risky. Focusing on active online buyers would produce accurate assessment of a website quality, thus ensuring highly reliable data for e-marketers. Further, this study involved those between 18 – 25 years who are also part of Y-Generation consumers. Expanding this research to cover various groups of Y-Generation would be interesting to research because this generation is expected to actively involved in online purchasing. For e-marketers, assessing their websites quality from different consumers in different countries is also important because online shopping is about borderless world.

References

[1] Haque, A. and Khatibi, A., (2005). E-shopping: Current Practices and Future Opportunities towards Malaysian Customer Perspective. Journal of Social Sciences, vol. 1, Issue 1, pp. 41-46.

[2] Cheng, S.Y., Tsai, M.T., Cheng, N.C., Chen, K.S., (2012). Predicting Intention to Purchase on Group Buying Website in Taiwan, Virtual Community, Critical Mass and Risk, Online Information Review, Vol.36 (5), pp. 698-712.

[3] Oriental Daily News (2009).

[4] International Data Corporation (IDC) Asia-Pacific

[5] Mohd. Shoki, M.A., Ng Sze Yan, Norhayati Z., Kamaruzaman A.R, Khalid I. (2014) Online Purchasing Behaviour of Malaysia's Young Consumers. Recent Trends in Behavioural Science. Taylor and Fracis Group. pp: 167-173.

[6] Shergil, G.S., and Chen, Z. (2005). Web-based shopping: consumers' attitudes towards online shopping in New Zealand, *Journal of Electronic Commerce Research*, Vol. 6, No. 2, pp. 79-94

[7] Ruchi, Ashish and Gupta. (2010) 'To study the impact of website factors on consumer online buying behaviour in India', IJMMR, Volume1, Issue 1, pp. 1-16.

[8] Chayapa, K., and Cheng, L.W., (2011). Online Shopper Behavior: Influences of Online Shopping Decision. Asian Journal of Business Research, Volume 1 Number 2. 66 – 74.

[9] Tamara Walsh (2015). TMFSocialME. How Online Shopping Is Changing the Rules for Luxury Retail Stocks

[10] Zeithaml, V.A., Parasuraman, A., and Malhotra, A. "e-service quality (2000): definition, dimensions and conceptual model," Working paper, Marketing Science Institute, Cambridge, MA.

[11] Parasuraman, A. Zeithaml, V.A., Malhorta, A. (2005). E-S-QUAL, A Multiple-Item Scale for Assessing Electronic and Revolution in Web Site Functionalities. Decision Sciences, 35(3), 423-455.

[12] Loiacono, E. T., Watson, R. T. & Goodhue, D. L. (2002). "WEBQUAL™: A Measure of Web Site Quality," in Evans, K. & Scheer, L. (Eds), *Marketing Educators' Conference*: Marketing.

[13] Loiacono, Watson, and Goodhue, (2000), WEBQUAL: A Website Quality Instrument, Univ. of Georgia.

[14] Wolfinbarger, M., & Gilly, M. C. (2003). Etailq: Dimensionalizing, Measuring, and Predicting E-tail Quality. Journal of Retailing, 79(3), pp.183-198.
[15] Yoo and Donthu (2001), An Examination of selected Marketing Mix Elements and brand Equity. *J Acad Mark Sci*. pp.195-211
[16] Ranganathan, C., S. Ganapathy (2002). Key dimensions of business to consumer web sites. Information and Management , pp.457-465.
[17] Cyr . (2008). Modeling Web Site Design Across Cultures: Relationships to Trust, Satisfaction, and E-Loyalty, *Journal of Management Information Systems*, Vol. 24, No. 4: pp. 47-72
[18] Mohd. Shoki, M.A., Ng Sze Yan, Norhayati Z., Kamaruzaman A.R, Khalid I. (2014) Online Purchasing Behaviour of Malaysia's Young Consumers. Recent Trends in Behavioural Science. Taylor and Fracis Group. pp: 167-173.
[19] Lee, Gwo-Guang & Lin, Hsiu-Fen. (2005). Customer perceptions of e-service quality in online shopping. International Journal Retail and Distribution Management. 33(2), pp. 161-176.
[20] Kim, M. J., Chung, N., & Lee, C. K. (2011), The effect of Perceived Trust on Electronic Commerce, Shopping Online for Tourism Products and Services in South Korea. Tourism Management, Vol 32, pp. 256-265.
[21] Syed Sham Alam. (2008). Young Consumers Online Shopping: An Empirical Study, *Journal of Internet Business*, pp.81-89.
[22] Malhotra, N. K., Kim, S. S., & Agarwal, J. (2004), Internet Users' Information Privacy Concerns (IUIPC): The Construct, the Scale, and a Causal Model. ISR, Vol 15(4), pp. 336-355.
[23] Griffith, D.A., Kramph, R.F., & Palmer, J.W. (2001). The Role of Interface in Electronic Commerce: Consumer Involvement with Print versus Online Catalogs. International Journal of Electronic Commerce, 5(4), pp.101-114.
[24] Durkan, P.M. Durkin and J. Gillen,(2008). Exploring Efforts to Engender On-Line Trust, International Journal of Entrepreneurial Behaviour & Research, Vol. 9, No. 3, pp. 93-110.
[25] Djahantighi, F.S. and E. Fakar, E. (2010). Factors Affecting Customer's Trends for Reservation Foreign Hotels via Internet in Iran, International Bulletin of Business Administration, pp. 6-14.
[26] Ajzen, I. (2002), Attitudes, Personality, and Behavior, Open University Press, Milton-Keynes.
[27] Boudhayan Ganguly, Satya Bushan Dash, Dianne Cyr, Milena Head (2010). The effects of website design on purchase intention in online shopping: the mediating role of trust and the moderating role of culture. Int. J. Electronic Business, Vol. 8, Nos. 4/5, 2010. Pp: 302-330
[28] Park Haesun and Leslie Stoel. (2005). A model of socially responsible buying or sourcing decision-making processes. *International Journal of Retail and Distribution Management*. No.33. pp. 235–248.
[29] Gefen, D. (2002) Reflections on the Dimensions of Trust and Trustworthiness among Online Consumers, The DATA BASE for Advances in Information Systems, pp. 38-53.
[31] Roca, García and de la Vega, 2009). The importance of perceived trust, security and privacy in online trading systems Information Management & Computer Security, Vol. 17 Iss: 2, pp.96 - 113
[32] Niranjanamurthy, M., and Dharmendra, C. (2013). The study of E-Commerce Security Issues and Solutions, International Journal of Advanced Research in Computer and Communication Engineering, Vol. 2, Issue 7.
[33] Ural, T. (2009). İşletme ve Pazarlama Etiği. Ankara: Detay Yayıncılık.
[34] Wolfinbarger, M., & Gilly, M. C. (2001). Shopping Online for Freedom, Control, and Fun, California Management Review, 43 (2), pp. 34-55.
[36] Hill R.R. (1998). Analytical comparison of optimization problem generation methodologies. *Proceedings of 30th conference on winter simulation*. Pg 609-616.
[37] Ruchi Nayyar (2010) Ruchi, Ashish and Gupta. (2010) 'To study the impact of website factors on consumer online buying behaviour in India', IJMMR, Volume1, Issue 1, pp. 1-16.
[39] Sharma and Khattri, 2013). Study of Online Shopping Behaviour and Its Impact on Online Deal Website. Asian Journal of Management Research 3, Issue 2, 394 – 405.
[40] ALMowalad , Areeg and Putit , Lennora (2013) Factors influencing Saudi Arabian women's shopping behavior in online purchase activities, UiTM.
[41] Fifi Yusmita, Nik Kamariah Nik Mat, Mulyagus Usman Muhammad, Yuhainis Mohd Yusoff, Fitrisal azhar, Saeed Behjati (2012). Determinants of Online Purchasing Behavior in Nanggroe Aceh Darussalam. American Journal of Economics June 2012, Special Issue: 153-157 DOI: 10.5923/j.economics.20120001.34

IOP Conf. Series: Materials Science and Engineering **131** (2016) 012015 doi:10.1088/1757-899X/131/1/012015

Adaptive Neuro-fuzzy approach in friction identification

Muhammad Zaiyad Muda @ Ismail

Fakulti Kejuruteraan Mekanikal, Universiti Teknologi MARA, Shah Alam, Selangor, Malaysia.

zaiyad13@yahoo.com.my

Abstract. Friction is known to affect the performance of motion control system, especially in terms of its accuracy. Therefore, a number of techniques or methods have been explored and implemented to alleviate the effects of friction. In this project, the Artificial Intelligent (AI) approach is used to model the friction which will be then used to compensate the friction. The Adaptive Neuro-Fuzzy Inference System (ANFIS) is chosen among several other AI methods because of its reliability and capabilities of solving complex computation. ANFIS is a hybrid AI-paradigm that combines the best features of neural network and fuzzy logic. This AI method (ANFIS) is effective for nonlinear system identification and compensation and thus, being used in this project.

1. Introduction

Friction is the tangential reaction force that occurs when there is relative motion between surfaces [1]. Despite its importance in many applications, it also affects the attempts of designers in making an efficient system or machines. In industry, the uses of DC motors are quite common, for example in motion control, positioning and actuation system. Friction identification and compensation comes in place when there is need to improve the current motion accuracy especially when it involves precise machining and manufacturing processes.

This study includes the understanding of various number of classical friction models, namely Coulomb, static and Stribeck friction. DC motor mathematical model is briefly explained, which will be used to identify required data needed for friction modelling using ANFIS. Artificial Intelligent approach has been introduced in control system for friction identification due to complexity in friction models [2]. Generally, the objective of system identification is to estimate the dynamic model structure by means of observing the input and output data [3].

2. Friction model

Friction models are described by plotting the friction force and velocity [4].Among the classical friction models that frequently studied are Coulomb, static and Stribeck friction. Another friction model known as viscous damping is the most utilized model; however, it is rarely found [5]. Coulomb friction, also known as dry friction [6], is a constant opposing torque for nonzero velocities [7]. The magnitude of kinetic friction according to Coulomb model depends on the magnitude of normal force and the direction of slippage, not the speed.

The phenomenon of Stribeck friction occurs due to the use of fluid lubrication, where the magnitude of friction is inversely proportionate to the velocity [8]. It occurs at the low velocity, and normally encountered in the lubricated servo-controlled machines. In the other hand, static friction is the friction when sticking [9]. The maximum frictional force is normally happen at the small distance from beginning point. The maximum static friction coefficient represents the maximum value of the friction force [10]. Figure 1 shows the various types of friction model.

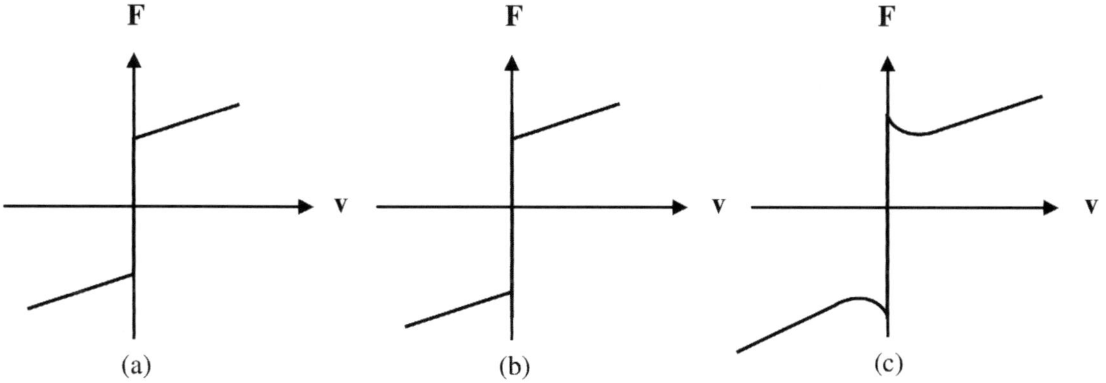

Figure 1. Friction models: (a) Coulomb; (b) Static friction; (c) Stribeck.

3. DC motor mathematical model

DC motor is one of the most frequently used electromechanical systems. It consists of both electrical and mechanical elements [11]. Figure 2 shows a simplified DC motor model, and Figure 3 shows a block diagram for a DC motor that is obtained by using Laplace Transform.

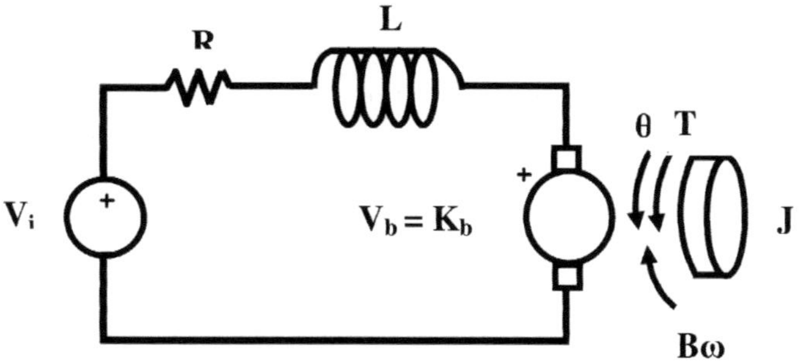

Figure 2. DC motor model

The mathematical model of DC motor is described as follows:

$$V_{in}(t) - Ri(t) - L\frac{di}{dt} - K_b\,\omega(t) = 0 \tag{1}$$

$$J\frac{d\omega}{dt} + B\omega(t) - K_t\, i(t) = 0 \tag{2}$$

Hence the transfer function is given by:

$$\frac{\Omega(s)}{V_{in}(s)} = \frac{K_t}{(Js+B)(Ls+R)+K_bK_t} \tag{3}$$

IOP Conf. Series: Materials Science and Engineering **131** (2016) 012015 doi:10.1088/1757-899X/131/1/012015

The DC motor parameters are described as follows:
V_{in} = input voltage
R = motor resistance
L = motor inductance
K_b = back electromotive-force constant
K_t = motor torque constant
J = motor inertia
B = motor friction

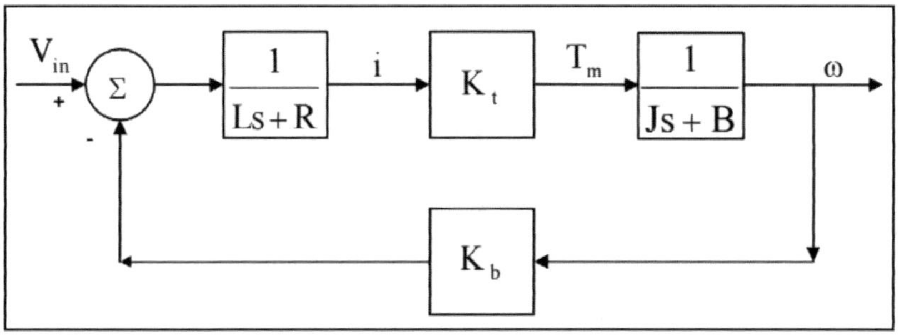

Figure 3. DC motor block diagram.

In determining the value of the motor frictional torque (T_m), the motor current (I_m) that has been measured in the experiment is multiplied with the motor torque constant (K_t) due to the equation:

$$T_m = K_t I_m \tag{4}$$

4. Methodology

Figure 4 shows the flowchart that summarizes the overall experiment. It starts with development of Simulink model, and then followed by data collection. The important data in this experiment are angular velocity and motor current. Next, the data is trained by using ANFIS toolbox that is available in MATLAB. The friction modeling is evaluated based on the difference between the actual and trained data.

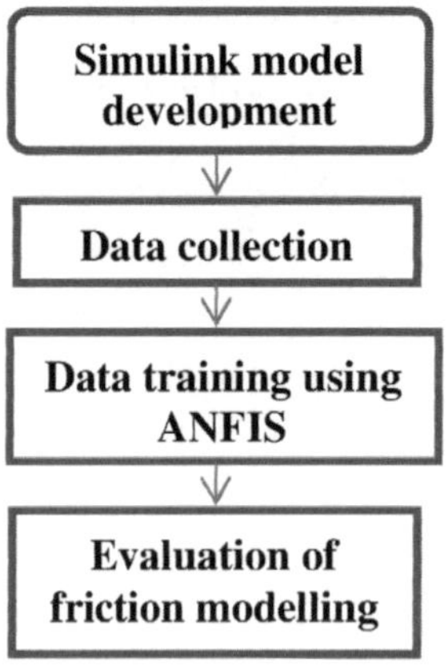

Figure 4. Experiment flowchart.

4.1 Experimental setup

This experiment involves the use of the Modular Servo System (MSS) as shown in Figure 5. Three important components in this system are DC motor, tachogenerator and encoder. In the Modular Servo System, several modules are mounted at the metal rail and coupled with small clutches. The modules are arranged in the chain. The rotation angle of the DC motor shaft is measured using an incremental encoder.The tachogenerator is connected directly to the DC motor and used to generate a voltage signal proportional to the angular velocity.

Figure 5. Modular Servo System configuration.

Then, in order to generate the data required for friction identification, MATLAB Sofware which includes Simulink and Simscape is used.

4.2 Friction identification experiment

In this experiment, the objective is to find out the relationship between the angular velocity of the DC motor with the friction torque where the voltage is used as variable. The first step for this experiment is the system modelling using the MATLAB software, Simulink and Simscape. Figure 6 shows the model used in friction identification experiment. The model is developed by referring to the standard DC motor configuration available in mathworks, and then modified to suit the need of this project.

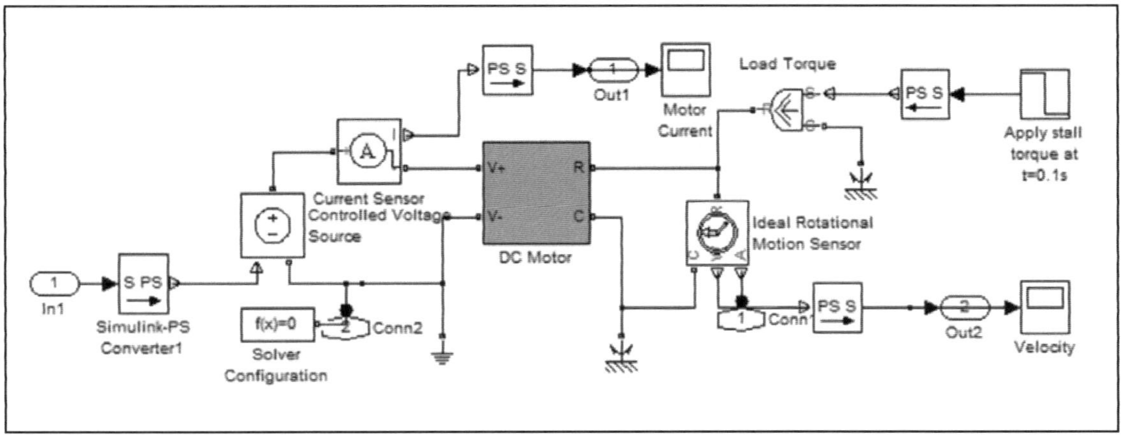

Figure 6. Friction identification model in MATLAB.

IOP Conf. Series: Materials Science and Engineering **131** (2016) 012015 doi:10.1088/1757-899X/131/1/012015

5. Result

The voltage input in the model is set to a certain values before it starts to run. After that, the result of the motor current and angular velocity is recorded in a table so that the process of analysing the data will be easier. Furthermore, in the model system, the fixed parameter such as the motor inductance, resistance, motor inertia and so on is applied in the system in order to get the results. The data is studied in order to understand the relationship between the velocity and the frictional torque.

Figure 7 shows the relationship between the velocity and frictional torque.

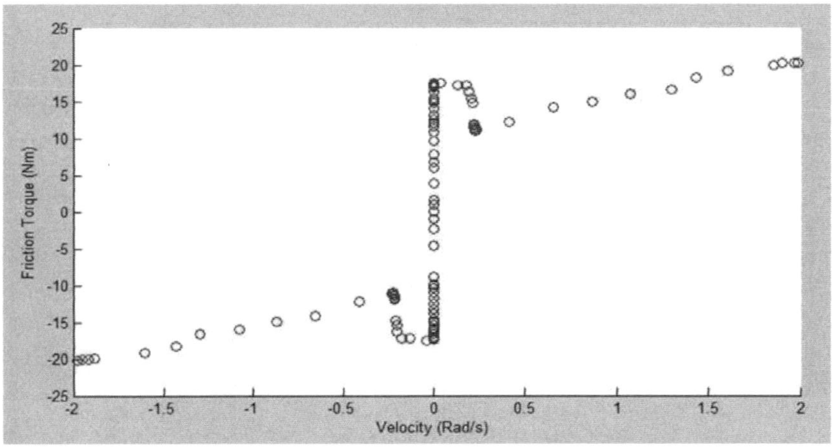

Figure 7. Frictional torque versus velocity graph.

The angular velocity and frictional torque data are trained using the ANFIS method. These data went through several training process with 100 numbers of epochs. This process only allows 0.0001 error tolerance provided by the trained data.

Based on the Figure 8, the training error during the data training process is 9.6428 based on the 100 epochs reading.

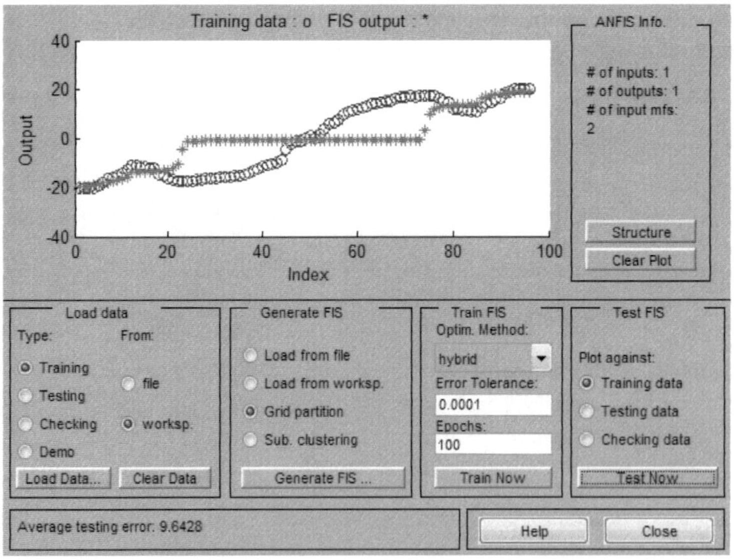

Figure 8. Frictional torque and angular velocity data trained using ANFIS.

6. Conclusion

The study carried out has successfully verified the experimental results of the friction identification using the Adaptive Neuro-Fuzzy Inference System (ANFIS). The difference between the real data and trained data is 9.6428%, which is acceptable since the target is below 10%. The friction model that has been built in this project can be used in the second phase of the project, which is friction compensation. The performance of DC motor is expected to be improved once the ANFIS based friction model is included into the system.

7. References

[1] Virgala I and Kelemen M 2013 International Journal of Mechanics and Applications 3(1) 26-30

[2] Tijani I B, Rini A and Momoh J E S 2011 Advances in Mechatronics 43-68

[3] Tolgay K and Sawsan A 2013 Proceedings on the 2013 International Conference on Systems, Control, Signal Processing and Informatics 233-240

[4] Chin-Jer L, Her-Terng Y and Yun-Cheng T 2013 IEEE/ASME Transactions on Mechatronics 18(4) 1385-1396

[5] Richard C and Cutkosky M R 1999 *Proceeding of the 1999 ASME IMECE* 14-19

[6] Popov V L 2010 Contact Mechanics and Friction

[7] Ciliz M K and Tomizuka M 2007 Engineering Applications of Artificial Intelligence 20 898-911

[8] Marton L 2006 *Acta Polytechnica Hungarica* 3(3) 45-58

[9] Olsson H, Astrom K J, Canudas de Wit C, Gafvert M and Lischinsky P 1997 European Journal of Control 1-37

[10] Sergio S M and Ricardo C 2013 Mathematical Problems in Engineering Volume 2013 1-8

[11] Virgala I, Frankovsky P and Kenderova M 2013 American Journal of Mechanical Engineering Volume 1 1-5

Acknowledgements

The author would like to thank Ministry of Education Malaysia and Universiti Teknologi MARA (UiTM) Shah Alam, Selangor for their support. This project is supported by the Ministry of Education Malaysia (MOE) under the Fundamental Research Grant Scheme (FRGS) (600-RMI/FRGS 5/3 (77/2013)).

Identifying Critical Success Factors for TQM and Employee Performance in Malaysian Automotive Industry: A Literature Review

Aimie Nadia Dedy, *Norhayati Zakuan, Ahamad Zaidi Bahari, Mohd Shoki Md Ariff , Thoo Ai Chin
Faculty of Management,
Universiti Teknologi Malaysia, 81310 UTM Skudai, Johor, Malaysia

Muhamad Zameri Mat Saman
Faculty of Mechanical,
Universiti Teknologi Malaysia, 81310 UTM Skudai, Johor, Malaysia

E-mail: norhayatimz@utm.my

Abstract. TQM is a management philosophy embracing all activities through which the needs and expectations of the customer and the community and the goals of the companies are satisfied in the most efficient and cost effective way by maximizing the potential of all workers in a continuing drive for total quality improvement. TQM is very important to the company especially in automotive industry in order for them to survive in the competitive global market. The main objective of this study is to review a relationship between TQM and employee performance. Authors review updated literature on TQM study with two main targets: (a) evolution of TQM considering as a set of practice, (b) and its impacts to employee performance. Therefore, two research questions are proposed in order to review TQM constructs and employee performance measure: (a) Is the set of critical success factors associated with TQM valid as a whole? (b) What is the critical success factors should be considered to measure employee performance in automotive industry?

Keywords – Total Quality Management Automotive Industry, Employee performance, and Malaysia and Structural Model

1. Introduction

Nowadays in new high technology, the competition has become a major factor in the organization to succeed. Many companies have tried to increase competition not only meet customer needs even more than that. This can only be achieved by the organization through improved employee performance, reduced costs, improved customer satisfaction, improved product performance and sustained efforts towards the organization's goals. In order for companies to survive and grow in the future, it is essential that they deliver high quality goods and services. Total quality management is one of the ways of managing organization to achieve excellence.

Total quality management is one of the ways of managing organization to achieve excellence. Total means everything in organization and quality means in degree of excellence. When refer to management it means by organizing, controlling, planning, directing to achieve certain goal. In management practices total quality management play important role to providing competitive advantage in organization. "In order to enhance the competitive position and improve business performance, companies worldwide, large and small, manufacturing and services, have applied the principles of total quality" (Aarabi et al., 2012).

In Malaysian automotive industry there are a lot of quality practices to make sure that they meet the standard quality in producing their products. Many organization faced the same problem when they want to produce the product which are they do not follow the standard of quality that customer want. Because to ensure that company can compete in market is high. Other than that, many organizations do not have an employee that very expertise to control the quality their production

The purpose of this paper is to review a relationship between TQM and employee performance. It is an attempt to contribute to the extant literature by incorporating TQM practices on employee performance relationship within the context of an emerging automotive industry.

2. Literature Review

TQM can be defined in many ways; difference source can result in difference definition. The definition that selected from several journals is "TQM has been defined as a set of organizational strategies, practices, and tools for organizational performance improvement" (Kaynak, 2003) and TQM advocates argue that it cannot be successful without employee involvement (Deming, 1982). The proposed TQM practices in this study adopted mainly from previous work by Zakuan et al., (2010) which has been conducted in Malaysian automotive industry.

2.1. Critical Success Factors for TQM

Based on literature review and preliminary study, the authors have identified six important elements of TQM practices namely: customer focus, leadership, training, teamwork, communication, and top management. Different researchers have adopted different TQM constructs for testing its effect on company performance. These practices are top management leadership, training and education, employee relations, quality assurances, supplier quality, product design, process management, strategic planning, customer focus, information technology and analysis, people management.

Most of the TQM success factors dealing with organization and their people have been reviewed in previous studies such as teamwork, reward and recognition, customer focus, organizational trust, extensive training, high level of communication, management commitment at all levels, employee involvement, empowerment and organizational culture (Narasimhan and Nair2005). Table 1 show that the factor of TQM from the past researcher.

Table 1: The factors of TQM

Factor of TQM	Authors	Total
Top management commitment	Sila and Ebrahimpour , 2002, Saraph et al.1989, Flynn et al. 1994, Ahire et al. 1996, Rao et al. 1999,	5
Leadership	Lewis et al. 2009, Evans and Jack, 2004 , Sila and Ebrahimpour 2002, Cemal 2012, Sadikoglu 2010, Rahman et al. & Lewis et al. 2009, Grandzol and Gershon 1998, Quazi and Padibjo 1998, Conca et al. 2004,	9
Teamwork	Guimaraes et al. 2007, Lewis et al. 2009, Sila and Ebrahimpour 2002, Rahman et al. & Lewis et al 2009, Black and Porter 1996,	5
Training	Guimaraes et al. 2007, Lewis et al.2009, Sila and Ebrahimpour 2002, Cemal zehir 2012, Esin sadikoglu 2010, Rahman et al. & Lewis et al. 2009, Saraph et al. 1989, Ahire et al. 1996, Rao et al. 1999	9
Customer focus and satisfaction	Guimaraes et al. 2007, Lewis et al. 2009, Evans et al. 2004 and Dean et al. 2004, Sila and Ebrahimpour 2002, Cemal zehir 2012, Esin sadikoglu 2010, Rahman et al. & Lewis et al 2009, Flynn et al. 1994, Black and Porter 1996, Ahire et al. 1996, Grandzol and Gershon 1998, Quazi and Padibjo 1998, Conca et al. 2004,	12
Communication	Guimaraes et al. 2007, Lewis et al. 2009, Evans et al. & Dean et al. 2004, Rahman et al. & Lewis et al.2009, Black and Porter 1996,	5
Process management	Evans et al. & Dean et al. 2004, Esin sadikoglu 2010, Saraph et al. 1989, Flynn et al. 1994, Quazi and Padibjo 1998, Rao et al. 1999, Conca et al .2004	7
Product design	Cemal zehir 2012, Rahman et al. & Lewis et al. 2009, Saraph et al. 1989, Flynn et al. 1994, Ahire et al. 1996, Rao et al. 1999,	6
Employee involvement	Guimaraes et al. 2007, Evans et al. & Dean et al. 2004, Esin sadikoglu 2010, Ahire et al. 1996, Rao et al. 1999	5
Continues improvement	Evans et al. & Dean et al. 2004, Sila and Ebrahimpour 2002, Esin sadikoglu 2010, Conca et al. 2004	4

Based on the previous studies through reviewing and synthesis of TQM literature, the author has identified 6 critical success factors of TQM as follows: (i) Top Management (ii) Leadership (iii) Communication (iv) Customer focus (v) Teamwork (vi) Training

(i) Top Management Commitment

According to Zakuan et al. (2008) top management commitment provides the motivating factors and all the resources for organization and controlling activities within the organization. Meanwhile, Hoang and Igel (2005) noted that top management commitment, employee involvement, and employee empowerment are practises that mainly relate to human aspects in the companies while process management, information and analysis system, and strategic planning are considered as mechanistic elements of TQM.

(ii) Leadership

Leadership is the ability to adapt the setting so everyone feels empowered to contribute creatively to solving the problems. Leaders must understand culture and recognise those elements that cannot be changed. They must be able to create an environment where they can empower others to act both independently and interdependently. They must provide a vision that focuses on quality and meeting customer's expectations. In other words, top management need transformational leadership skill. On the other hand, the role of middle management in initiating and institutionalising small, incremental improvements has been noted by Frohman (1997).

(iii) Communication

Dale (2003) noted that organizations are totally reliant on communication, which is defined as the exchange of ideas, messages, or information by speech, signals, or writing. Without communication, organizations would not function. If communication is diminished the entire organization suffers. When communication is thorough, accurate, and timely, the organization tends to be vibrant and effective.

(iv) Customer focus

According to Deming (1986) customer focus is the most important part of production, means producing and delivering products and services that fulfil customers' present and future needs and expectations. It also refers to exceeding customers' expectations in order to ensure long-term organizational success and survive.

(v) Teamwork

According to Quazi et al., (2002) teams are appropriate when there is a need for coordination of activity, when major breakthroughs in performance are required. It has been reported that teams are very useful for integration of activities, generating production efficiencies and for providing innovative approaches to production issues (Eisenhardt and Tabrizi 1995).

(vi) Training

The review of literature corroborates the importance of training as an important factor for successful TQM implementation (Palo & Padhi 2003; Quazi, Hong & Meng 2002). Training is considered a vehicle for implementing and reinforcing quality practice (Reed, Lemak and Mero 2002). Effective training and employee involvement have also been found to be important for initiating quality management practice in Indian context (Joseph et al 1999).

2.2 Performance measure for employee performance

Performance can be defined as a social construct which is often blurred in use. Employee performance can be explained as the set of behaviours that are relevant to the target of the organization or the organizational unit in which a staff works, (N.M. Zakuan et al., 2008, Ahmad, 2013). In this study, two performance measure for employee performance has been identified which are job satisfaction and workplace environment.

i. Job satisfaction
According to Kaynak (2003), had defines Job Satisfaction "as an individual's general attitude toward" their job. In fact, Total Quality Management covers every aspect on the way of life and operation that is conducted in an organization. This is particularly true for employee involvement and job performance that is directly affected by TQM on employee involvement because of established participative activities and job performance because of the conditions created necessary for good performance.

ii. Workplace environment
Sadikoglu and Zehir, (2010) stated that the furniture dimensions and arrangement, spatial layout, noise level, temperature, the total amount and accessibility of file and work storage, and the height and density of workplace partitions have a great impact on job satisfaction and performance, for both the individual and the team.

3. The proposed conceptual structural model

A conceptual model guides the author to determine the result that will be analysing the relationship between TQM and employee performance. In other words, conceptual structural model involves identifying the network of relationship among variables as shown in Figure 1. Several studies show a relationship between TQM in each of its forms and organizational performance (Kaynak, 2003). It has been argued that implementation of TQM practices will enhance business performance

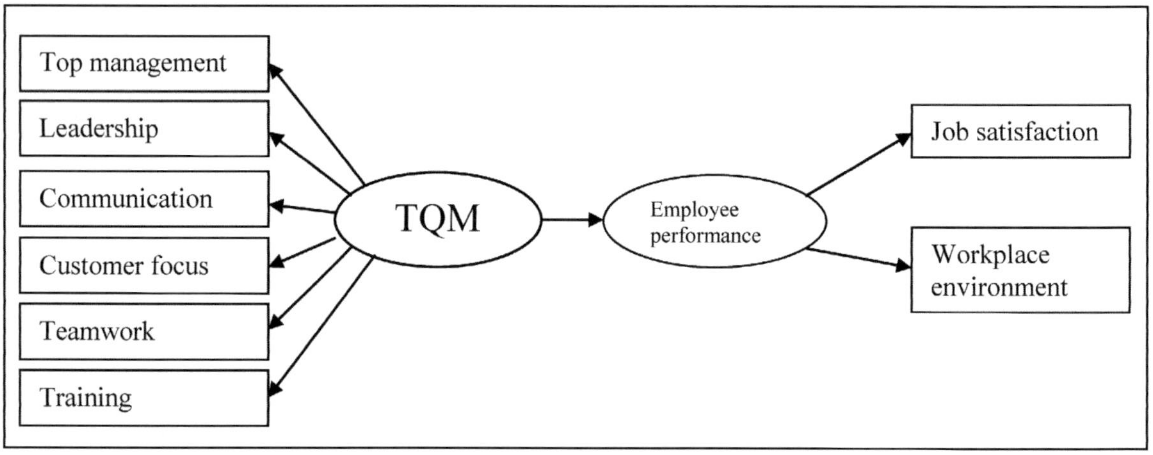

Figure 1: The proposed conceptual model

In this model, all possible relationships between each of six TQM dimensions considered as the independent variables which is top management, training, leadership, teamwork, communication, and customer focus and employee performance considered as the dependent variables.

4. Conclusion

The main aim of this paper is to review the relationships between total quality management and their impact on employee performance in Malaysian automotive industry. Even though many studies have been performed to identify for successful implementation TQM practices, it is found that there are still lack of study to investigate the relationships between TQM implementation on employee performance, especially in Malaysian car manufacturer and their suppliers. Thus, this research attempts to contribute to the development of total quality management and its effect to employee performance knowledge particularly in Malaysia.

To carry out this research, six TQM critical success factors and two employee performance measures have been investigated extensively as presented in the literature. Based on this review, a conceptual structural model that links six TQM critical success factors and employee performance measures which are employee satisfaction and workplace environment has been proposed.

Acknowledgement

The authors would like to thank the Faculty of Management, Universiti Teknologi Malaysia, Johor Bahru, Malaysia.

References

Aarabi, M., Saman. M.Z.M., Wong, K.W., Azadnia, A.H. and Zakuan, N. (2012). A comparative study on critical success factors (CSFs) of ERP systems implementation among SMEs and large firms in developing countries. *International Journal of Advancements in Computing Technology 4(9).* pp. 226-239

Ahire, S.L., Golhar, D.Y., Waller, M.A. (1996). Development and validation of TQM implementation constructs. *Decision Science*. 27(1), pp 23-56.

Ahmad, M. F., Zakuan, N., Jusoh, A., Tasir, Z. & Takala, J. (2013). Meta-analysis of the relationship between TQM and business performance. *IOP Conference Series: Materials Science and Engineering.* 1 ed. Vol. 46, 012020

Deming, W.E. (1982). *Quality, productivity and competition position.* MIT Institute for Advanced Engineering Study, Cambrige.

Evans, J.R., and Jack, E.P., (2003). Validating key results linkages in the Baldrige performance excellence model. *Quality Management Journal.* 10(2), pp. 7-24.

Fallahnejad1 M &, Shamsaddini Lora. E. (2015). Framework for connection between Total Quality Management and Innovation Processes. *Cumhuriyet University Faculty of Science Science Journal* 36 (3)

Flynn, B.B., Schroeder, R.G. and Sakakibara, S. (1994). A framework for quality management research and an associated measurement instrument. *Journal of Operations Management.* 11, pp. 339-366.

Jaafreh, A. B., & Al-abedallat, A. Z. (2012). The Effect of Quality Management Practices on Organizational Performance in Jordan: An Empirical Study. *International Journal of Financial Research*, *4*(1), 93–109.

Kaynak, H. (2003). The relationship between total quality management practices and their effects on firm performance. *Journal of Operations Management*, *21*(4), 405–435.

Narasimhan, R. and Nair, A. (2005). The antecedent role of quality, information sharing and supply chain proximity on strategic alliance formation and performance. *International Journal of Production Economics.* 96**,** pp. 301-313.

N.M. Zakuan, S.M. Yusof, T. Laosirihongthong (2008). Reflective review of relationship between total quality management and organizational performance, *Proceeding of the 4th IEEE International Conference on Management of Innovation & Technology (ICMIT2008).* pp. 444-448

Quazi, H.A. and Padibjo, S. R. (1998). A journey toward total quality management through ISO 9000 certification – a study on small and medium sized enterprises in Singapore. *International Journal of Quality and Reliability Management*. 15(5), pp. 489-508.

Rao, S.S., L.E. Solis, and Raghu-Nathan. (1999). A framework for international quality management research: Development and validation of a measurement. *Total Quality Management.* 10(7), pp. 1047-1075.

Sadikoglu, E. and Zehir, C. (2010) Investigating the effects of innovation and employee performance on the relationship between total quality management practices and firm performance: An empirical study of Turkish firms. *International Journal of Production Economics*, 127, 13–26.

Saraph, J.V., Benson, P.G. and Schroeder, R.G. (1989). An instrument for measuring the critical factors of quality management. *Decision Science*. 20, pp 810-829.

Sila, I. and Ebrahimpour, M., (2002). An investigation of the total quality management survey based research published between 1989 and 2000: A literature review. *International Journal of Quality & Reliability Management.* 19(7), pp. 902-970.

Zakuan, N.M., Yusof, S.M., Laosirihongthong, T., and Shaharoun, A.M. (2010) Proposed relationship of TQM and organisational performance using structured equation modelling. *Total Quality Management & Business Excellence*.

An Analysis of the Impact of Total Quality Management on Employee Performance with mediating role of Process Innovation

Aimie Nadia Dedy, *Norhayati Zakuan, Siti Zaleha Omain, Kamaruzzaman Abdul Rahim, Mohd Shoki Md Ariff , Zuraidah Sulaiman
Faculty of Management,
Universiti Teknologi Malaysia, 81310 UTM Skudai, Johor, Malaysia

Muhamad Zameri Mat Saman
Faculty of Mechanical,
Universiti Teknologi Malaysia, 81310 UTM Skudai, Johor, Malaysia

E-mail: norhayatimz@utm.my

Abstract. TQM practices are important to provide customer satisfaction by improved product performance and sustained towards the organizational goal. The objective of this study was to analyse the relationship between TQM practices, process innovation and employee performance. In this study, six critical success factors of TQM have been identified namely customer focus, leadership, training, teamwork, communication, and top management. Based on thorough literature review, employee performances have been measured by two constructs which are job satisfaction and workplace environment. The study used methodology of quantitative approach. The questionnaires for this study were randomly distributed to 102 employees in the selected car manufacturer companies. The respondents were choosing from management team, supervisor, technician and others worker. Results of the study support the proposed hypotheses that there are significant relationship between TQM practices, process innovation and employee performance.

Keywords – Total Quality Management, Car Manufacturer Industry, Employee Performance, Malaysia and Regression Analysis

1. Introduction

TQM has been thoroughly discussed in the past literature and can be seen as one of a management philosophy characterized by its principles, implementation, and strategies that emphasizes upon continuous improvement in quality, increased empowerment of employees, top management support, employee enhancement, teamwork, transformational leadership, rewards and incentives, and suppliers management. Ahmad et al., (2013), noted that most of the reviews on TQM have been undertaken in the electronic and manufacturing sectors and have later spread to service industry.

Recently there are many problems faced by the organization in terms of quality management among employees. For example, they need to compete in the market to make the company remain competitive. In Malaysia automotive industry, many companies overlook the fierce competition from other countries. This resulted may impact to the businesses in Malaysia.

The research question of this paper is does there is relationship between TQM practices, process innovation and employee performance? Therefore, to answer the research question, the purpose of this paper is to analyze the relationship between TQM, process innovation and employee performance. It is important for all organizations to have knowledge about quality management. Given the significance of automotive industries to the Malaysian economy, the researcher decided to evaluate the impact of TQM on employee performance with mediating by process innovation. In order to survive in a competitive market place, total quality management practices implementation is one of the key issues that can help align organization's to stay competitive.

2. Literature Review

TQM has been widely recognized and successfully implemented in many small and large organizations, giving them the edge in international as well as local competitiveness through the production of high quality products to satisfy customer needs (N.M Zakuan et al., 2008). Quazi and Padibjo (1998) stated that TQM efforts in the United States of America and Japan highlighted the increasing importance of TQM and its impact on profitability. This is supported by Lakhal *et al.* (2006), who noted that organizations with TQM systems in place consistently exceeded industry standards for return on investment. Six critical factors of TQM practices namely: customer focus, leadership, training, teamwork, communication, and top management have been identified in this study

Employee performance can be defined as the set of competencies that are related to the goals of the organization or the organizational unit in which an employee's works, (Aarabi et al., 2012). In this study, author will measure employee performance based on two elements which are job satisfaction and workplace environment.

2.1 Hypothesis Development

In this model, all possible relationships between each of six TQM dimensions considered as the independent variables which are top management, training, leadership, teamwork, communication, and customer focus and employee performance considered as the dependent variables as shown in Figure 1.

TQM practices were positively related with employees' work-related attitudes, such as job involvement, job satisfaction, career satisfaction, and organizational commitment. TQM practices also encourage employees' participation, promote empowerment, and recognize that employees play an important role in achieving the organizations', objectives and treat employees as primary resource. (Evans and Jack, 2004), Thus hypothesis 1 being develop as below:

Hypothesis 1: TQM has a direct and positive relationship with employee performance

TQM practices were positively related with employees' work-related attitudes, such as job involvement, job satisfaction, career satisfaction, and organizational commitment.TQM practices also encourage employees' participation, promote empowerment, recognize that employees play an important role in achieving the organizations', objectives and treat employees as primary resource. (Norliza and Muhd Hasmi, 2006), Thus hypothesis 3 being develop as below:

Hypothesis 2: TQM has a direct and positive relationship with employee performance

Employee performance is positively related to process innovation. Employees refer to the non-management employees of the organisation and the role they play in affecting innovation management. This factor takes into account the various personal characteristics associated with employees (Bharadwaj and Menon, 2000) and the motivation of employees to become innovative (Mostafa, 2005). Thus hypothesis 2 being develops as below.

Hypothesis 3: process innovation has a direct and positive relationship with employee performance

Kaynak, 2003 stated that in a total quality setting, employee involvement in decision making service is highly valued, encouraged, and rewarded. In such work environment, employee will work harder to improve the process, product or service quality. Incremental improvement of the existing process can come from learning curve, which means that an experienced worker will lead to decrease in cost per unit over time, (Anderson et al 1994). Thus Process innovation was to carry out about the mediator impact to total quality management. It means to know what would happen to TQM is innovation practise will be apply in the process. And also to study the relationship between innovation and employee performance which to what will happen to employee performance is would be increase or decrease. Implementation of innovations gives an opportunity to modernize the structures of manufacturing and services, to improve products and technologies, to raise their international competition, (Laura and Bronius, 2005). Thus hypothesis 4 being develops as below.

Hypothesis 4: the implementation of TQM increases with a mediating role of process innovation

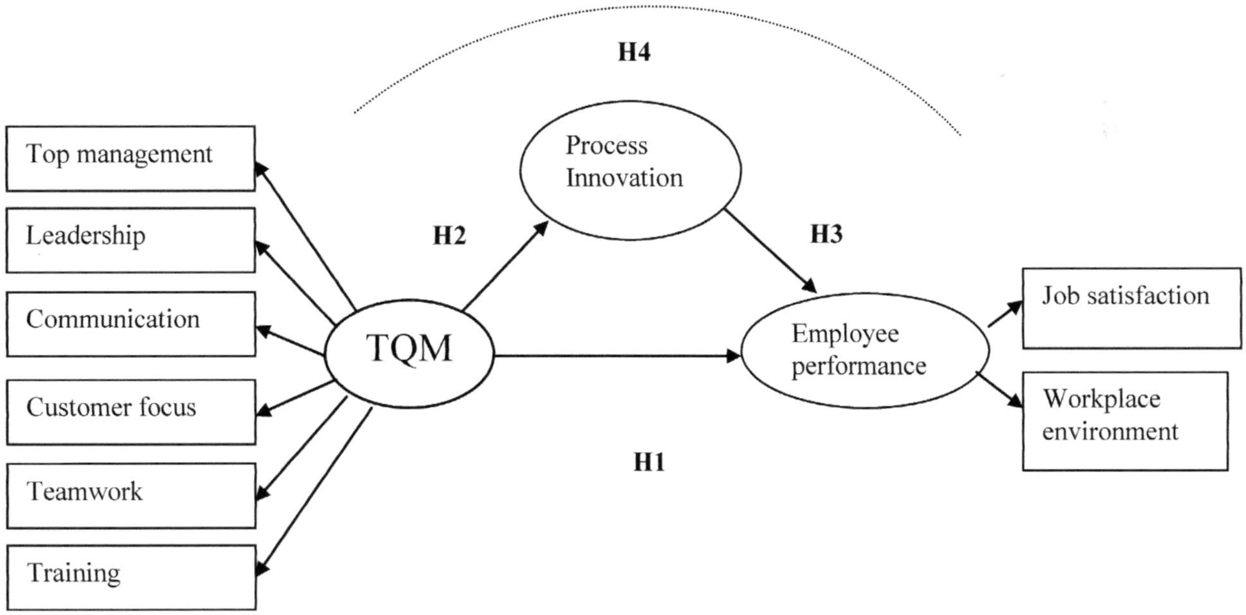

Figure 1: The conceptual model

3. Research Method

Chang (2002) describes survey as questioning the respondents and the recording of their responses to be used as data for analysis. A survey technique is considered the most economical among methods available for data collection due to its ability in performing efficient data collection (Rao et al., 1999), which had been used by the author.

The questionnaires for this study were randomly distributed to 102 employees in the selected car manufacturer companies. The respondents were choosing from management team, supervisor, technician and others worker. The data used in this research was through questionnaire to obtain quantitative data to testing of statistical of the hypotheses.

4. Results

4.1 Analysis of Mean score for Total Quality Management

Table 1 list out the mean, level of significant and rank of dimension for TQM. Customer focus represented the highest significant level with mean score 4.253 compare with others which was ranked at first place and determined as the most critical success factor for TQM practises. Next, it was followed by the leadership with 4.071 mean score, training with 3.961 mean score and communication with 3.978 means score. However, teamwork and top management were ranked as fifth and six places as its mean score was lower compared to other dimensions with the mean score 3.655 and 3.766 respectively. In short, customer focus is determined as the most critical success factor while team work as the less critical success factor based on the mean score.

Table 1: Mean Score Test for Total Quality Management

Dimension	N	Mean	Level of significance	Rank
Top management	102	3.766	High	5
Leadership	102	4.071	Very high	2
Customer focus	102	4.153	Very high	1
Training	102	3.961	High	3
Communication	102	3.978	High	4
teamwork	102	3.655	high	6

4.2 Analysis of means score for employee performance

Table 2 indicates, employee satisfaction and working environment showed high level of significance based on the range of means score. However, employee satisfaction has the highest mean score with 3.848 and followed by the working environment with 3.834 means score. Hence, employee satisfaction is identified as the critical success factor in the employee performance based on the ranking.

Table 2: Mean Score Test for Employee Performance

Dimension	N	Mean	Level Of Significance	Rank
Employee satisfaction	102	3.848	High	1
Working environment	102	3.834	High	2

4.3 Analysis of relationship between TQM and EP

In this section, linear regression was conducted to analyse the effect of TQM (independent variable) on the EP (dependent variable). From Table 3, there was a positive effect of TQM on EP which was significant at 0.01 level, (β-0.809, t=11.179, Sig. =0.000). It also showed that R square for this regression model was 0.555; it means that 55.5 per cent of the variance in employee performance was explained by the total quality management. Thus, it can be concluded that there is a significant relationship between TQM and EP which represent the TQM practises could affect EP.

Table 3: Linear Regression between TQM Practises and EP

Linear regression				
	Employee performance			
	Beta (β)	**Std. Error**	**t-value**	**p-value**
(constant)	.739	.280	2.638	.000
TQM	.809	.072	11.179	.000
F statistic	124.970			
R square	.555			

Note: Significant at p <0.01

4.4 Analysis of relationship between TQM and Process innovation

To identify the relationship between TQM and PI, a linear regression was conducted to determine the effects of TQM (independent variable) on the PI (dependent variable). From the finding in Table 4, there indicated positive effect of TQM on PI which was significant al 0.01 level, (β=0.953, t=11.637,Sig.=0.000). Besides, R square for this regression model was 0.575. It means 57.5 percent of the variance in process innovation was explained by the total quality management. In short, it can be concluded that there is a significant relationship between TQM and EP which show that the TQM could affect PI.

Table 4: Linear Regression between TQM Practises and PI

Linear regression				
	Process innovation			
	Beta (β)	**Std. Error**	**t-value**	**p-value**
(constant)	**.056**	**.317**	**.177**	**.000**
TQM	**.953**	**.082**	**11.637**	**.000**
F stat	**135.414**			
R square	**0.758**			

Note: Significant at p <0.01

4.5 Analysis of relationship between PI and employee performance

Table 5 presents the finding of the regression analysis between PI (independent variable) and EP (dependent variable). The result showed a positive effect of PI on EP which was significant at 0.01 level, (β=0.560, t=8.518, Sig.=0.000). The R square for this regression model was 0.420, it means that 42 percent of the variance in employee performance was explained by the process innovation. Thus, it can be concluded that there is a significant relationship between PI and EP which also proved that the PI could affect EP.

Table 5: Linear Regression between PI and EP

Linear regression				
	Employee performance			
	Beta (β)	**Std. Error**	**t-value**	**p-value**
(constant)	**1.763**	**.248**	**7.118**	**.000**
PI	**.560**	**.066**	**8.518**	**.000**
F stat	**72.559**			
R square	**.648**			

Note: Significant at p <0.01

4.6 Analysis of Sobel test for Process innovation mediator

There are three prerequisites must be fulfilled before conducting Sobel test. Firstly, independent variables should have significant effect on mediator. Second, there must be a significant between mediator variable and dependent variable. Lastly, independent variables must have significant effect on dependent variables without independent variables must have significant effect on dependent variables without including the influences of mediator. According to the finding on Table 4, TQM was a statically significant predictor of PI with ($p<0.01$). Next, process innovation was a statically significant predictor of EP with ($p<0.01$). Moreover, Table 3 also showed that TQM was a statically significant predictor of EP with ($p<0.01$). Thus, it can be explained that there might consist of partial mediating effect.

Next, to compare the coefficients between TQM and EP to determine whether there is a partial mediator. From the Table 6, it showed that the coefficient between TQM and PI with EP is $\beta = 0.648$ and the coefficient between TQM and EP was equal to $\beta = 0.809$. By comparing two unstandardized coefficients, there showed a decreasing of coefficient. So, it can be concluded there was a partial mediator effect with the existing of PI mediator. Sobel test was used in this section to prove that whether mediation effect is statically significant on TQM to EP. A statically significant level of $p<0.05$ and $z>1.97$ is defined as an existing of mediation. Based on the results the test statistics is equal to 2.8525202 ($z>1.97$), with p- value equal to 0.00433741 ($p<0.05$). Thus, it means that the relationship between TQM and EP is mediated by PI and predictability of dependent variables (EP) by means of independent variables (TQM) will be more significant.

Table 6: Linear Regression between PI and EP

Multiple regression				
	Employee performance			
	Beta (β)	**Std. Error**	**t-value**	**p-value**
(constant)	**.729**	**.276**	**2.639**	**.010**
TQM	**.648**	**.110**	**5.914**	**.000**
PI	**.169**	**.087**	**1.941**	**.000**
F stat	**66.096**			
R square	**.572**			

Note: Significant at $p<0.01$

Based on the previous analysis results, Table 4.7 indicated the results of the hypothesis presented in this study. All hypotheses were accepted by the following findings of the analysis conducted on the variables.

Table 4.10: The results of the hypothesis presented in this study

Hypotheses	**Result**
H1: TQM has a direct and positive relationship with EP	Supported
H2: TQM has a direct and positive relationship with PI	Supported
H3: PI has a direct and positive relationship with EP	Supported
H4: The implementation of TQM increases with a mediating role of PI in automotive industry	Supported

5. Discussions

5.1 Results on TQM Practises Dimensions Measure

Based on the results in Tables 1, it showed that customer focus was distinguished as the most critical success factor for TQM practises with the highest mean score 4.153. This outcome can be proved by the Deming (1982), customer focus is the most important part of production, means producing and delivering products and services that fulfil customers' present and future needs and expectations. It explained to exceeding customers' expectations in order to ensure long-term organizational success and survival. It is necessary for every organization to create a customer focused culture.

There are numerous reasons why more and more companies are choosing to become customer oriented. The simplest of these is that it gives them a better understanding of what the customers want. In order to reach of success, companies need to be more focused on delighting customers, rather than on making profits. Many respondents very agree that they increased personal contacts between the organization and customer. Used customer requirement as the basis for quality and also support activity to improve customer satisfaction. Also always alert and update to the needs and feedback from customers. With all this activity they can make sure they can achieve the exceeding customer expectations.

5.2 Results on the EP Dimensions Measure

Based on the result, employee satisfaction obtained the highest score with 3.848 and followed by working environment with 3.834 mean score. Hence employee satisfaction is identified as the critical success factor in the organizational performance based on the ranking. This is particularly true for employee involvement and job performance that is directly affected by TQM on employee involvement because of established participative activities and job performance because of the conditions created necessary for good performance. According to Romano (2002), had defines employee satisfaction "as an individual's general attitude toward" their job. In fact, Total Quality Management covers every aspect on the way of life and operation that is conducted in an organization.

5.3 Results of Regression between TQM Practises and EP

From the findings of linear regression, the result shows there was a significant and positive effect of TQM practises on EP. "TQM practices were positively related with employees' work-related attitudes, such as job involvement, job satisfaction, career satisfaction, and organizational commitment. TQM practices also encourage employees' participation, promote empowerment, and recognize that employees play an important role in achieving the organizations goals, its objectives and treat employees as primary resource."

5.4 Results of Regression between TQM and PI

The relationship between TQM and process innovation are positively related because both have the same role in improving the quality of product or service of the company. Flynn (1994) stated satisfied and experienced employees have historical knowledge and insight on their external customers' needs and requirements and the skills to innovate new products to satisfy the customers and improve competitiveness of the firm.

5.5 Results of Regression between PI and EP

The relationship between process innovation and employee performance is positively related to process innovation. Employees refer to the non-management employees of the organisation and the role they play in affecting innovation management. This factor takes into account the various personal characteristics associated with employees (Bharadwaj and Menon, 2000) and the motivation of employees to become innovative (Mostafa, 2005).

5.6 Results on Mediating Role of PI

Thus Process innovation was to carry out about the mediator impact to total quality management practices. It means to know what would happen to TQM is innovation practise will be apply in the process. And also to study the relationship between process innovation and employee performance which to what will happen to employee performance is would be increase or decrease. Implementation of innovations gives an opportunity to modernize the structures of manufacturing and services, to improve products and technologies, to raise their international competition, (Laura and Bronius, 2005). Hence, it will make company better in industry and gained more profit from their product.

6. Conclusion

Finally, this research has indicated that the customer focus was distinguished as the most critical success factor for TQM practises. Whereby, employee satisfaction is identified as the critical success factor for the employee performance based on the ranking.

The second objective is to examine the relationship between TQM practises, PI and EP in automotive industry. It was also found that there was positive correlation among dimensions of TQM practises. Besides, there was also a strong correlation appeared between dimensions of TQM practises and EP. Furthermore, the direct positive relationship was proven to be existed between the TQM with PI, TQM with EP and PI with EP. It has fulfilled the first three hypotheses in this research.

The third objective is to investigate the effect of process innovation (PI) mediator toward total quality management (TQM) to employee performance in automotive industry. In this research, the sobel test for mediating role of PI proved that there was an indirect effect through PI and it could stronger the effect of TQM towards EP. As a result, all the objectives of this study have been achieved.

Acknowledgement

The authors would like to thank the Faculty of Management, Universiti Teknologi Malaysia, Johor Bahru, Malaysia.

References

Aarabi, M., Saman. M.Z.M., Wong, K.W., Azadnia, A.H. and Zakuan, N. (2012). A comparative study on critical success factors (CSFs) of ERP systems implementation among SMEs and large firms in developing countries. *International Journal of Advancements in Computing Technology 4(9).* pp. 226-239

Ahmad, M. F., Zakuan, N., Jusoh, A., Tasir, Z. & Takala, J. (2013). Meta-analysis of the relationship between TQM and business performance. *IOP Conference Series: Materials Science and Engineering.* 1 ed. Vol. 46, 012020

Chang, T.L. (2002). *Six Sigma: A framework for small and medium sized enterprises to achieve total quality*. PhD Dissertation.: Cleveland State University.

Deming, W.E. (1982). *Quality, productivity and competition position.* MIT Institute for Advanced Engineering Study, Cambrige.

Evans, J.R., and Jack, E.P., (2003). Validating key results linkages in the Baldrige performance excellence model. *Quality Management Journal.* 10(2), pp. 7-24.

Lakhal, L., Pasin, F., and Limam, M. (2006). Quality management practice and their impact on performance. *International Journal of Quality & Reliability Management*. 23(6), pp. 625-646.

N.M. Zakuan, S.M. Yusof, T. Laosirihongthong (2008). Reflective review of relationship between total quality management and organizational performance, *Proceeding of the 4th IEEE International Conference on Management of Innovation & Technology (ICMIT2008).* pp. 444-448

Quazi, H.A. and Padibjo, S. R. (1998). A journey toward total quality management through ISO 9000 certification – a study on small and medium sized enterprises in Singapore. *International Journal of Quality and Reliability Management*. 15(5), pp. 489-508.

Rao, S.S., L.E. Solis, and Raghu-Nathan. (1999). A framework for international quality management research: Development and validation of a measurement. *Total Quality Management.* 10(7), pp. 1047-1075.

Romano, P. (2000). ISO 9000: what is its impact on performance? *Quality Management Journal.* 7(3), pp. 38-56.

Graphene Nanoribbons (GNRs) for Future Interconnect

Rahmat Saptono Duryat

Universitas Indonesia, Kampus UI Depok, Jawa Barat 16424, Indonesia

e-mail: rahmat.saptono@ui.ac.id

Abstract. Selecting and developing materials for the future devices require a sound understanding of design requirements. Miniaturization of electronic devices, as commonly expressed by Moore Law, has involved the integration level. Increase of the level has caused some consequences in the design and selection of materials for interconnection. The present paper deals with the challenge of materials design and selection beyond the nanoscale limit and the ability of traditional materials to cope with. One of the emerging materials, i.e. Graphene, will be reviewed with particular reference to its characteristics and potentials for future interconnection.

1. Introduction

Interconnection is one of the important areas in electrical packaging, the science and art of connecting electronic circuit components to reliably perform some design functions and constraints involving handing and protection for assembly processes. Basically, the term refers to the process and technique of making electrical connections between the bond pads of the chip and a lead-frame, substrate, or even another chip[1].

Materials selection and development for electrical packaging application is governed by a set of design requirements. A clear understanding of the functions and constraints is a prerequisite. In general, the long established functions of IC packaging are: (1) to provide electrical links between the devices on the chip and the macroscopic environment, (2) to dissipate extra thermal energy generated by the device during operation, (3) to protect sensitive electrical connection on the chip from chemical degradation and contamination, (4) to provide mechanical support to handle a small and delicate chip, and (5) to provide a sufficient electrical interface so that the IC performance is not considerably degraded by the package design[2].

One of the critical stages in material selection and development is translating the functional requirements into a set of material specifications. Traditionally, materials are employed based on their physical properties and performances at any design levels involving their manufacturability and reliability to perform designed functions within given constraints. Key attributes for conventional packaging materials involve (1) strength, (2) electrical conductivity, (3) thermal conductivity, (4) thermal coefficient of expansion, and (5) manufacturability[3].

While dimension is becoming smaller and smaller, reliability issues, with particular reference to electro-migration, come to appear[4,5]. This stress-induced atomic transport phenomenon is driven by an electric field and related to the flow of charge. In electronic packaging, reliability is a complex issue involving mechanical, thermal, and electrical constrains.

Interconnection technology cannot be separated from semiconductor technology[6]. They are associated with the evolution of device designs and materials. In semiconductor technology, interconnection deals with the wiring system that distributes clock and other signals to the various functional blocks of a CMOS integrated circuit and also provides necessary power and ground connections[7,8]. The development and trend of semiconductor technology are basically governed by the ability of industry to reduce the minimum feature sizes that are employed to manufacture integrated circuits[9]. The feature scaling involves integration level, cost, speed, power, compactness, and functionality. One of the most frequently cited scaling trends is in integration level, which is commonly expressed by Moore's Law[9]. Basically, it is assumed that the number of processor on an integrated circuit doubles in about 2 years.

The present paper addresses the challenges of interconnect designs and materials beyond the nano-scale limit and the capacity of conventional materials to cope with it. One of the emerging materials for interconnects will be reviewed with particular reference to its characteristics, properties, and performances related to the future applications.

2. Challenge Beyond the Nanoscale Limit

Increase of integration levels has caused some consequences to the interconnection design and materials. As an illustration, multiprocessor component on-chip technology has now become a technological trend, meaning that tens to hundreds components need to be integrated in one chip and have to communicate each other during the execution of applications[10]. In general, the higher the degree of integration, the larger the number of components has to be integrated in one single chip, and so the smaller the feature scale and the more complex the interconnection design are.

From the perspective of design, it is assumed that at a particular point, when the gap ratio between interconnect delay and gate delay increases significantly, the performance and reliability of the system is no longer dictated by the elementary devices but the interconnections. When the bottleneck exists in the communication, the interconnect-centric design flow is becoming essential, meaning that the interconnect design and optimization must be well involved in every single stage of design process[11].

To some extent, the optimization and synchronization of the interconnect design at local and global level can compensate the effect of scaling and hence maintain the performance and reliability of the overall system. Optimizing the geometry and architecture may provide a workable solution at global level. The geometrical path length reduction solution involves extreme parallelisms, multi core, and 3-D solutions. These solutions, however, will also provide further challenges as well as new opportunities in interconnect technology.

From the viewpoint of materials, miniaturization has also caused important consequences. Unlike processor and power that normally take benefits from dimensional scaling, interconnect performance is degraded. Despite the underlying physics of interconnect materials[12,13] and the effect of dimensional changes[14,15] had been extensively studied, their technological implications have not become a major concern until the dimensional scaling reaches a particular point. In addition to the reliability or electro-migration issue in the first stage of miniaturization, the intrinsic performance of the existing interconnects becomes affected when the feature scaling approach the physical limit of the materials. These issues are even more important and critical as the technology continuous to go beyond the nanoscale limit (<100 nm node).

3. Limit of Conventional Copper Interconnect Materials

Srivastava and Baneerje[16] addressed briefly the issue of the interconnect performance (resistivity and interconnect delay) and reliability (current carrying capacity) beyond 90 nm entering the nanotechnology. Increase of resistivity with the dimensional scaling is depicted in Figure 1.

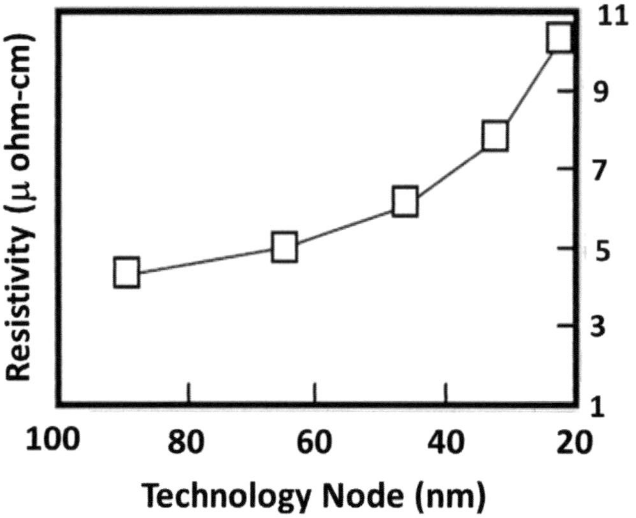

Figure 1. Resistivity of Cu beyond 90 nm node interconnect at T = 120°C (After ref.[16])

It was assumed that the phenomenon is related to the surface and internal boundary of the conductor, which become more significant as the cross-sectional dimension approaches the mean free path of electrons (about 40 nm at room temperature). It can also be seen that the effects are becoming more significant as the volume fraction ratio of (surface and internal) boundary to the cross section increases with the scaling.

Scaling analysis on material properties is revealed in ref.[17] based on previously developed compact analytical model related to the technology nodes projected by ITRS. Scaling of Cu resistivity for the ITRS intermediate wires at 300 K is depicted in Figure 2.

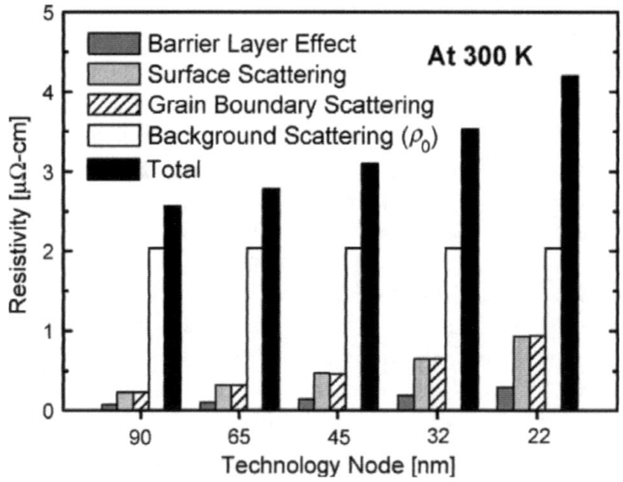

Figure 2. Scaling of Cu resistivity for the ITRS intermediate wires at 300 K (After ref.[17])

The calculated resistivity indicated the significance increase chiefly due to surface and boundary scattering as expected. Referring to Im et.al[17], the contribution of surface and boundary scattering to the total resistivity is generally the same and consistently increases with the scaling. Effect of barrier layer was also included to the calculation as its fraction was assumed become more significant with the cross section scaling. It was assumed that the background scattering due to phonon, electrons, and impurities provided a constant contribution to the resistivity, which is independent at a given temperature.

More comprehensive study of the resistivity of copper wires with lateral dimensions of 100 nm and smaller was reported by W. Steinhögl et.al[18]. Surface and boundary scattering had been identified as the root-causes of the resistivity decrease and the distinction between the two contributions were confirmed in this study. Feldman et.al[19] studied the dependence of resistivity on surface profile in nanoscale metal films and wires and concluded that roughness at short wavelengths less than 100 nm dominates scattering, and that primarily specular scattering should be achievable for root-mean-square roughness below about 0.7 nm. Feldman et.al[20] studied in detail the cause of scattering and confirmed the effects of internal lattice structures.

For local connections, surface and internal boundary scattering will also affect the Resistance-Capacitance (RC) delay as depicted in Figure 3.

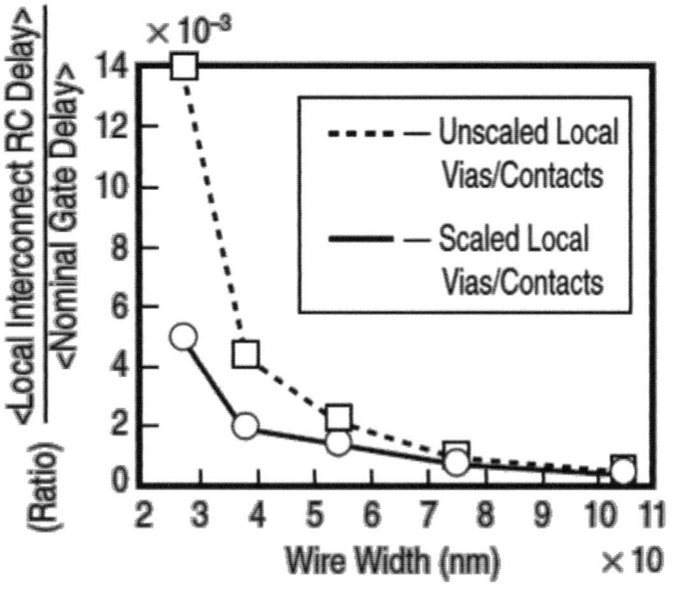

Figure 3. Ratio of local interconnect RC delay to nominal gate delay with the scaling (After ref.[16])

Solutions for RC delays may be applied to extend the physical limit of materials. Decreasing capacitance or lowering the dielectric constant of the material, which is achieved by introducing porosity or changing its chemical, can compensate reduced resistivity. This, however, has to be done at the expense of Elastic Modulus that is related to the reliability issue. Croes et.al[21] has studied the effect of dimensional and porosity scaling on the reliability of interconnects. With respect to stress-induced voiding, it was shown that despite the copper stress decreased with the introduction of low-κ materials, the gradient of stress increased. Since stress gradient is the driving force for voiding, it is assumed that the stress-induced becomes more severe and that voiding in a via becomes an actual issue. More comprehensive review on electro-migration issues at submicron level interconnects are found in ref.[22,23].

Constraint on conventional copper interconnect materials has driven research and development in both devices and materials. ITRS has specified three different areas for emerging interconnect solutions involving (1) Cu-extension, (2) Cu-replacement, (3) Native device interconnects[8]. Traditionally, global interconnect delay has been identified as the "grand challenge" to which the efforts have been concentrated. The future of interconnect technology, however, is more likely dependent upon the technological breakthrough in materials technology.

4. Potential and Characteristics of Graphene Nanoribbons (GNRs)

One dimensional Carbon-based nanomaterials have been identified as emerging research materials for both Cu-replacement and native-device interconnects[8]. Recent theoretical studies, which involve modeling and performance analysis of candidate materials in comparison to the conventional interconnect materials, have confirmed the potential of carbon-based nanomaterials for future interconnect devices[24,25]. Basic parameter of Graphene is depicted in Table 1[26].

Table 1 Basic Parameters of Graphene (After ref.[26])

Parameters	Value
C-C bond length, a_0	1.4 Å
Lattice constant	2.46 Å
Hopping Amplitude	
Nearest Neighbor, t	2.8 eV
Next-nearest, t'	0.1 eV
Third-nearest, t"	0.07 eV
Fermi Velocity, υ_F	1.1 x 10^6 m/s

Properties of the carbon-based nanomaterials related to VLSI interconnects application have been presented in ref.[25]. The properties of GNRs and Copper are compared in Table 2.

Table 2 Key Properties of Cu and GNRs (After ref.[25])

Properties	Cu	GNRs
Maximum current density (A/cm2)	10^7	$>10^8$
Melting Point (K)	1357	3800
Density (g/cm^3)	8.94	2.09-2.33
Tensile Strength (GPa)	0.22	-
Thermal Conductivity (x10^3 W/m.K)	0.385	3-5
Temperature Coefficient of Resistance (x 10^{-3}/K)	4	-1.47
Mean Free Path (nm) at Room Temperature	40	10^3

It is learned from Table 2 that 1-D Carbon-based nanomaterials exhibit several advantages compared to the existing interconnect materials involving: (1) electrical conductivity, (2) thermal conductivity, and (3) current carrying capacity. From the theoretical analysis it has been assumed that the candidate materials can reduce interconnect delay (by up to about 60% for global interconnect) and power consumptions (by up to about 50% for global interconnect)[25]. These quantitative comparisons give an insight on why it is promising.

Following the exploration of Carbon Nano Tubes (CNTs), Graphene had come to be an emerging material for research. The milestone is when a reliable preparation and study of monocrystalline graphitic layers of few atoms thick, including a single-layer Graphene, was firstly reported[26]. Graphene is basically a flat single layer of carbon atoms closely packed into a 2-D honeycomb lattice, which is a basic building block for graphitic materials of other dimensionalities involving fullerenes (0D), carbon nanotubes (1D), and graphite (3D)[28]. Graphene represents a conceptually new class of materials of one atom thickness which, due to its outstanding crystal quality and electronic properties, has opened a new gateway for low dimensional physics and novel applications[28,29]. Since then, the structures, properties, processes, and potential applications of Graphene have been studied extensively[26,29-36].

GNRs are basically 1-D strips of 2-D Graphene of a particular width and pattern of edge, which can be simply considered as unrolled CNTs. GNRs is patterned from Graphene and specified based on the pattern of the ribbon edges. Referring to Figure 4, two most common types of GNRs are specified as "Arm Chair" A-GNRs and "Zigzag" Z-GNRs.

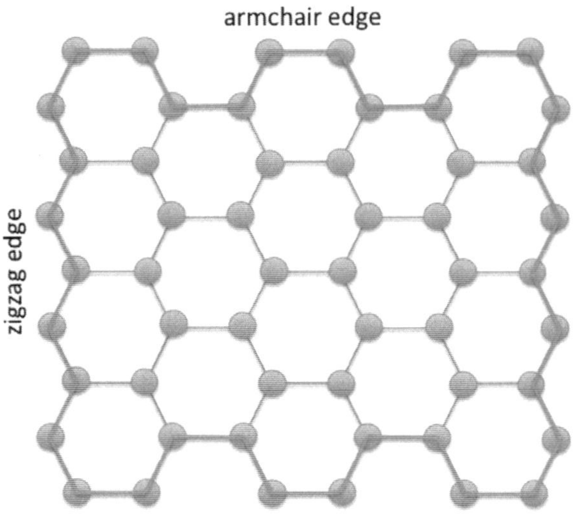

Figure 4. A piece of a honeycomb lattice displaying both zigzag and armchair edge

Generally speaking, GNRs exhibit mechanical, electrical, and thermal properties comparable to that of CNTs as they share similar basic structure. Characteristic of GNRs is the edge structure that specifies boundary conditions and hence dictates physical properties[37]. Additionally, the edges can also be manipulated by functionalization or doping. Key advantage of GNRs over CNTs is in the fabrication process, in which GNRs can be patterned from Graphene in more controllable way.

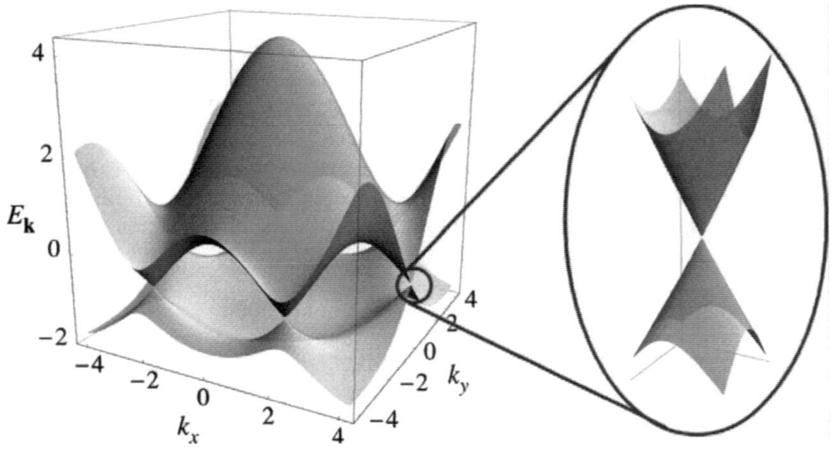

Figure 5 Band structure (3-Dimensional E-k Diagram) of Graphene (After ref.[30])

Unique properties of GNRs as well as CNTs can be explained from the band structure of their 'mother' materials, Graphene. Li et.al[25] have provided a clear explanation about the relationship between the important properties related to the interconnect application and the band structure of

Graphene. Figure 5 shows a Three Dimensional E-k diagram calculated based on tight binding approximation, which is described as massless Dirac-Fermion. The energy spectrum is characterized by 6 pairs of conical valence and conduction bands met at a single point in momentum space located at the 6 corners of Hexagonal 2D BZ whose apexes are called Dirac points. The E-k relationship at low energy level is linear leading to zero effective mass for electrons and holes so called massless Dirac-Fermions. The location of the cross section of the cut lines in the k-space determines resulted band structure and hence (semi)metallic and semiconductor behavior of GNRs.

The electronic properties of GNRs are characterized by the geometry of the edge that specifies the fixed edge boundary conditions (compared to periodic boundary conditions around the tube circumference in CNTs). Metallic – Semiconductor behavior of GNRs is dictated by the width of ribbon. It was predicted that Armchair-GNRs are metallic when N = 3M-1, where M is integer and N is the number of atom across the width of the ribbon[37]. Electrical transport is strongly affected by edge scattering. Rozhkov[38] showed that edge disorder can be introduced to tune the band gap of an armchair nanoribbon back to zero gap and proposed it as a mean for band gap tuning.

Neto et.al[30] have revealed the different behavior of electrons in Graphene in metal. It was assumed that electrons are insensitive to disorder and electrons-electrons scattering. Li et.al[25] revealed that the long mean free path of GNRs at low bias is due to weak mechanical and suppressed optical scattering at room temperature. High mobility of electrons in Graphene is related to its mean free path. It can be estimated from conductivity measurement of suspended specimen. High mobility of electrons in Graphene ranging from 2000 cm^2/V to 200,000 cm^2/V was revealed in ref.[31]. Since mobility is independent of temperature in the range of 10-100°C it can be inferred that the dominant scattering mechanism is initially associated to the Graphene defects. Improving sample preparation as well as removing substrates improved the mobility significantly. It was also revealed that the both electrons and holes mobility are nearly the same.

Thermal conductivity of has been extensively study due to its importance in energy dissipation and thermal management of micro and sub micro devices and interconnect. Guo et.al[41] studied the thermal conductivity of GNRs with different edge shapes as a function of length, width, and strain using non equilibrium molecular dynamics method. It was revealed that thermal conductivity is sensitive to the edge shapes, widths, and strains.

High value of strength of GNRs can be explained from strong sp2 hybrid bonds[25]. Indirect measurement of mechanical properties of monolayer Graphene had been reported by Lee et.al[42] based on a Nanoindentation/AFM-based method introduced by Frank et.al[43]. Young Modulus, E = 1 TPa, Failure Stress σ_f = 0.13 TPa, and Failure Strain, ε_f = 0.25 were used as datum for the load bearing capacity of the GNRs that is scarcely measured experimentally due to difficulty in sample preparation. Molecular Dynamic Analysis[44] and Atomic non-linear FE analysis of GNRs[45] had exhibited results within the order of magnitudes. Effects of temperature and temperature variation rates had also studied by the use of molecular dynamic simulations[46]. The results were reported more sensitive to geometry factors (size and edge) with higher L/w ratio. Giant plasticity (superior permanent deformation and very high flow stress) of GNRs was expected to occur at ultra-narrow dimension of several atomic width as simulated with molecular dynamics[47]. Characterizations of statics and dynamic mechanical properties both experimentally and numerically have been extensively studied, but still, there is a lot of space for exploration.

5. Closure

Increase of integration levels in device miniaturization has challenged the interconnect design and materials. Interconnection-centric design is now more important as the critical performance of the system is more dictated by the interconnection. On the other hand, multifaceted dimensional constraints and the ability of long established materials to carry the function seem to approach the limit. The future of interconnects will more depend on the break trough of technology according to the grand strategy of replacement and native device. One of the emerging materials is Graphene that can be patterned into GNRs in more controllable way. The new class of materials exhibits basic advantages in term of electrical and thermal conductivity as well as current carrying capacity in addition to its load bearing capacity. The unique properties of this material are related to its bond and band structures, which are particularly affected by its geometric boundary conditions. Atomic and molecular model could be effectively employed to simulate the behavior of this material in response to the electrical, mechanical and thermal load related to the reliability issues.

References

[1] K.D. Cluff and M.G. Pecht, "Electronic Packaging Technologies" in Mechanical Engineering Handbook, edited by Frank Kreith, Boca Raton: CRC Press LLC, 1999
[2] W.D. Callister, Fundamental of Materials Science and Engineering 5th Edition, New York: John Wiley and Sons Inc., 2001
[3] W.C. Fackler, "Electronic Packaging" in Mechanical Engineer's Handbook 2nd ed., edited by Myer Kutz. New York: John Wiley & Sons, Inc. 1998
[4] J.R. Lloyd 1999 *Microelectron. Eng.* **49** 51-64
[5] D.G. Pierce and P. G. Brusius 1997 *Microelectron. Reliab.* **37**(7) 1053 -1072
[6] G.C. Schwartz and K.V. Srikrishnan, "Preface" in Handbook of Semiconductor interconnection Technology 2nd ed., edited by Geraldine C. Schwartz and Kris V. Srikrishnan, Boca Raton: Taylor and Francis Group LLC, 2006.
[7] International Technology Road Map for Semiconductor 2010 Update Overview
[8] International Technology Road Map for Semiconductor 2009 Edition Interconnect
[9] G.E. Moore 1965 *Electron.* 114–117
[10] S. Pasricha and N. Dutt 2008 *IPSJ Trans. Sys. LSI Des. Method.* **1** 2-17
[11] J. Cong 2001 *Proceed. IEEE* **89**(4) 505-528
[12] B. Segall 1961 *Phys. Rev.* **124**(6) 4294-4297
[13] B. Segall 1962 *Phys. Rev.* **125**(1) 109-122
[14] S. Ciraci and I.P. Brata 1986 *Phys. Rev. B* **33**(6) 4294-4297
[15] I.P. Brata, S. Ciraci, G.P. Srivastava 1986 *Phys. Rev. B* **34**(12) 8246-8257
[16] N. Srivastava and K. Banerjee 2004 *JOM* **56**(10), 30-31
[17] S. Im, N. Srivastava, K. Banerjee, and K.E. Goodson 2005 *IEEE Trans . Electron. Dev.* **52**(12), 2710-2719
[18] W. Steinhögl, G. Schindler, G. Steinlesberger, M. Traving, and M. Engelhardt 2005 *J. App. Phys.* **97** 023706
[19] B. Feldman, R. Deng and S.T. Dunham 2008 *J. App. Phys.* **103**(11) 113715-1-4.
[20] B. Feldman, S. Park, M. Haverty, S. Shankar, and S.T. Dunham 2010 *Phys. Status Solidi B* **247**(7) 1791–1796
[21] K. Croes, C. J. Wilson, M. Lofrano, G.P. Beyer, Z. Tokei 2011 *Microelectron. Eng.* **88** 614–619
[22] C.M. Tan, A. Roy 2007 *Mat. Sci. Eng. R* **58** 1–75
[23] H. Ceric and S. Selberherr 2011 *Mat. Sci. Eng. R* **71** 53–86
[24] H. Li, C. Xu, N. Srivastava, K. Banerjee 2009 IEEE Trans. Elec. Dev. **56**(9) 1799-1821.
[25] H. Li, C. Xu and K. Banerjee 2010 *IEEE Des. Test Comp.* **27**(4) 20-31
[26] A.V. Rozhkov, G. Giavaras, Y.P. Bliokh, V. Freilikher, F. Nori 2011 Phys. Rep.
[27] K.S. Novoselov, A.K. Geim, S.V. Morozov, D. Jiang, Y. Zhang, S.V. Dubonos, I.V. Grigorieva, A.A. Firsov 2004 *Science* **306** 666-669
[28] K. Geim and K.S. Novoselov 2007 *Nat. Mat.* **6** 183-191
[29] A.H. Castro Neto 2010 *Mat. Tod.* **13**(3) 12-17
[30] A.H. Castro Neto, F. Guinea, N.M. R. Peres, K. S. Novoselov and A. K. Geim 2009 *Rev. Mod. Phys.* **81** 110-155
[31] A. Soldano, A. Mahmood and E. Dujardin 2010 *Carbon* **48**(8) 2127-2150
[32] M.J. Allen, V.C. Tung and R.B. Kaner 2010 *Chem. Rev.* **110**(1) 132-45
[33] D.S.L. Abergel, V. Apalkov, J. Berashevich, K. Ziegler and T. Chakraborty 2010 *Adv. Phys.* **59**(4), 261–482

[34] M. Terronesa, A.R. Botello-Méndezb, J. Campos-Delgadoc,, F. López-Uríasd, Y.I. Vega-Cantúd, F. J. Rodríguez-Macíasd, A.L. Elías , E. Munoz-Sandovald, A.G. Cano-Márquezd, J-C Charlierb, H. Terronesbet 2010 *Nano Today* **5**(4) 351-372
[35] Y. H. Wu, T. Yu and Z. X. Shen 2010 *J. App. Phys.* **108** 071301
[36] V. Singh, D. Joung, L. Zhai, S. Das, S.L. Khondaker, S. Seal 2011 *Prog. Mat. Sci.*
[37] K. Nakada, M. Fujita, G. Dresselhaus, M.S. Dresselhaus 1996 *Phys. Rev. B* **54**(24) 17954-17958
[38] N. Gorjizadeh and Y. Kawazoe 2010 *J. Nanomat.* 1-7
[39] Z. Xu and M.J. Buehler 2009 *Nanotech.* **20** 375704 1-8
[40] A.V. Rozhkov, S. Savel'ev and F. Nori 2009 *Phys. Rev. B* **79** 125420
[41] Z. Guo, D. Zhang, and X-G Gong 2009 *App. Phys. Lett.* **95** 163103 1-3
[42] C. Lee, X. We, J.W. Kysar, and J. Hone *Science* **321** 385
[43] I.W. Frank, D.M Tanenbaum, A.M. Zande, and P.L. McEuen 2007 . *Vac. Sci. Technol B* **25**(6) 2668.
[44] H. Bu, Y. Chen, M. Zou, H. Yi, K. Bi a, Z. Ni 2009 *Physics Letters A* **373** 3359–3362
[45] S.K. Georgantzinos, G.I. Giannopoulos, D.E. Katsareas, P.A. Kakavas, N.K. Anifantis 2011 *Comp. Mat. Sci.* **50** 2057–2062
[46] L. Xiong and Y. Gao, "Temperature Effect on the Tensile Mechanical Properties of Graphene Nanoribbons" in Proceedings of the Sixth International Conference on Nonlinear Mechanics, edited by Z. Zhou, Lancaster: DEStech Publications, 2013.
[47] A.S. Kochnev and I.A. Ovid'ko 2015 *Rev. Adv. Mater. Sci.* **43** 77-82.

IOP Conf. Series: Materials Science and Engineering **131** (2016) 012019 doi:10.1088/1757-899X/131/1/012019

Development in Geared Turbofan Aeroengine

A. L. Mohd Tobi[1,*], A. E. Ismail[1]

[1]Structural Integrity and Monitoring Research Group,
Faculty of Mechanical and Manufacturing Engineering,
Universiti Tun Hussein Onn Malaysia, Batu Pahat, Johor, Malaysia.

[*]abdlatif@uthm.edu.my

Abstract. This paper looks into the implementation of epicyclic gear system to the aeroengine in order to increase the efficiency of the engine. The improvement made is in the direction of improving fuel consumption, reduction in pollutant gasses and perceived noise. Introduction of epicyclic gear system is capable to achieve bypass ratio of up to 15:1 with the benefits of weight and noise reduction. Radical new aircraft designs and engine installation are being studied to overcome some of the challenges associated with the future geared turbofan and open-rotor engine.

1. Introduction

Year 2001 marks new direction of aviation industry in Europe where the direction of the industry is shaped based on the target set by the Advisory Council for Aviation Research and Innovation in Europe (ACARE) as stated in the Vision 2020 documents [1]. The pressure to be a socially responsible aviation industry, which maintains the balance between economy and the ecosystems within an ethical framework, has been put forward with the aim to reduce fuel consumption (which contributes to CO2 emission), nitrogen oxides (NOx) and noise emission.

The target is set to reduce fuels consumption by 50%, NOx reduction by 80% for landing and takeoff and 50% reduction of perceived noise (relative to the year 2000 aircraft). To ensure continuity to the effort, Flightpath 2050 has been published in 2011 [2] to bring the effort beyond the year 2020 and towards year 2050. Further reduction of fuel consumption to 75%, NOx reduction to 65% and perceived noise emission to 90% has been set (**Table 1**). This will allow major aeroengine manufacturer to plan the development in the long run.

Table 1. Fuel-burn and emissions reduction goals set by ACARE, after [3].

Category	**ACARE**	
	Vision 2020	**Flightpath 2050**
Fuel	50%	75%
	Relative to year 2000 aircraft	
NOx	80%	90%
	Relative to year 2000 aircraft	
Noise	50%	65%
	Relative to year 2000 aircraft	

One way of achieving the set target in vision 2020 is by increasing the efficiency of the engine through the increase in its propulsion efficiency. A percentage of 20% of reduction in fuel consumption will be coming from engine improvement, another 20% from aircraft improvement and the remaining 10% from air traffic management (**Figure 1**) [4]. For the NOx reduction, 72% is expected to come from engine improvement and the remaining 8% target is expected to come from aircraft improvement and air traffic management. For the noise emission, the majority improvement is expected to come from the engine itself.

Aeroengine improvement can be achieved via increasing its propulsion efficiency, mechanical transfer efficiency and thermal efficiency. **Figure 2** shows the current engine efficiency and its limit of theoretical efficiency. The direction of future aeroengine is expected to take the route of increasing propulsive efficiency. This work will look into epicyclic gearbox application in aeroengine as a way to improve propulsion efficiency of aeroengine. This is align in the direction of improving fuel consumption, reduction in pollutant gasses and perceived noise

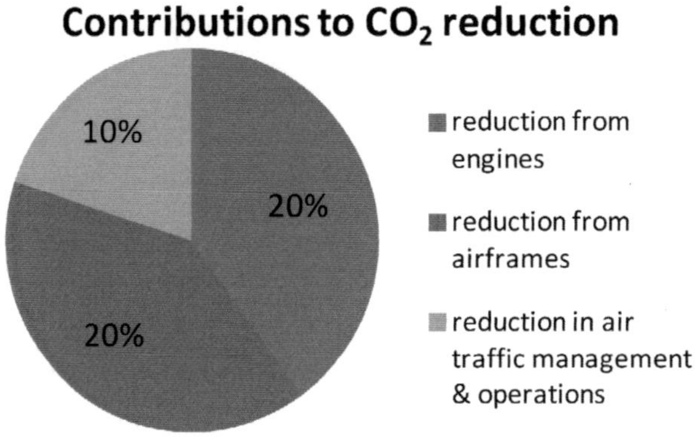

Figure 1. Contribution of engines, airframe, and air traffic management towards the target of 50% reduction in CO2 emission based from year 2000 aircraft, after [4].

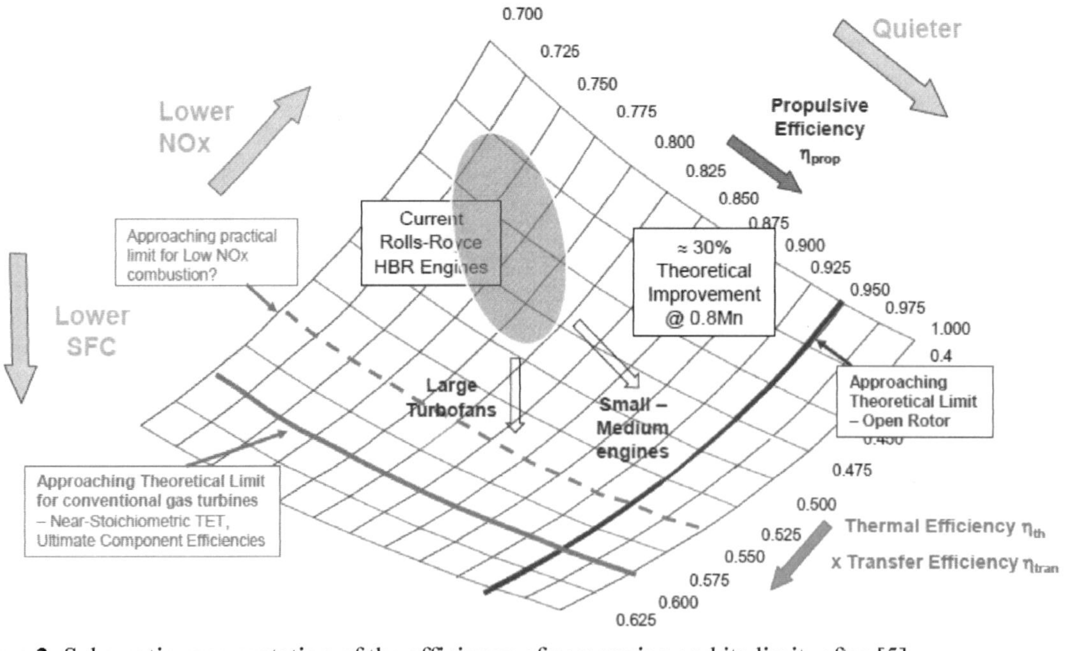

Figure 2. Schematic representation of the efficiency of aeroengine and its limit, after [5].

IOP Conf. Series: Materials Science and Engineering **131** (2016) 012019 doi:10.1088/1757-899X/131/1/012019

The report will be laid out as follows. The next section will present the architecture of epicycle gearbox. Current development and implementation of geared aeroengine is presented in Section 3. Geared Turbofan™ (GTF) engine from Pratt & Whitney (PW) and UltraFan™ engine from Rolls-Royce (RR) will also be presented in the section. The report then will be followed by the discussion section on challenges and issues related to geared turbofan aeroengine. The work is then summarised in the conclusions section.

2. Epicylic gearbox

At present, the architecture of turbofan engine fan blade is rotated by the low pressure (LP) spool energy extracted by the LP turbine blade. Bypass ratio is the ratio of air passing over the core of the engine. High bypass ratio will increase the propulsive efficiency. At the moment, typical engine have 80% of bypass ratio. This can be achieved by high volume intake of air larger diameter of fan. As the fan blade size increases, the tip linear velocity will increase, thus causing noise and efficiency reduction, whilst slowing the speed will reduce the optimum speed of the LP compressor.

One way to overcome this problem is by having a gear system which reduces the shaft's speed from the turbine to fan through the implementation of epicycle gear system. Not only that, fewer compressor stages are needed due to increase in efficiency. This equates to weight savings, fewer parts to maintain and quieter engine [6].

Epicyclic gear system consists of 3 gears being assembled as a system. They consist of a central sun gear, one or more planet gear and a ring gear as shown in **Figure 3**. All the planet gears are hold by planet carrier frame and rotate concentrically with the ring and sun gear.

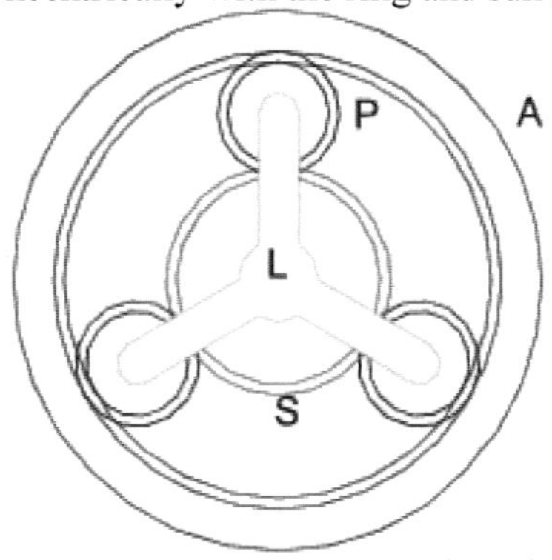

Figure 3. Typical Epicyclic Gear Arrangement, Sun gear (S), planet gear (P), planet carrier (L), and ring gear (A), after [7].

There are 3 different arrangements of epicyclic gear system. They are summarised in **Table 2**. The advantages and disadvantages of the epicyclic gear system are listed below.

Advantages:

- Load sharing between several gears.
- Compact and efficient arrangement for space and weight saving due to inline arrangement.
- Concentric rotation of input and output shaft provides less bending moment reaction from rotation radial forces.

Disadvantages:

- Complex design.
- Assembly of meshing gear is limited to certain location of teeth.
- Requires additional gear for non-inline assembly.

Table 2. 3 types of arrangement for epicyclic gear system, reproduced from [8].

Arrangement	Fixed gear	Input gear	Output gear	Gear ratio	Range of ratio input to output	Advantages/ Challenges
Star	Carrier frame	Sun gear	Ring gear	(N_R/N_S)	2:1 - 11:1 (opposite direction)	Torque balances due to counter rotating
Planetary	Ring gear	Sun gear	Carrier frame	$(N_R/N_S)+1$	3:1 - 12:1 (same direction)	Stress carried by joints from carrier to planet gear
Solar	Sun gear	Ring gear	Carrier frame	$(N_S/N_R)+1$	1.2:1 – 1.7:1 (same direction)	Stress carried by joints from carrier to planet gear
N_S=number of sun teeth, N_R=number of ring teeth (per gear)						

3. Current development

3.1. Geared Turbofan

Epicyclic gear system in star arrangement is used for the speed reduction from the LP turbine blade to the fan blade in Pratt & Whitney Geared Turbofan™ PurePower® engine arrangement. The LP shaft attached to the sun gear acts as driver and the ring gear is attached to the fan blade. The planet carrier holds 5 planet gears and doesn't rotate relative to the engine nacelle (**Figure 4**). Speed reduction of about 3:1 (9000 rpm from turbine to 3200 rpm on the fan) is achieved and resulted optimum working condition on the LP turbine blade and the fan blade [9].

Figure 4. Epicyclic gear box setup with 5 planet gears in Pratt & Whitney Geared Turbofan™ engine, after [10].

Figure 5 shows the difference between the conventional turbofan and the GTF arrangement. The main differences which can be seen are; i) the increase of the fan blade diameter allowing higher bypass ratio, ii) weight reduction due to less stages of LP compressor LP turbine blade, and iii) the existence of gear system in between the LP compressor and the fan [10]. The differences in turns have decreased the system complexity, reduced the overall length of the engine, saved cost, and improved efficiency. The larger fan with slower speed will be less noise pollutant to the airport and surround community [10].

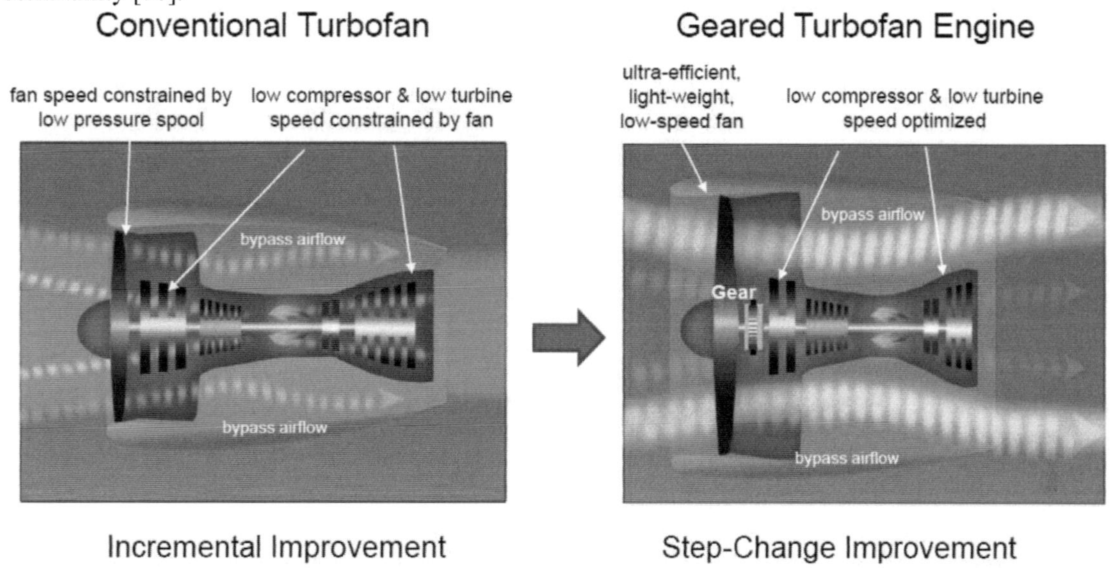

Figure 5. Comparison between conventional turbofan and geared turbofan architecture, after [10].

Table 3 summarises the GTF engine ranges developed by PW showing the bypass ratio which increases up to 12:1. The engine configuration can be set to match the requirement from 70 – 230 seating aircraft for corporate and narrow body aircraft market. It also shows the increase in the fan diameters. The thrust can be scaled in the range from 10,000 to 40,000 lbs. The engine is claimed to improve 16% of fuel consumption, 3,000 tonnes reduction of CO2 per year, halves NOx emission and noise reduction of 77% [11].

3.2. *UltraFan™*

Rolls-Royce is expected to come out with a geared turbofan engine called UltraFan™ engine in 2025 (**Figure 6**). The technology will be based from their scalable three spools Advance engine which is expected to increase the bypass ratio from 11:1 to 15:1 [13]. The UltraFan™ engine which equipped with variable ditch fan thrust reverse will provide weight save in nacelle to overcome bigger nacelle for longer fan blade. LP turbine will be removed to make it to two and half spool arrangement, thus reducing more weight (**Figure 7**). The aim is to provide better efficiency and lighter engine to compete with the GTF from PW in the narrow body and single aisle market [14]. The UltraFan™ engine will also be used for the development of open-rotor engine by Rolls-Royce to meet the Flightpath 2050 target [15].

Table 3: PW GTF PurePower® Engine Family range, reproduced after [12].

Engine Program	**PW1200G**	**PW1500G**	**PW1100-JM**	**PW1400G**
Engine Type	Geared Turbofan™ (GTF) Engine with scaled engine core; current models from 10,000 to 40,000 pounds of thrust	Geared Turbofan™ (GTF) Engine with scaled engine core; current models from 10,000 to 40,000 pounds of thrust	Geared Turbofan™ (GTF) Engine with scaled engine core; current models from 10,000 to 40,000 pounds of thrust	Geared Turbofan™ (GTF) Engine with scaled engine core; current models from 10,000 to 40,000 pounds of thrust
Aircraft Family	Mitsubishi Regional Jet	Bombardier CSeries	Airbus A320neo	Irkut MC-21
Aircraft Models	MRJ70, MRJ90	CS100, CS300	A319neo, A320neo, A321neo	MC-21-200, MC-21-300, MC-21-400
Passenger capacity	70-96	100-145	124-220	130-230
Engine Models (thrust in pound-force)	PW1215G:15,000lbs PW1215G:17,000lbs	PW1521G:21,000lbs PW1524G:23,300lbs	PW1124G:24,000lbs PW1127G:27,000lbs PW1133G:33,000lbs	24,000-33,000lbs
Architecture	1-G-2-8-2-3	1-G-3-8-2-3	1-G-3-8-2-3	1-G-3-8-2-3
Bypass Ratio (BPR)	9:1	12:1	12:1	12:1
Fan Diameter	56 inches	73 inches	81 inches	81 inches
Entry Into Service (EIS)	2014	2013	October 2015	2016

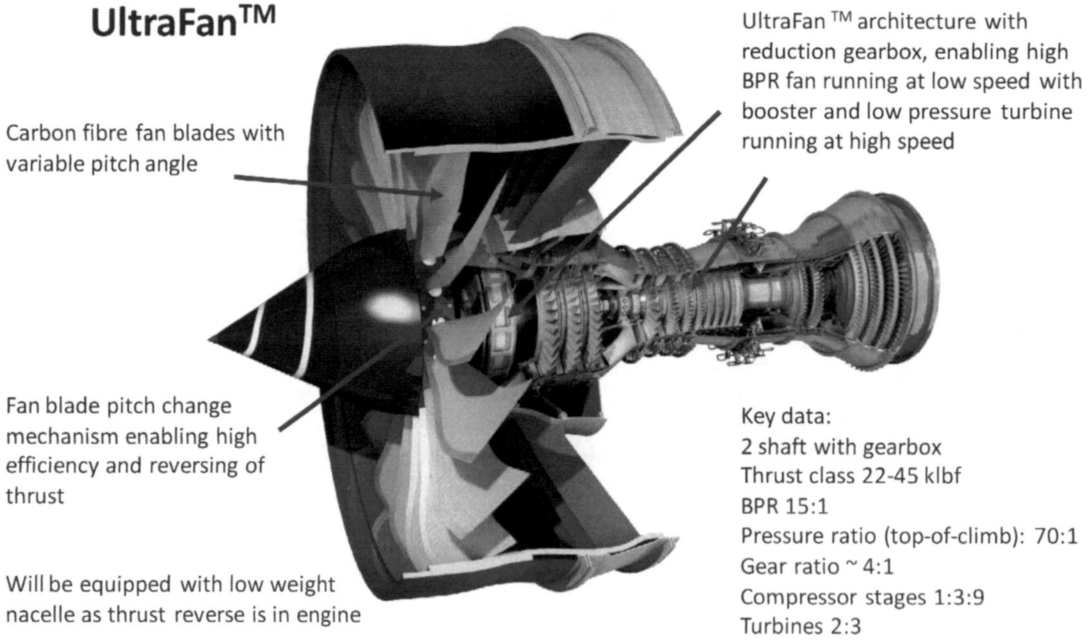

Figure 6. UltraFan™ engine architecture and key data based on Rolls-Royce materials, after [16].

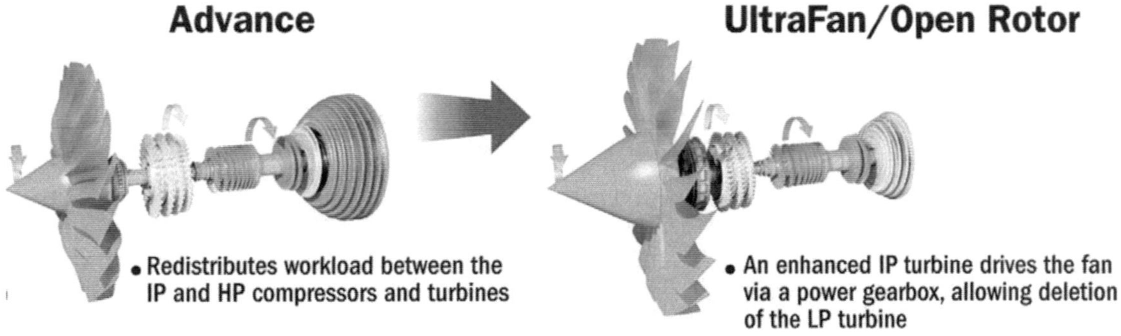

Figure 7. Schematic view of the development of Rolls-Royce's UltraFan™ engine based on the Advance engine, after [17].

4. Discussions

4.1. Engine installation

One of the challenges with the high bypass ratio is the increase in the size of the engine. The conventional under wing engine installation causes ground clearance difficulty (**Figure 8**). An unconventional installation of the engine might be required to overcome this problem. This in turns will require radical new aircraft design. Not only that, the performance of the engine for below or above wing installation is going to be different.

There are few possible designs that have been studied to scale the geared engine from turbofan to open-rotor concept. Lockheed Martin's boxed-wing and MIT double bubble design are among one of the few designs looked at (**Figure 9**) [18, 19]. Recently, PW has proposed a radical engine installation of reverse and tilted engine configuration. The propulsion and the gas generator are separated, tilted and on reverse flow direction to each other (**Figure 10**) [20]. All the new revolutionise designs are due to accommodate the installation of the high bypass ratio engine.

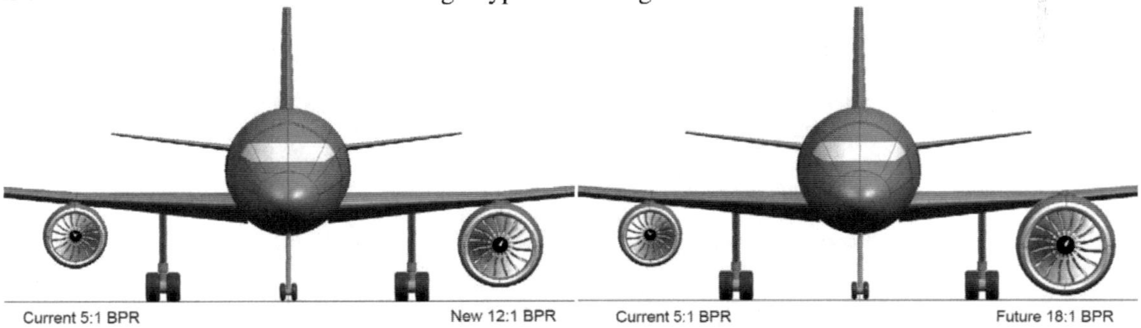

Figure 8. Schematic view of increase in engine size relative to its under wing installation position of conventional engine, after [10].

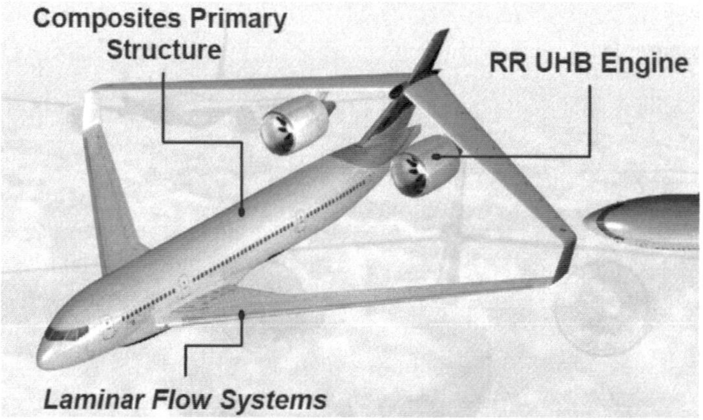

(a)

(b)

Figure 9. New concept of aircraft design for open-rotor and turbofan high bypass ratio engine: a) Lockheed Martin's box-wing concept [18], and b) MIT's double bubble concept [19].

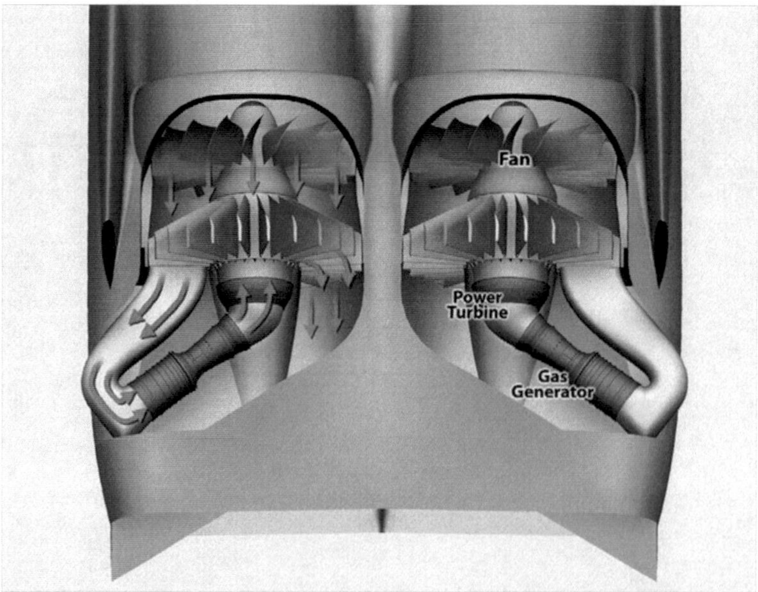

Fig. 10: PW's proposed innovative reversed, separated, and tilted propulsion concept, after [20].

4.2. Mechanical reliability and drag

The introduction of gear system has added additional mechanical components reliability issue. This will add to different dynamic behaviour due to another shaft which turns at different speeds. The interaction is further complicated due to the reduction of compressor and turbine stages. The increase in engine size introduces further drag to the engine. Increase in thrust might be offset by the drag of bigger engine and possibly lower cruising speed (at same thrust) and shallow climbing angle.

5. Conclusions

The following key points are highlighted for the technology advancement and challenges in geared aeroegine:

- Larger fan with smaller core is required to increase the bypass ratio of the engine.
- Geared turbofan could potentially increase bypass ratio to 15:1 in order to increase the propulsive efficiency of aeroengine.
- Unconventional design of aircraft is being designed to overcome the limitation of engine installation.
- Reliability performance of geared turbofan is still unknown.

Acknowledgments

Authors would like to thank Universiti Tun Hussein Onn Malaysia and Ministry of Higher Education Malaysia for providing financial assistance for the project conducted through the grant of RACE 1441.

References

[1] European Commission. (2001). European Aeronautics: A Vision for 2020, Office for Official Publications of the European Communities, Luxembourg, ISBN 92-894-0559-7.

[2] European Commission (2011). Flightpath 2050: Europe's Vision for Aviation. Publications Office of the European Union, Luxembourg, ISBN:978-92-79-19724.

[3] W.R. Graham, C.A. Hall, M. Vera Morales. (2014). the potential of future aircraft technology for noise and pollutant emissions reduction. Transport Policy. 34 , 36–51.

[4] R. Parker. (2011). Green Aeroengines: Aero-engine technology to reduce the impact of aviation on the environment. Unpublished presentation at: *Green Aviation*, January 6, 2011, Imperial College, London, United Kingdom. Accessed at https://workspace.imperial.ac.uk/greenaviation/Public/speakerspresentations/Parker.pdf

[5] J. Whurr. (2013). Future Civil Aeroengine Architectures & Technologies. Unpublished presentation at: *The 10th European Conference on Turbomachinery Fluid Dynamics and Thermodynamics ETC 10,* April 15, 2013, Lappeenranta University of Technology, Finland. Accessed at http://www.etc10.eu/mat/Whurr.pdf

[6] Leeham Co EU. (2014). *GE engines faces challenge from PW, RR GTF technology.* Available: http://leehamnews.com/2014/04/07/ge-engines-faces-challenge-from-pw-rr-gtf-technology/. Last accessed 9th May 2015.

[7] R. Beardmore. (2013). *Epicyclic Gears.* Available: http://www.roymech.co.uk/Useful_Tables/Drive/Epi_cyclic_gears.html. Last accessed 9th May 2015.

[8] D.W. Dudley. (1962). *The design, manufacture, and application of gears.* New York: McGraw Hill.

[9] J. Shapiro. (2008). *Green Technology: Jets Gear up to fly Greener.* Available: http://machinedesign.com/news/green-technology-jets-gear-fly-greener. Last accessed 19th May 2015.

[10] M. Winter. (2013). *A View into the Next Generation of Commercial Aviation.* Available: https://www.aiaa.org/uploadedFiles/Events/Other/Corporate/2013_-_Aerospace_Todayand_Tomorrow/Winter.pdf. Last accessed 9th May 2015.

[11] The Flying Engineer. (2013). *Pratt and Whitney PW1100G Geared Turbofan Engine.* Available: http://theflyingengineer.com/flightdeck/pw1100g-gtf/. Last accessed 9th May 2015.

[12] Pratt & Whitney. (2015). *PurePower Engine Family Specs Chart.* Available: https://www.pw.utc.com/Content/PurePowerPW1000G_Engine/pdf/B-1-1_PurePowerEngineFamily_SpecsChart.pdf. Last accessed 9th May 2015

[13] Rolls-Royce Plc. (2014). *Three-shaft engine design.* Available: http://web.archive.org/web/20061016141514/http://www.rolls-royce.com/civil_aerospace/technology/threeshaft.jsp. Last accessed 9th May 2015.

[14] R. Wall. (2014). *Rolls-Royce Unveils Engine for Future Boeing, Airbus Planes.* Available: http://www.bloomberg.com/news/articles/2014-02-26/rolls-royce-unveils-new-engine-for-future-boeing-airbus-jets. Last accessed 9th May 2015.

[15] I. Goold. (2014). *Rolls-Royce Advances Toward UltraFan.* Available: http://www.ainonline.com/aviation-news/air-transport/2014-07-14/rolls-royce-advances-toward-ultrafan. Last accessed 9th May 2015.

[16] Leeham Co EU. (2014). *Rolls-Royce plans for new single-aisle, twin-aisle airplane engines.* Available: https://leehamnews.com/2014/03/10/rolls-royce-plans-for-new-single-aisle-twin-aisle-airplane-engines/ Last accessed 9th May 2015.

[17] G. Norris. (2014). *Rolls-Royce Details Advance and UltraFan Test Plan.* Available: http://aviationweek.com/commercial-aviation/rolls-royce-details-advance-and-ultrafan-test-plan?NL=AW-05&Issue=AW-05_20140825_AW-05_683&YM_RID=%27email%27&YM_MID=%27mmid%27&sfvc4enews=42&cl=article_1. Last accessed 9th May 2015.

[18] E.M. Greitzer et al. (2011). N+3 Aircraft Concept Designs and Trade Studies, Final Report. NASA/CR-2011-216794.

[19] K.C. Martin. (2012). *NASA environmentally responsible aviation.* Oral Presentation at the 50th AIAA Aerospace Sciences Meeting. Accessed at https://www.aiaa.org/uploaded Files/About-AIAA/Press-Room/Key_Speeches-Reports-and-Presentations/2012/Martin-Lockheed-AVC-AIAA-GEPC2.pdf

[20] G. Norris & G. Warwick. (2015). *A Reversed, Tilted Future for Pratt's Geared Turbofan?.* Available: http://aviationweek.com/technology/reversed-tilted-future-pratt-s-geared-turbofan. Last accessed 9th May 2015.

Part Transportation Improvement in Warehouse of an Automotive Factory

Tiwaporn Choojan[1] and Parames Chutima[1,2]

[1]Industrial Engineering Department, Chulalongkorn University, Bangkok 10330, Thailand
[2]Regional Centre for Manufacturing Systems Engineering, Chulalongkorn University, Bangkok 10330, Thailand

E-Mail: cparames@chula.ac.th

Abstract. This research is focus on the problems cause the assembly line stop in a Japanese automotive manufacturer in Thailand. The problem arose from not being able to supply parts to the assembly line in time and supplying wrong parts to the assembly line. Methods to resolve these problems involved the followings: 1) design new tag labels, 2) improve operations method and 3) redesign the parts storage layout and supply routes. The result showed that the time for transferring parts from the warehouse to the preparing area and the overlapped routes were reduced. Moreover, the supply accuracy from the preparation area to the assembly line was improved.

1.Introduction

This case study concerns about the in-plant parts transportation of a well-known Japanese automotive manufacturer in Thailand. This company receives parts from suppliers and assembles them all together to produce various models of cars, e.g. sedan, SUV, etc. Its daily and monthly production rates depend on customers' demand. The revenue of the company earns from the number of cars being produced according to the delivery plan. If cars are not produced in time, several costs such as storage, overtime and parts expediting costs will be incurred to the company.

The major cause of the delayed production arises from the assembly line stoppage. The incident that always raises such problem is supplying parts to the assembly line are not properly organized and sent in a timely manner. Historical data on the line stoppage between January and July 2015 showed that the number of delays from part supplying shortage occurred 13 times (193 minutes) and wrongly supply parts occurred 6 times (20 minutes). Both of these ineffective operations are the logistics department's responsibility.

Two logistics operation areas include the warehouse using for storage parts and the parts preparation area for picking parts according to various models of cars to send to the assembly line. The logistics department's duties comprise as follows: 1) receive parts from supplier (Step a) at receive area by operator #1, 2) move parts part from truck port to warehouse (Step b) by operator#2, 3) sorting part separate to preparation area (Step c), put away part on shelf and storage part in warehouse (Step d) by operator #3 after production order part them pick part by pick order sheet at warehouse (Step e), 4) move part that pick already from warehouse to preparation area (Step f) and put away part to flow rack(Step g) by operation#4, 5) pick system show data part that must to picking parts according to various models of cars (Step h) (pick system was arrangement software) ,part are picked by operator#5 at preparation area, and 6) move part that pick already from preparation area to assembly line (Step i) by operator#6 (Fig. 1).

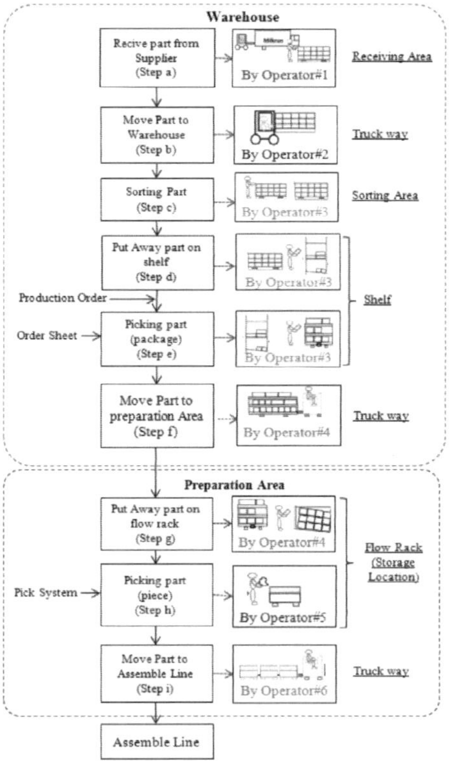

Figure 1. The logistics department's duties.

From the historical data analysis, it is noticeable that the causes of parts supplying late to the assembly line came from the followings:

1. Wrongly placing parts in the part preparation area (put away part to flow rack or storage location (Step g) by operation#4).
2. Input incorrect data into the parts arrangement software causing operators not being able to prepare necessary parts or prepare wrong parts send to the assembly line (parts according to various models of cars (Step h) by Operator#5).
3. Pickup parts less than necessary or pickup mistaken parts (parts according to various models of cars (Step h) by Operator#5)
4. Late delivery parts to the assembly line (move part from preparation area to assembly line (Step i) by operator#6).

From the aforementioned problems, the Lean technique is employed to identify the inefficient working steps and find the appropriate procedure for productivity improvement. Generally, the Lean technique is a tool to improve working process, methods to manufacture, and working environment so that operators can work better and comfortably. As a result, operators can increase their efficiency by using proper materials, machines and instruments as well as specifying the suitable parts removing method [1-2]. In addition, production process chart, operations chart, Why-Why analysis and ECRS (**E**: **E**liminate, **C**: **C**ombine, **R**: **R**earrange, **S**: **S**implify) are also applied in this study [3-5].

2. Problem Analysis

The detail analyses of two aforementioned problems were conducted by brainstorming of engineers and shop floor supervisors. The root causes of these two problems can be explained as follows:

2.1 Parts Supply Delay

The Why-Why analysis was employed to identify the root causes of the parts supply delay problem (Step i of Fig. 1). The result was shown in Table 1.

Table 1. Why-Why analysis.

Why-Why	Parts supply delay problem
Why1	Operator#6 delays moving part from preparation area to assembly line
Why2	Tow-truck driven by Operator#6 cannot move parts to the assemble line
Why3	Tow-truck driven by Operator#6 waits for an available truck way to gain access into the preparation area
Why4	Traffic jam in the truck way of the preparation area
Why5	Another tow-truck (Operator#4) utilizes the same truck way simultaneously due to Operator#4 spends too much time in keeping parts in the storage location
Why6	Operator#4 spends too much time in finding the storage location for each part

Table 1 shows the problem analysis of part supply delay by using the Why-Why technique to identify the main root causes of the problems. The problem was that Operator#6 had to delay moving parts from preparation area to assembly line because the tow truck used to move parts to the assembly line was blocked by another tow truck (tow truck's operation#4). According to Fig. 1, operator#4's duties were to move parts from warehouse to preparation area (Step f) and put away parts to storage locations (Step g). Also, he had to operate the process of keeping parts on the truck way which he often spent too much time in keeping parts in the storage location. After analysis the put away process, it was found that Operator#4 spent too much time in finding the storage location for each part. So, the root cause of the part supply delay problem came from Operator#4 spent too much time in finding the storage location for each part on tuck way that utilizes the same truck way with Operator#6. This effected the movement of parts to the assemble line by Operator#6 (Fig. 2).

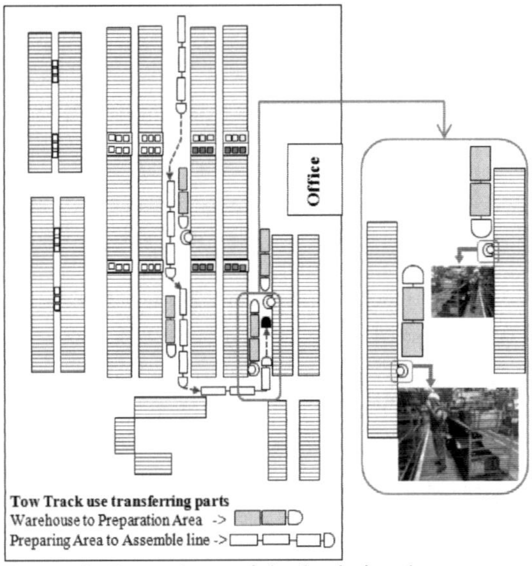

Figure 2. Operations of the logistics department.

2.2. Storing Parts into Wrong Locations

The Why-Why analysis was also used to identify the root causes of the storing parts into wrong location problem (Step g of Fig. 1). The result was shown in Table 2.

Table 2. Why-Why analysis.

Why-Why	Storing parts into wrong location problem
Why1	Wrong part number (Operator#5)
Why2	Part number at the storage location and part label attached on the package are different
Why3	Operator#4 places the package at the wrong location
Why4	Operator#4 uses his/her own experience to find proper locations to keep parts

From Table 2, it is clear that the root cause of the storing parts into wrong location problem comes from Operator#4 uses his/her own experience to find proper locations to keep parts. Visual matching between the part label and part number location was prone to be problematic.

3. Process Improvement

To increase the efficiency of the logistics operations, three improvement methods were developed and implemented as follows.

3.1. Design new tag label

To alleviate the problem caused by Operator#4 does not know where to place parts in the storage location, currently location label undefined the location code (Fig. 3). Therefore, the location code and the revised form of location label were established. A location label is newly designed to indicate the storage location. First, it is necessary to denominate all of the storage locations by defining ten digits

IOP Conf. Series: Materials Science and Engineering **131** (2016) 012020 doi:10.1088/1757-899X/131/1/012020

or alphabets which signifies the preparing area code, shelf number as well as parts position sequence (Fig. 4). The example is shown in Fig. 5. The information regarding the locations where parts have to be placed was added into the parts location label affixed on the flow rack. This information can help Operator#4 find the correct parts storage location more easily [6]. In addition, the problem about the input incorrect data into the parts arrangement software was resolved by the verification of the line supervisor.

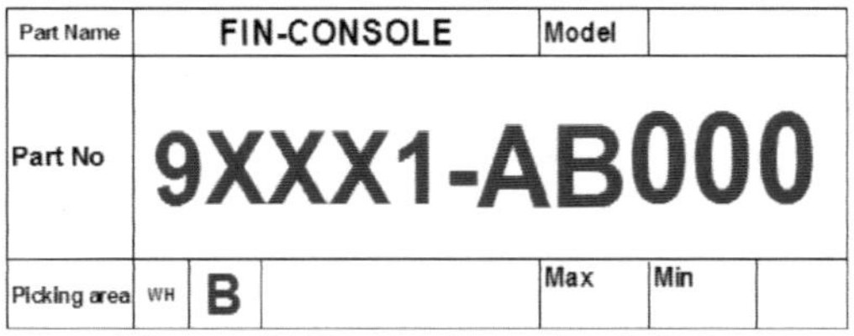

Figure 3. Current Location labels.

Digits or Alphabets	X	-	XX	-	XX	-	X	XX
Meaning	1		2		3		4	5

Meaning	Example
1. Zone	A,B,C
2. Station number	01,02,03
3. Shelf number	01,02,03
4. Level	A: Level1, B: Level 2
5. Position	01,02,03

Figure 4. Meaning of Location Code.

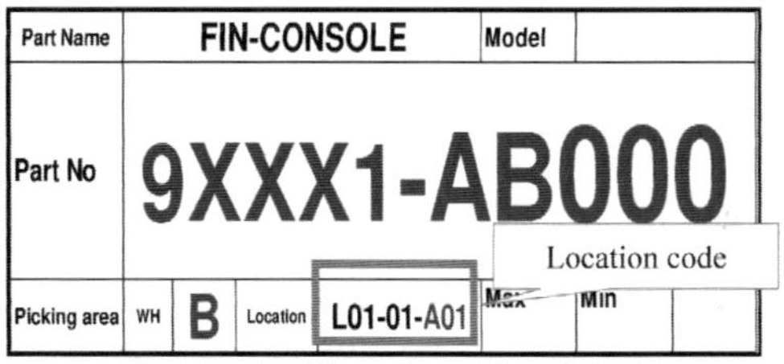

Figure 5. New Location labels.

3.2. Improving operations method

The new operations method aims to reduce time to find parts storage locations for keeping parts in the preparation area was developed by adapting the ECRS technique of the Lean concept [7]. The number of part numbers per storage cycle in the preparation area was reduced. The process of putting away parts in the warehouse (Step d) and the process of picking parts by the order sheet (Step e) by Operator#3 were eliminated. Therefore, after receiving parts from suppliers, parts are sorted and then sent them directly to the preparation area (Fig. 6).

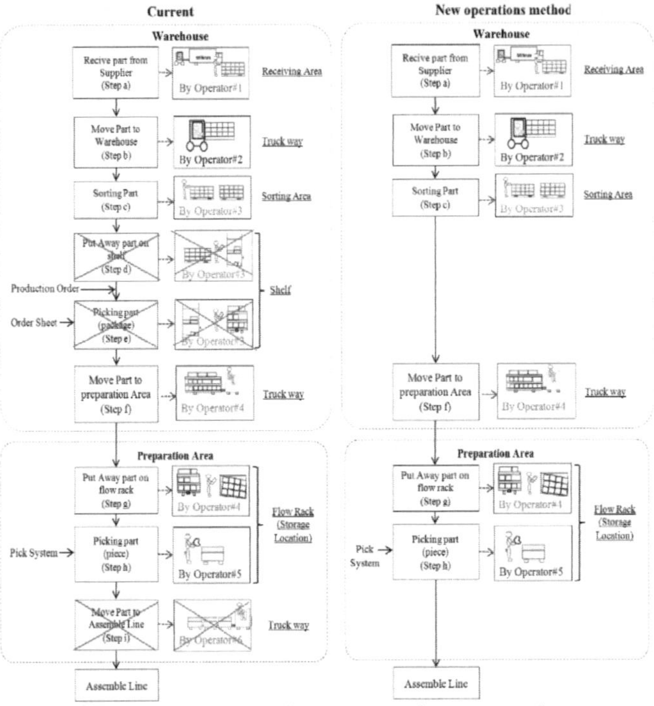

Figure 6. Current and new operations procedures.

3.3. Design a new layout and supplying routes

To decrease the number of tow trucks that operate work in the preparation area, some parts that needed to be prepared in the preparation area were transferred to be prepared in the area of the assembly line instead (Figs. 6-7). The benefits were to increase the space in the preparation area and decrease the number of tow trucks moving into the preparation area.

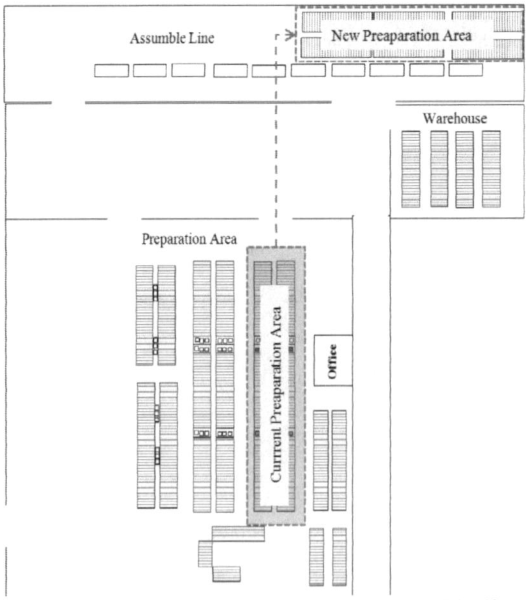

Figure 7. Transfer some parts to the assembly line.

IOP Conf. Series: Materials Science and Engineering **131** (2016) 012020 doi:10.1088/1757-899X/131/1/012020

4. Results and Discussion

The results obtained after the implementation of the new workflow found that the logistics operations steps was reduced from nine to six by reducing steps in storing parts into the warehouse and arranging parts as stated in the parts order sheet. To decrease part quantity per cycle of Operator#4, he has to reduce the movement of parts from the warehouse to the preparation area (Step f) from 86 packages to 64 packages (-22 packages) and decrease time in finding the storage location for each part from 45 times to 13 times (-32 times) (1 part number find the storage location 1 time). The target operating time of logistics department was 442 minutes per shift, and the operating time of transferring part was just 331 minutes per shift which was just 74.9% of the 100% target. After the process was improved, the operating time of transferring part increases to 439 minutes per shift which is 99.2% of the 100% target. Therefore, the efficient of the part transferring of Operator#4 was improved by 24% (from 74.9% to 99%) as shown in Table 3.

Table 3. Compare results between current and new operations methods.

Detail	Current	New operations methods
Parts quantity per cycle (Package)	86	64
Part number	45	13
Operating Time per Shift (minute)	331	439
Efficient of parts transferring of Operator#4	74.90%	99.20%
	24%	

5. Conclusion

This research aims to improve the transportation inside the logistics operating area by applying the Lean concept as a tool. The research started with the collection of the current data and analysing workflow by using the Why-Why technique to identify the main root causes of the problems. Having analysed the operations procedure in detail, the new workflow was designed using the ECRS technique. The results showed that the transferring time and time loss for keeping parts were reduced substantially. The problem about storing mistake parts into the preparing area was also resolved. Consequently, the parts transferring can be operated on time and wrong parts supply to the assembly line was also reduced.

References

[1] S S Mahapatra and S R Mohanty 2007 Lean manufacturing in continuous process industry :An empirical study *Journal of Scientific & Industrial Research* **66** 19-27.

[2] T Melton 2005 The benefits of Lean manufacturing. What Lean Thinking has to Offer the Process Industries *Chemical Engineering Research and Design* **83(A6)** 662–673.

[3] K Wajanawichakon and C Srimitee 2012 ECRS's Principles for a Drinking Water Production Plant *IOSR Journal of Engineering* **2(5)** 956-960.

[4] D Sindhuja, N M Gandhi, P Madhumathi 2012 Redesigning of Horn Assembly Line Using Ecrs Principles *International Journal of Engineering and Innovative Technology (IJEIT)* **1(3)**.

[5] S Karawatthanaworrakul, S Tongthavornsuwan and V Tangwarodomnukun 2015 *The Journal of Industrial Technology* **11(2)**.

[6] C Bartram, A Bones and D K Sobek II 2014 Efficiency Improvement within Supplies Inventory at a Critical Access Hospital *Proceedings of the 2014 Industrial and Systems Engineering Research Conference*.

[7] P Ongkunaruk and W Wongsatit 2014 An ECRS-based line balancing concept: a case study of a frozen *Business Process Management Journal* **20(5)**.

Defective Reduction in Automotive Headlining Manufacturing Process

Saranya Rittichai[1] and Parames Chutima[1,2]

[1]Industrial Engineering Department, Chulalongkorn University, Bangkok 10330, Thailand
[2]Regional Centre for Manufacturing Systems Engineering, Chulalongkorn University, Bangkok 10330, Thailand

E-Mail: cparames@chula.ac.th

Abstract. In an automobile parts manufacturing company, currently the headlining process has a lot of wastes resulting in a high cost of quality per year. In this paper, the Six Sigma method is used to reduce the defects in the headlining process. Cause-and-effect matrix and failure mode and effect analysis (FMEA) were adopted to screen the factors that affect the quality of headlining. The 2^{k-1} fractional factorials design was also use to determine the potential preliminary root causes. The full factorial experiments was conducted to identify appropriate settings of the significant factors. The result showed that the process can reduce the defects of headlining from 12.21% to 6.95%

1. Introduction

Six Sigma is a method with highly disciplined for reducing the variability of the process to ensure customer satisfaction and cost reduction of the organization. It is one of the most effective improvement strategies for overall operation excellences [1]. Six Sigma is the statistical term of a process that produces defects fewer than 3.4 defects per million opportunities [2]. Six Sigma provides a systematic steps known as DMAIC (i.e. Define, Measure, Analyze, Improve, and Control) in a formalized problem solving process [3]. The DMAIC process can improve any type of process in any organization to improve its efficiency and effectiveness [4].

The case study company was established in the early 2013s to manufacture various automotive parts to domestic and international customers. The automotive headlining process is focused in this study since it produces most defects (12.21% or 122,082 DPPM) when comparing with another process. Its value is estimated around 3 million baht per year. Currently, the main production problem is the dimension of the part which do not meet the customer specifications. Hence, this research aims to reduce the defective parts in the automotive headlining process by using the Six Sigma approach.

2. Define Phase

The Define phase is the starting point of the DMAIC Six Sigma approach. This phase help define the problem of interest in detail and thus making the way ahead clearer in terms of actions required to be taken. The prolong analysis and discussion are carried out with the project charter team members to identify the main problem which impacts significantly on the defective rate. The historical manufacturing defect data from January 2015 to May 2015 showed that the proportion of wastes were classified as follows: part dimension out-off standard 45.30%, feeble parts 12.90%, peeling parts

9.30%, creasing parts 9.10%, bulging parts 6.80%, bloated parts 6.20%, wrinkle parts 5.70% and other 4.50% (Fig. 1). It is clear that the part dimension out-off standard is the highest proportion; therefore this problem is subject to be improved in this research.

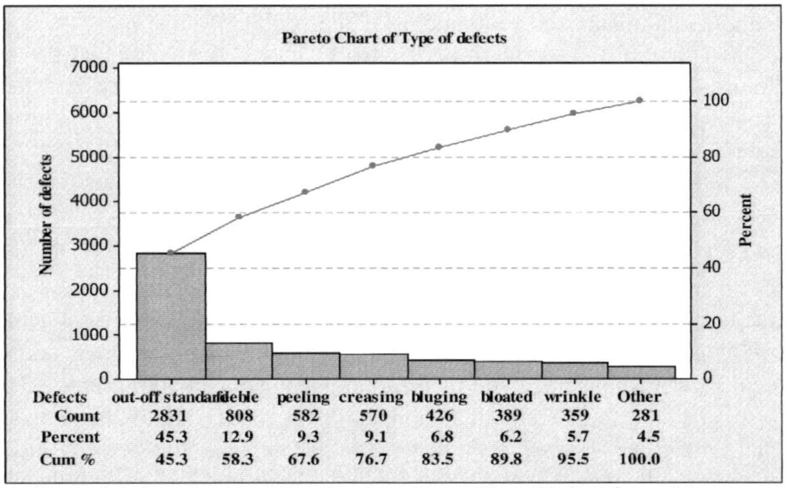

Figure 1. Pareto chart of defect (source: Production data of headlining from 1 January – 31 May 2015)

3. Measure Phase

This phase involves two main steps, i.e. measurement system analysis (MSA) and root-cause analysis.

3.1. Measurement System Analysis

The analysis of gauge R&R with attribute data was used to measure the goodness of current measurement system [5]. This approach was used to analyze the repeatability and accuracy of operators who perform self-inspection on the finished parts. The test results showed that three operators have the percentage of repeatability equal to 95%, 97% and 97%; whereas the accuracy in measuring is 95%, 97% and 97%, respectively (Fig. 2). As a result, the overall effective repeatability and accuracy are equal to 92.5% and 92.5%, respectively. It was clear that all operators were failed by the testing criteria of the MSA which were set as 100%. Therefore, the company had to sent all operators to the retraining course (i.e. QC parts inspection course). The training program was conducted into two parts, i.e. theoretical and practical. In the theoretical part, operators were trained to understand about the basic background and limitation of parts. For practical training, a QC expert provided insight knowledge of how to check parts properly and effectively. After completing the course, the attribute MSA was reconducted to these operators again and the testing results showed that all of them passed the given standard (Fig. 3).

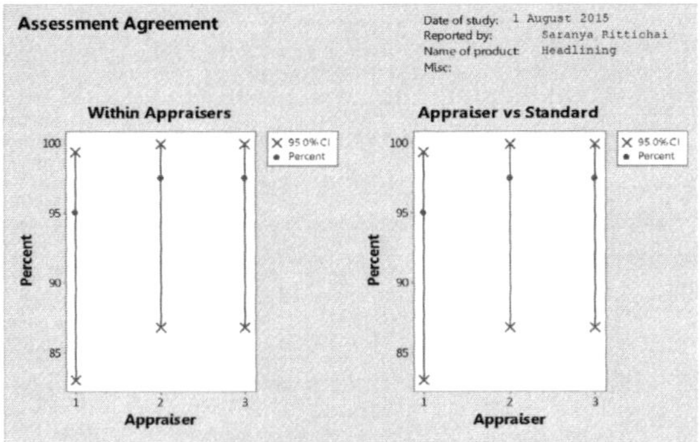

Figure 2. Attribute test performance chart (Before-training)

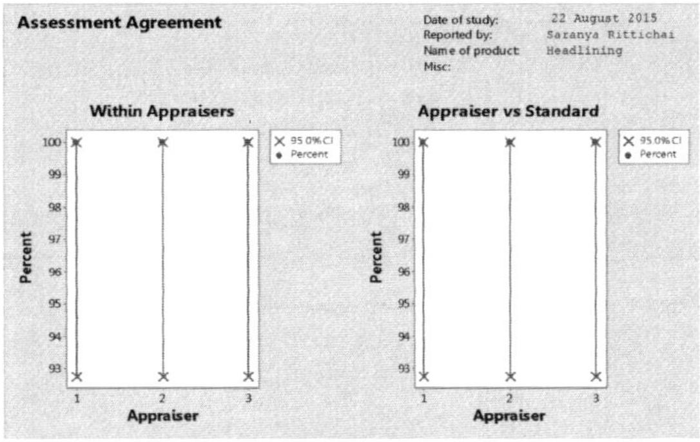

Figure 3. Attribute test performance chart (After-retraining)

3.2. Root-Cause Analysis

The brainstorming meetings by a knowledgeable team who specializes in the production process of the headlining process were conducted to articulate the relevant potential factors that can affect quality of headlining. The cause-and-effect diagram with 4M (i.e. Man, Machine, Material and Method) was applied to this process to indicate the root-causes of the problems. For the part dimension out-off standard problem, 12 potential factors were concluded (Fig. 4). The causes of the problem can be summarized as follows. First, operators do not have enough skill to place part correctly on the platform. This makes parts shaken while cutting their edges by the water jet cutting machine. In addition, it also causes operator fatigue. Second, there is no method to specify the proper air pressure level and the cycle time of cleaning platform which cause parts located not tightly adhere to the platform. Third, the machine's components are damaged and the type of the holder on the machine platform is not appropriate which causes misalignment of parts. Forth, material deformation and inappropriate type of cleaning tools cause scraps remaining on the platform.

These factors were then used to analyse the relationship between the cause and its associate effects using the cause-and-effect matrix. A priority rate of 10 point was given for each problem that causes part dimension out-off standard problem and the score ranges between 1-10 points. These scores were sorted from large to small by the Pareto chart and the factors with significant impacts (score > 50 point) were selected for further analysis. As a result, the relevance factors were reduced to 8 (Fig. 5). These factors are further screened by using the Failure Mode and Effects Analysis (FMEA). The results showed that 4 factors with the highest scores were identified (Fig. 6). These factors were chosen for the next experiments.

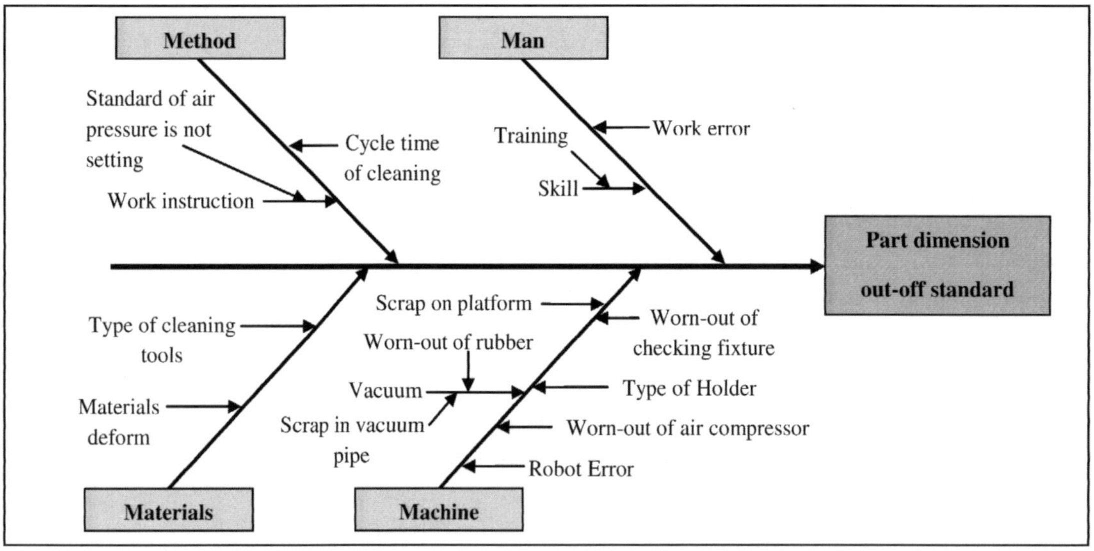

Figure 4. Cause-and-effect diagram of the part dimension out-off standard

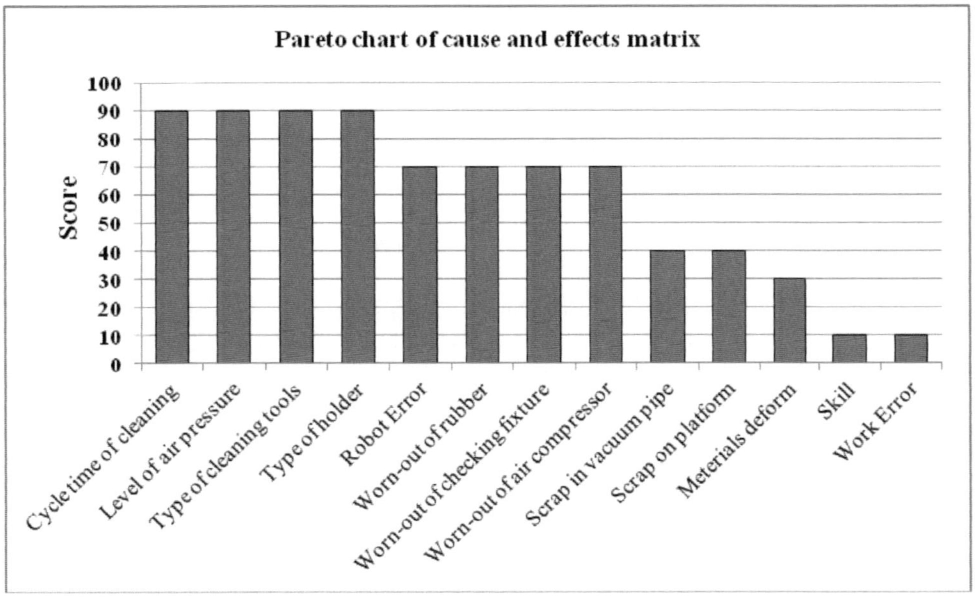

Figure 5. Pareto chart of the cause-and-effects matrix

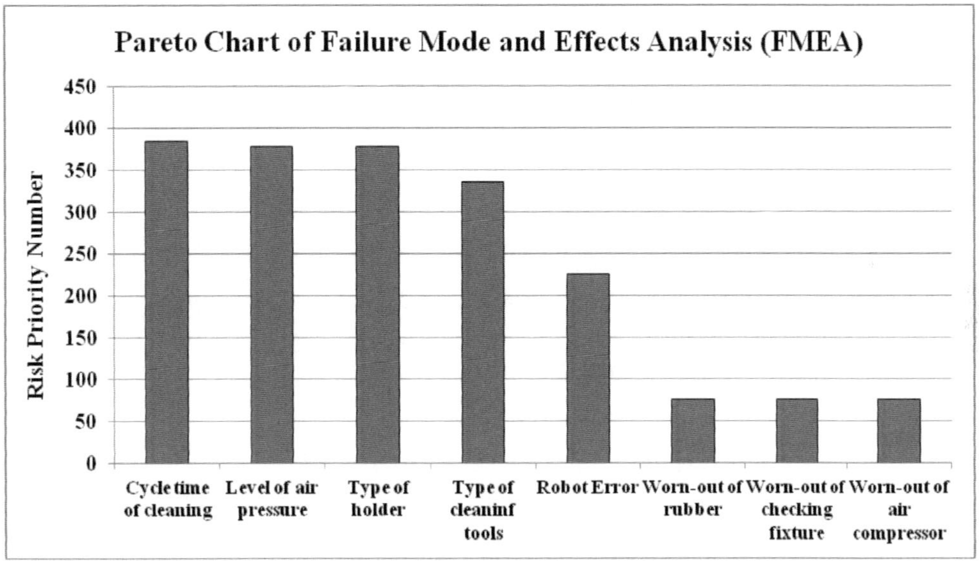

Figure 6. Pareto chart of the FMEA

4. Analysis Phase

In this phase, the experimental design method was adopted to find the factors that show statistically significant effects to the part dimension out-off standard defect. Four factors obtain from the previous phase were experimented including air pressure, types of the holder on the platform, types of cleaning tools and cycle time of cleaning. The 2^{k-1} fractional factorial design [6] with 4 factors and 2 levels of each factor were adopted to test whether or not the main effects and their interactions were significant (Table 1).

Table 1. Factors and levels in the 2^{k-1} fractional factorial design

No.	Factors	Units	Symbols	Levels of factor	
				Low (-1)	High (1)
1	Air pressure	Bar	A	5	10
2	Type of holder	-	B	Old	New
3	Type of cleaning tools	-	C	Old	New
4	Cycle time of cleaning	Time/Day	D	1	4

The experimental data were analyzed by using the Minitab software [7]. From the Pareto chart of effects (Fig. 7), the factor effect with its value greater than 0.02053 is a significant factor at 95% confident interval equal. The Pareto chart of effects showed that the main effects A and B have P-values less than 0.05 (significant level α), whereas factors C and D and interactions (AD, AB and AC) were not significant. Therefore, it can be concluded that the main effects A and B are significant at 95% confident interval. As a result, all significant factors need further investigation to find their appropriate level settings.

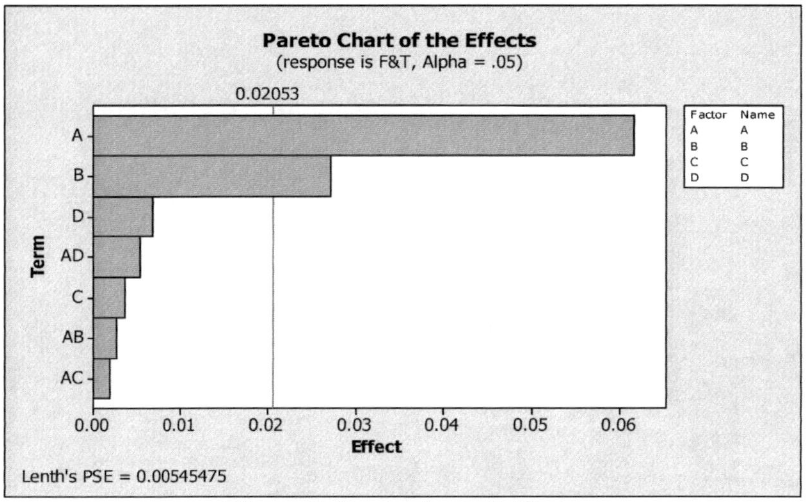

Figure 7. Pareto chart of effects

5. Improvement Phase

In this phase, more levels of the significant factors derived from the 2^{k-1} factional factorial experiments were tested to find their appropriate levels that can reduce defects in the headlining process. The general full factorial design was used in this experiment. Three levels of air pressure and two levels of holder on the platform were randomly run with 2 replicates (Table 2).

Table 2. Factors and level in full factorial design

Factors	Units	Symbols	Levels of factor		
			Low (-1)	Medium (0)	High (1)
Air pressure	Bar	A	5	8	10
Type of holder on platform	-	B	Old	-	New

The statistical analysis showed that both main effects obtained the P-values less than 0.05 (Fig. 8). Therefore, it can be concluded that the main effects affected the defective at 95% confident interval. Since the main effects were used as a mean to select the proper setting for factors, the factors A was set at 10 Bar (high level of factor A) and B was set with the new type holder on platform (high level of factor B) (Fig.9).

General Linear Model: F&T versus A, B

Factor	Type	Levels	Values
A	fixed	3	5, 8, 10
B	fixed	2	Old, New

Analysis of Variance for F&T, using Adjusted SS for Tests

Source	DF	Seq SS	Adj SS	Adj MS	F	P
A	2	0.0097543	0.0097543	0.0048771	72.84	0.000
B	1	0.0012569	0.0012569	0.0012569	18.77	0.005
A*B	2	0.0001175	0.0001175	0.0000588	0.88	0.463
Error	6	0.0004018	0.0004018	0.0000670		
Total	11	0.0115304				

S = 0.00818293 R-Sq = 96.52% R-Sq(adj) = 93.61%

Figure 8. Results of ANOVA

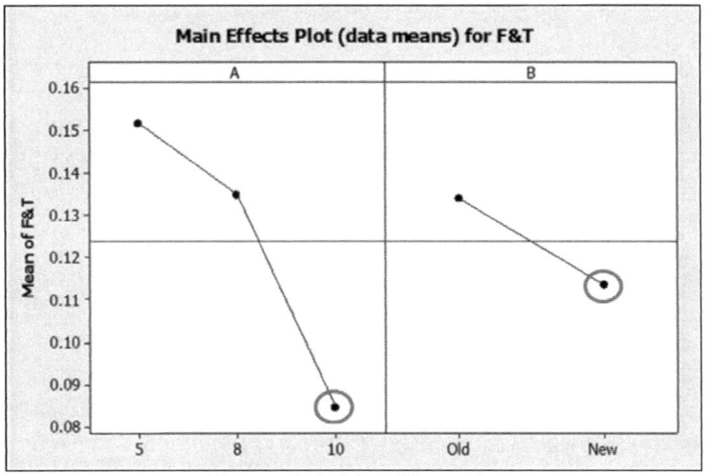

Figure 9. Main effects plot

6. Control Phase

In this phase, all settings of the significant factors derived from the previous phase were implemented in the process and the data relating to the defect occurred in the automotive headlining process were collected for 30 days. It was found that the quantity of defects was reduced substantially (Fig.10).

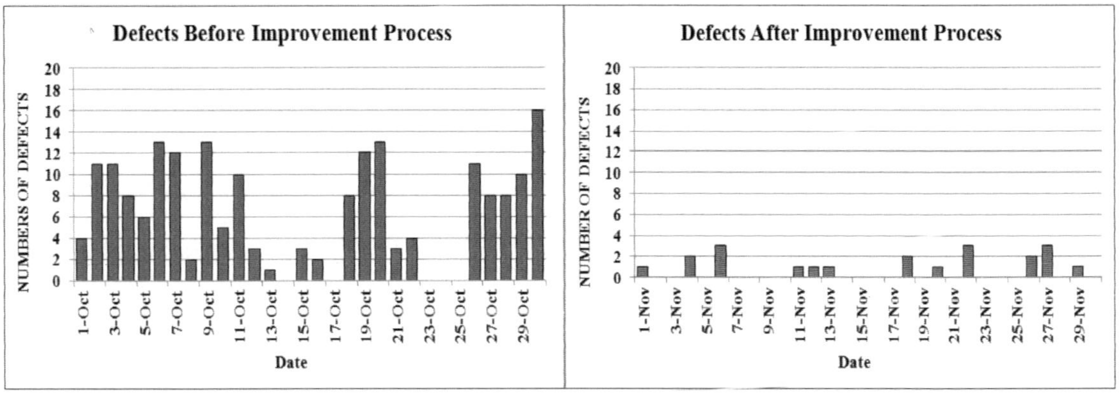

Figure 10. Comparison of defects

To maintain the quality of the headlining process after the improvement, the work instruction for operational control in the headlining process was written. In addition, the operators were trained to understand the importance of controlling these parameters to ensure that the performance improvements were maintained.

7. Conclusion

The objective of this research was to applied the Six Sigma method to reduce wastes that occurred in the automotive headlining manufacturing process. Five steps of the Six Sigma were systematically utilized so as to find the proper parameter settings of the significant factors. The experimentation results indicated that the defect was decreased when the air pressure of the water jet cutting machine is increased and when the new part holder type was used. Therefore, it can be concluded that the air pressure and holder on the platform factors had significant effects on the problem of interest. After applying all new settings to the process, the defect rate was reduced to 69,489 DPPM.

References

[1] D.A. Desai, P. Kotadiya, N. Makwana and S. Patel, 2015 Curbing variations in packaging process through Six Sigma way in a large-scale food-processing industry, Journal of industrial engineer international, **11(1)**, 119-129

[2] C. Gygi, N.D. Carlo and B. Williams 2005 *Six Sigma for dummies* (New Jersey: Wiley Publishing, Inc)

[3] T. Bertels 2003 *Rath & Strong's six sigma leadership handbook (*New Jersey: Wiley Publishing, Inc)

[4] H.C. Hung and M.H. Sung 2011 Applying six sigma to manufacturing processes in the food industry to reduce quality cost, Scientific Research and Essays, **6(3)**, 580-591

[5] C.T. Carroll 2013 *Six Sigma for powerful improvement* (New York: Taylor & Francis Group)

[6] D.C. Montgomery 2008 *Design and analysis experiments* (America: John Wiley & Sons, Inc)

[7] D.C. Montgomery and G.C. Runger 2010 *Applied statistics and probability engineers* (America: John Wiley & Sons, Inc)

Lean Six Sigma Application in Rear Combination Automotive Lighting Process

Thanwarhat Sodkomkham[1] and Parames Chutima[1,2]

[1]Industrial Engineering Department, Chulalongkorn University, Bangkok 10330, Thailand
[2]Regional Centre for Manufacturing Systems Engineering, Chulalongkorn University, Bangkok 10330, Thailand

[2]E-Mail: cparames@chula.ac.th

Abstract. The case study company produces various front and rear lightings for automobiles and motorcycles. Currently, it faces two problems, i.e. high defective rate and high inventory. Lean Six Sigma was applied as a tool to solve the first problem, whereas the other problem was managed by changing the production concept from push to pull. The results showed that after applying all new settings to the process, the defect rate was reduced from 36,361 DPPM to 3,029 DPPM. In addition, after the implementation of the Kanban system, the company achieved substantial improvement in lead time reduction by 44%, in-process inventory reduction by 42%, finished good inventory reduction by 50%, and finished good area increased by 16%.

1. Introduction

Automotive parts industry is vital to the growth of national economy. Part manufacturing companies search for management methodologies that allow them to improve their products and processes, decrease cost, increase profitability and customer's satisfaction. The case study company is a large-sized automotive lighting manufacturer in Thailand. The company produces high varieties of front and rear lightings supplying to automotive and motorcycle assembly companies.

Preliminary production survey indicates that the rear combination automotive lighting of the T9 product has highest defects (10,762 DPPM) comparing with other models resulting from high process wastes. This results in its current production cost being found substantially higher than its selling price. As a result, in this paper, the Lean Six Sigma method [1-7] is applied to reduce defects in the T9 manufacturing process.

The steps of Lean Six Sigma include the waste identification (7 wastes), as well as the defect waste reduction in process injection (5 phases of the Six Sigma approach, i.e. DMAIC standing for define, measure, analysis, improvement and control phases).

2. Define Phase

This research focuses on the production of the T9 model of the rear combination automotive lighting since its defect rate is highest. The production of the T9 model consists of three processes, i.e.

injection, surface, and assembly. From the historical defect reports between January 2015 and July 2015, it was found that the injection process had the highest defect rate of 36,361 DPPM comparing with the others. In addition, the injection process is the first process in which can cause several consequence effects to the two subsequence processes. Four major defects found in the injection process include silver line (39.8%), dust (19.3%), short shot (15%), and flash (7.4%). As a result, the application of Lean Six Sigma is emphasized on the highest defect proportion, i.e. silver line defect.

3. Measure Phase

This phase involves two main steps, i.e. waste measurement and problem identification analysis, which can be described as follows [6].

3.1 Waste Measurement

Based on the 7-wastes concept, wastes can be arisen from over production, unnecessary transportation, waiting, unnecessary inventory, defects, unnecessary motions, and inappropriate processing [2, 3]. The result of the waste assessment of the T9 model is shown in Table 1. It is obvious that the types of wastes that influence most to the T9 model include defects (37.70%), over production (31.39%), and unnecessary inventory (16.39%).

Table 1. Summary of Waste Measurement.

No.	Types of waste	Frequency (%)
1	Over production	31.39
2	Unnecessary transportation	4.11
3	Waiting	5.34
4	Unnecessary Inventory	16.39
5	Defects	37.70
6	Unnecessary motions	1.42
7	Inappropriate operation	3.65

Currently, the T9 model manufacturing process is run based on the push production concept [8] where the forecasted schedule is used to determine the production quantity to be produced in each month. The major steps of the T9 model manufacturing process are sketched as shown in Figure 1.

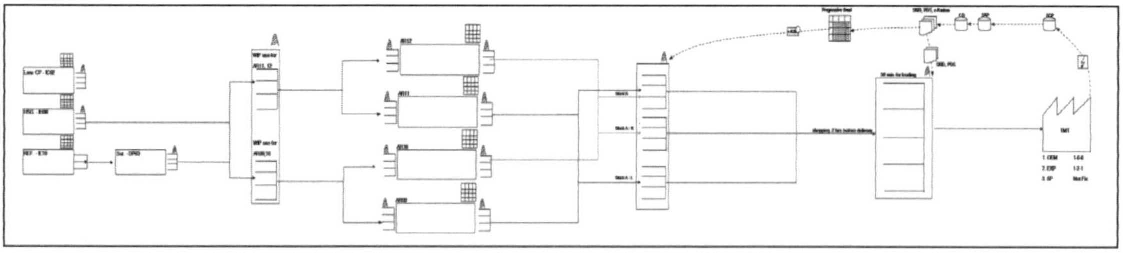

Figure 1. Major steps of the T9 model manufacturing process.
(Note: Manufacturing process is run based on the push production concept. The processes comprises injection, surface and assembly)

3.2 Problem Identification Analysis

A number of brainstorming meetings by a knowledgeable team who specializes in the production process of the rear combination automotive lighting were systematically conducted to indicate the relevant factors that can affect defects occurred in the injection process. The cause and effect analysis with 5M1E was applied in this process [3]. It was revealed that the number of tentative causes was 16 (Table 2). These factors were used for further analyzing the relationship between cause and related effect using the cause and effect matrix. By scoring these relationships (i.e. 0, 1, 3, and 9), these scores were sorted from large to small by the Pareto chart and the factors with significant impacts (high score is 9) were selected for further analysis. As a result, the relevance factors were reduced to 7 (Fig. 2).

Table 2. Cause and Effect Diagram with 5M1E.
(Source: Brainstorm meetings by a knowledgeable team who specializes in the production process of the rear combination automotive lighting.)

Item	Area cause	Process input	Total
1	Man	Operators lack of training	3
2		Operators lack of knowledge	1
3		Operators lack of experiences and skills	1
4	Material	Quality of plastic	3
5	Machine	Design of mold does not fit	9
6		Machine breakdown	9
7		Clamping force for injection molding	3
8		Screw and Barrel are corroded	3
9	Method	Time of bake plastic	9
10		Temperature of bake plastic	9
11		Speed of injection	9
12		Pressure of injection	9
13		Time of cooling part	9
14	Measurement	Measurement system instability	3
15	Environment	Humidity in working area	3
16		Temperature in working area	0

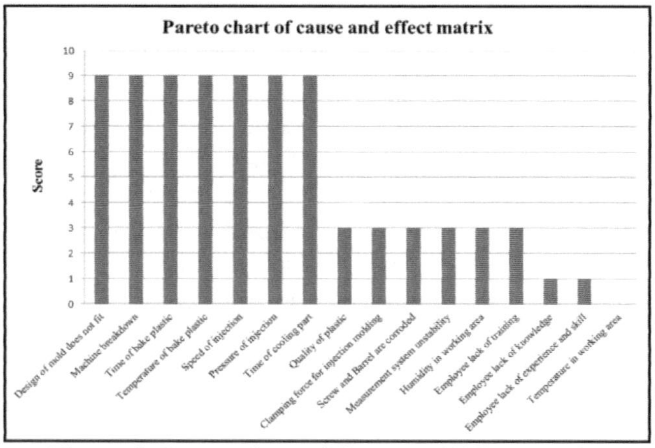

Figure 2. Pareto chart of Cause and Effect matrix.

The root causes with the highest score (9 points from Figure. 2) were selected for further analyze the factors that had significant impacts to the silver line defect. The risk priority number of each root cause was calculated (Table 3).

Table 3. The Risk Priority Number of Each Root Cause.

No.	Potential Failure	Risk Priority Number (RPN)
1	Design of mold does not fit	108
2	Machine breakdown	120
3	Time of bake plastic	486
4	Temperature of bake plastic	648
5	Speed of injection	720
6	Pressure of injection	648
7	Time of cooling part	448

Fig. 3 shows five potential factors that were concluded from the brainstorming meetings including speed of injection, pressure of injection, temperature of bake plastic, time of bake plastic and time of cooling part. The result shows that five factors have a high score. As a result, these five factors were selected for experiments in the next phase, which was 96% of the total score.

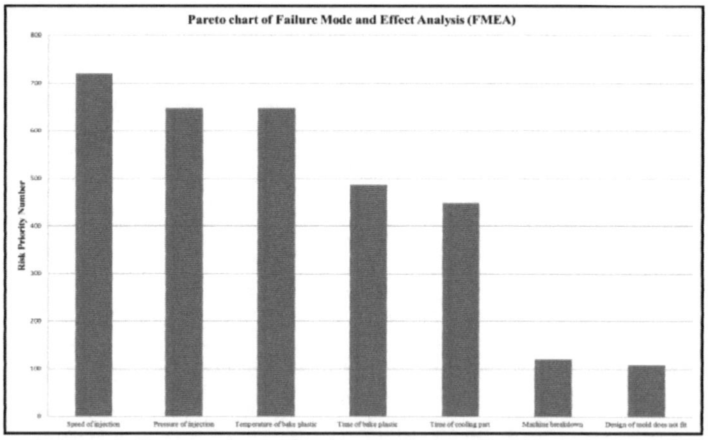

Figure 3. Pareto chart of Failure Mode and Effect Analysis.

4. Analysis Phase

The 2k-1 fractional factorial design [11] was employed to test whether or not these factors and their interactions have significant impacts on the silver line defect (Table 4). The levels of each factor were mainly chosen from the minimum and maximum adjustable values of the instruments used in the process. All 16 trials were conducted within a completely randomized experiment. The response variable in this study was the defect occurred in the injection process.

Table 4. Factors and Level of Factors in 2^{5-1} Fractional Factorial Design.

No.	Factors	Units	Symbols	Levels of factor	
				Low (-1)	High (1)
1	Speed of injection	percentage	A	10	20
2	Pressure of injection	kilograms/square centimeter	B	150	160
3	Temperature of bake plastic	Celsius	C	90	100
4	Time of bake plastic	hour	D	3	4
5	Time of cooling part	second	E	25	30

The experimental data were analyzed by using the Minitab software and the Pareto chart of effects (Fig. 4). The factor effect with its value greater than 0.1588 is a significant factor at 95% confident interval. The Pareto chart of effects showed that main effects A and B have the P-values less than 0.05 (significant level α), whereas factors C, D, E and interactions (AB, AC, AD, AE, BC, BD, BE, CD and DE) were not significant. Therefore, it can be concluded that the main effects A and B are significant at 95% confident interval. As a result, all significant factors need further investigation to find their appropriate level settings.

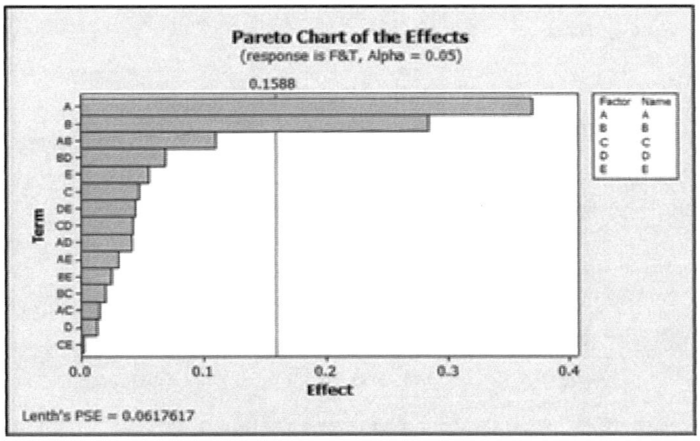

Figure 4. Pareto chart of effects.

5. Improvement Phase

The 2k-1 fractional factorial design [11] was employed to test whether or not these factors and their interactions have significant impacts on the silver line defect (Table 4). The levels of each factor were mainly chosen from the minimum and maximum adjustable values of the instruments used in the process. All 16 trials were conducted within a completely randomized experiment. The response variable in this study was the defect occurred in the injection process.

5.1 Experiment of relevant parameter

In this phase, more levels of the significant factors derived from the 2k-1 fractional factorial experiments were tested to find their appropriate levels that can reduce defects in the injection process of the rear combination automotive lighting. The general full factorial design was used in this experiment. Three levels of the speed of injection (A) and three levels of the pressure of injection (B) were randomly run with 2 replicates (Table 5).

Table 5. Factors and Levels of Factors in Factorial Experimental Design

No.	Factors	Units	Symbols	Levels of factor		
				Low (-1)	Medium (0)	High (1)
1	Speed of injection	percentage	A	10	15	20
2	Pressure of injection	kilograms/square centimeter	B	150	155	160

The statistics [12] in Figure. 5 shows that the main effects and two-way interaction have the P-values less than 0.05. Therefore, it can be concluded that the main effects and two-way interaction affected the defects in the injection process at 95% confident interval. Since the interaction effects were significant, the interaction effect plots were used as a mean to select the proper settings for the factors (Fig. 6). Hence, Factor A (speed of injection) was set at 20 percentage (high level of factor A), whereas Factor B (pressure of injection) was set at 155 kilograms/square centimeter (medium level of factor B).

General Linear Model: F&T versus Speed, Pressure

```
Factor    Type   Levels  Values
Speed     fixed       3  10, 15, 20
Pressure  fixed       3  150, 155, 160


Analysis of Variance for F&T, using Adjusted SS for Tests

Source          DF    Seq SS    Adj SS    Adj MS      F      P
Speed            2  0.060924  0.060924  0.030462  43.89  0.000
Pressure         2  0.022163  0.022163  0.011082  15.97  0.001
Speed*Pressure   4  0.037044  0.037044  0.009261  13.34  0.001
Error            9  0.006246  0.006246  0.000694
Total           17  0.126377


S = 0.0263439   R-Sq = 95.06%   R-Sq(adj) = 90.66%
```

Figure 5. Results of Analysis of Variance.

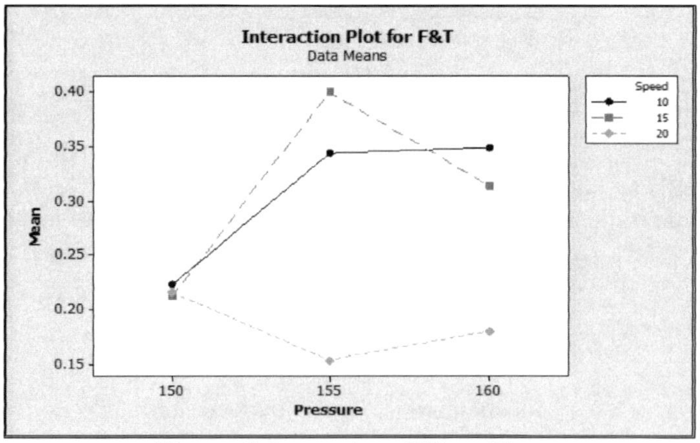

Figure 6. Interactions Effects Plot.

5.2 Pull concept implementation

Since the company applies the push production concept, the finished products are produced according to the forecasted demand without consideration of the actual customer's requirements. Consequently,

IOP Conf. Series: Materials Science and Engineering **131** (2016) 012022 doi:10.1088/1757-899X/131/1/012022

huge amount of WIP parts were found stagnated on the shop floor and the warehouse was fully packed with unnecessary finished goods. Because of the unorganized and less shop floor space, the congested material flow was noticeable in every production day.

To overcome this problem, the company decided to apply the pull production concept by implementing the Kanban system in the manufacturing process [9, 10]. It was also expected that this concept could bring about the reduction of inventory in the warehouse (Fig. 7).

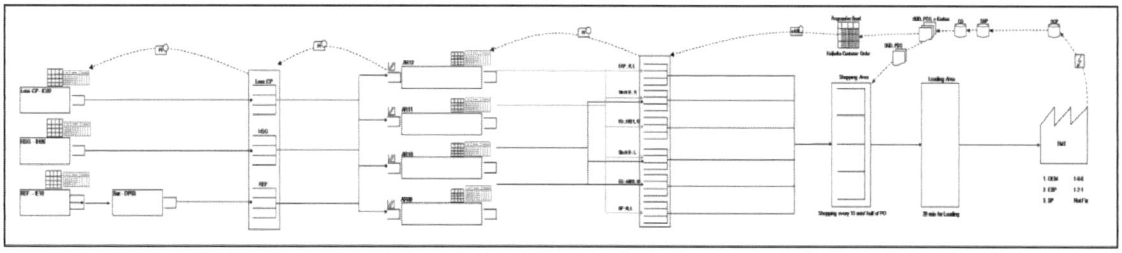

Figure 7. Development of the T9 model manufacturing process by Kanban system.

Before the Kanban pull system was realized on the shop floor, establishing of the mechanism and rule were necessary in order to assist the Kanban activity. The Kanban card was designed and issued to the production process (Figs. 8 and 9).

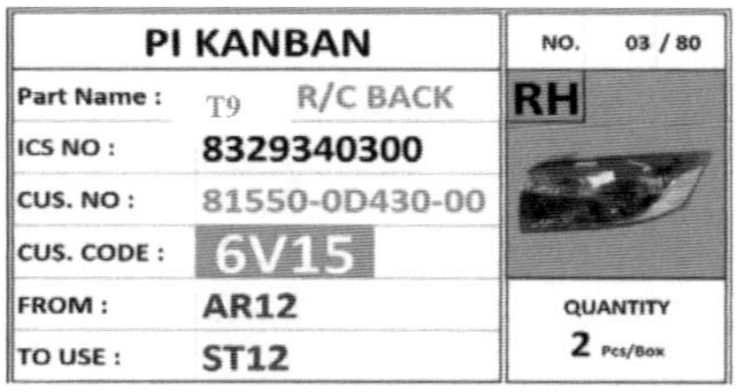

Figure 8. Example of Kanban cards: Production Instruction Kanban card

Figure 9. Example of Kanban cards: Part Withdrawal Kanban card with container

Fig. 8 shows that the production instruction Kanban card is a production order, which gives the instruction to the production process to produce the required product according to the needed quantity. In addition, the production withdrawal Kanban card (Fig. 9) is the authorization signal for the withdrawal of the product from the production process according to customer's needs.

The Kanban cards contain required information for the production process. To ensure that they are producing the right part in the right quantity at the right time according to customer's needs. The following information is contained in the Kanban card including customer information (such as customer's name), customer product name and type of model, product information (such as part name, part picture, and quantity per packing), production process address and storage area.

The Heijunka post used as the planning schedule to evenly distribute the product volume and variety throughout available production time was also established. Every Kanban card that was detached from the container during the withdrawal process was loaded into the Heijunka post. The card at the Heijunka post was used to withdraw the product from the production process to replenish the finished good that had been delivered to the customer. The frequency of withdrawal is scheduled according to the takt time. Therefore, Heijunka prevent the uneven loading of volume occurred in the production floor. To establish this post, few operational settings had to be determined, e.g. frequency of the withdrawal process and conveyance time, quantity of the Heijunka slot, maximum production quantity, etc.

6. Control Phase

In this phase, all settings of the significant factors derived from the previous phase were implemented in the process and the data relating to the defect occurred in the injection process of the rear combination automotive lighting were collected for 30 days. It was found that the quantity of defects was reduced substantially. The comparison between the number of defects before and after the improvement process was shown in Figs. 10 and 11.

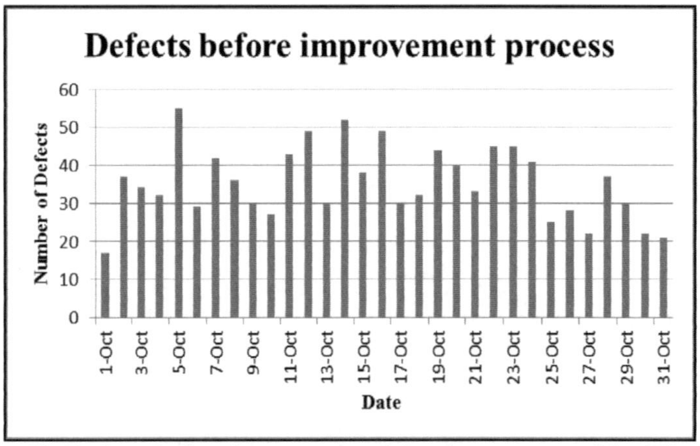

Figure 10. Defects before improvement injection process of the T9 model.

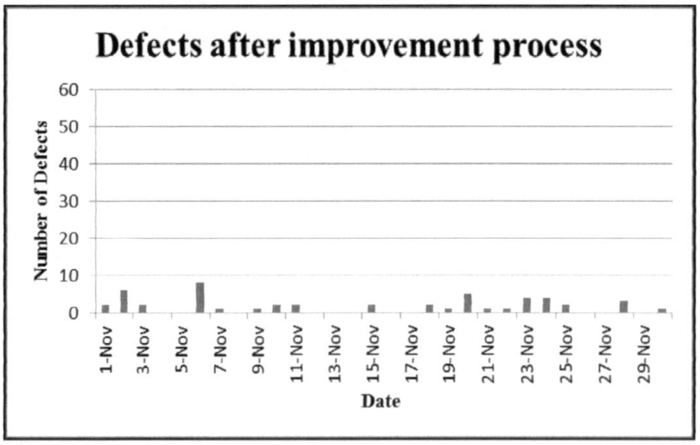

Figure 11. Defects after improvement injection process of the T9 model.

To maintain the quality of the injection process after the improvement, the work instruction for operational control in the injection process was written. In addition, the operators were trained to understand the importance of controlling these parameters to ensure that the performance improvements were maintained. In addition, to control these Kanban activities, a Kanban rule was established. The documented information related to all the guidelines and references was written and shared among all the production personnel that were involved in this activities. This helped the operators to manage and monitor the Kanban activities systematically.

7. Conclusion

The Lean Six Sigma method was applied to reduce wastes that occurred in the rear combination automotive lighting of the T9 manufacturing process. Currently, it faces two problems, i.e. high defective rate and high inventory. Five steps of the Six Sigma were systematically utilized so as to find the proper parameter settings of the significant factors. It was found that the speed of injection and pressure of injection had significant effects on the problem of interest. After applying all new settings to the process, the defect rate was reduced to 3,029 DPPM. In addition, after the implementation of the Kanban system, the company achieved substantial improvement in lead time reduction by 44%, in-process inventory reduction by 42%, finished good inventory reduction by 50%, and finished good area increased by 16%.

References

[1] Cabrita, M.D.R., J.P. Domingues, and J. Requeijo 2015 Application of Lean Six-Sigma methodology to reducing production costs: case study of a Portuguese bolts manufacturer *International Journal of Management Science and Engineering Management*, 1-9.

[2] Indrawati, S. and M. Ridwansyah 2015 Manufacturing Continuous Improvement Using Lean Six Sigma: An Iron Ores Industry Case Application *Procedia Manufacturing*, **4**, 528-534.

[3] Vinodh, S., S.G. Gautham, and A. Ramiya R 2011 Implementing lean sigma framework in an Indian automotive valves manufacturing organisation: a case study *Production Planning & Control*, **22(7)**, 708-722.

[4] Anderson, N.C. and J.V. Kovach 2014 Reducing Welding Defects in Turnaround Projects: A Lean Six Sigma Case Study *Quality Engineering*, **26(2)**, 168-181.

[5] Harjac, S.J., A. Atrens, and C.J. Moss 2008 Six Sigma review of root causes of corrosion incidents in hot potassium carbonate acid gas removal plant *Engineering Failure Analysis*,**15(5)**, 480-496.

[6] de Mast, J. and J. Lokkerbol 2012 An analysis of the Six Sigma DMAIC method from the perspective of problem solving *International Journal of Production Economics*, **139(2)**, 604-614.

[7] Erbiyik, H. and M. Saru 2015 Six Sigma Implementations in Supply Chain: An Application for an Automotive Subsidiary Industry in Bursa in Turkey *Procedia - Social and Behavioral Sciences*, **195**, 2556-2565.

[8] Naufal, A., et al., 2012 Development of Kanban System at Local Manufacturing Company in Malaysia–Case Study *Procedia Engineering*, **41**, 1721-1726.

[9] Rahman, N.A.A., S.M. Sharif, and M.M. Esa 2013 Lean Manufacturing Case Study with Kanban System Implementation *Procedia Economics and Finance*, **7**, 174-180.

[10] Kumar, C.S. and Panneerselvam, R. 2007 Literature review of JIT-KANBAN system *The International Journal of Advanced Manufacturing Technology*, **32**, 393-408.

[11] Montgomery. D.C. 2008 Design and analysis of experiments, U.S., John Wiley & Sons, Inc.

[12] Montgomery, D.C. and Runger, G.C. 2010. Applied statistics and probability for engineers, U.S., John Wiley & Sons, Inc.

Improvement of Automotive Part Supplier Performance Evaluation

Chalermkwan Kongmunee[1] and Parames Chutima[1, 2]

[1]Industrial Engineering Department, Chulalongkorn University, Bangkok 10330, Thailand
[2]Regional Centre for Manufacturing Systems Engineering, Chulalongkorn University, Bangkok 10330, Thailand

E-Mail:cparames@chula.ac.th

Abstract. This research investigates the problem of the part supplier performance evaluation in a major Japanese automotive plant in Thailand. Its current evaluation scheme is based on experiences and self-opinion of the evaluators. As a result, many poor performance suppliers are still considered as good suppliers and allow to supply parts to the plant without further improvement obligation. To alleviate this problem, the brainstorming session among stakeholders and evaluators are formally conducted. The result of which is the appropriate evaluation criteria and sub-criteria. The analytical hierarchy process is also used to find suitable weights for each criteria and sub-criteria. The results show that a newly developed evaluation method is significantly better than the previous one in segregating between good and poor suppliers.

1. Introduction

Nowadays, the business competition has become increasingly intensified in Thailand especially in the automotive industry which has grown and developed rapidly. One car consists of numerous different parts, which can be provided by various suppliers. This leads to high competition among automotive part suppliers who supply the same or similar parts to the automotive manufacturer. These suppliers are striving to gain a higher market share. To achieve the goal as such, they must pass several stringent criteria set by the automotive manufacturer. The case study company is a leading Japanese automotive assembly plant in Thailand. Currently, 243 suppliers pass the supplier selection process of the company. Therefore, they must have their operational processes evaluated on a monthly basis by several departments of the company, i.e. production, sales, and quality control departments. The evaluation process is not only conducted to assess the current performance of the suppliers, but also to further develop and improve the quality of the supplied parts. Furthermore, helping suppliers to reach the world class standard can mitigate potential future risks and problems that might affect the company, such as late orders and damaged goods, high turnover time and high part inspection time. As a consequence of the part supplier performance evaluation, the company can select the suppliers who can operate in line with its current competitive strategy. The overall supplier evaluation policy of the company weights equally in terms of delivery and quality of the parts supplied. Thus, the weight of part delivery has been set to 40%, the quality to 40% and the price to 20%. This weighting scheme forces the supplier to deliver high quality parts and also to be aware of the total lead time to ensure an on-time delivery. The part control department is responsible for the part supplier performance evaluation. A total of 12 evaluators have to evaluate 203 suppliers. These evaluators apply the same criteria in the part delivery performance measurement. It was found that the evaluation results are different based on their experience and self-opinion. As a result, the outcome of the evaluation may not reflect the actual delivery performance of the suppliers. In addition, the weights using in the evaluation criteria are not suitable. This study aims to improve the part delivery evaluation criteria to ensure the standardized evaluation process so that the results are trustworthy. The analytic hierarchy process (AHP) technique is applied to find suitable criteria and its associated weight. The measurement system analysis (MSA) is also exercised to calibrate the accuracy of the evaluators.

2. Literature Review

Due to the fact that the quality and performance of suppliers can affect the company, the part supplier performance evaluations are very important to any kind of business. Many researchers have studied and presented types and criteria which are different from each other, depending on the type of business and the strategy of the company.

2.1 Performance Evaluations

Evaluations are processes of gathering information from measurements and using evaluator opinion to determine the suitable scores through comparing the results to the standard. Supplier evaluations aim to reduce the risk when ordering goods, which assures that the supplier can meet customer needs and the results of the evaluation will be used to determine future purchase orders [1]. Evaluations help to choose efficient suppliers and avoid those who do not provide high quality products. The evaluations always ensure confidence with suppliers who pass all the criteria so that they maintain responsibility and succeed in filling customer needs [2]. Part supplier performance evaluations have diverse issues in determining the criteria. The primary three criteria are price, quality and lead time [3].

2.2 Multi-Criteria Decision Making (MCDM) Overview

MCDM is a method of balancing variables in the evaluation to produce the best outcome based on multiple criteria and their respective weights. To do this, finding clear methods and realistic weights can help the determination of the most effective and efficient supplier based on multiple criteria, not just only cost. Nowadays, there are many ways to deal with decision issues, which each has its own specialty. These can be used differently, depending on situation, appropriateness and specialization of the evaluator [4]. AHP is one of the most effective ways to deal with the complicated issues which are made easier to understand through simulation of human linguistics [5]. There are several ways to deal with determination issues by using MCDM, depending on the researchers, such as procurement and best supplier selection [6], location selection [7], environment risk assessment [8] and health care system [9].

2.3 Analytic Hierarchy Process

AHP was developed in 1970 by Thomas L. Saaty [10]. AHP is a technique that converts the feelings and opinions of evaluators into numerical data. A standard weighting of criteria is created based on the results of evaluations. AHP is a primary way to determine MCDM and is an effective and convenient process for creating a standard weight. AHP helps to determine the best suppliers based on evaluations. It can also be used to help with complicated multi-criteria decisions by comparing methods [11]. The primary technique of AHP involves presenting issue of hierarchy determination and comparing criteria through a statistical matrix. These results are then evaluated and the criteria are ranked and weighted. To obtain the best result, according to the study, the AHP technique is applied in many decision making situations, such as supplier performance evaluations [12, 13], supplier selection [14, 15], etc. The main benefits of AHP are that the results are formulated into numerical data, ranking criteria is simplified, and evaluation bias and opinion is reduced. AHP consists of three steps as follows: (1) Establishing the hierarchical structure, (2) Constructing the pairwise comparison matrix and (3) Calculating the consistency. More detail can be seen in [16]

3. Case Study

This study employs the example occurring in the automotive assembly plant. The company currently faces with the problem of part supplier performance evaluation. Top management found that the results of the evaluations were ineffective enough to reflect the actual part supplier performance. The reasons come from inappropriate weights being used in the evaluation criteria and non-standardized self-opinion of the evaluators. To analyse this problem, this research divided the suppliers into 3 groups depending on the importance of the supplied parts. AHP was then applied to these groups to specify the appropriate weights of the criteria. These criteria are further used to improve the delivery performance of the suppliers. Through doing this, a new standardized weights can be determined so that trustworthy and consistency results can be achieved.

4. AHP Implementation

AHP is a reliable and suitable method for decision procedures and help decision-makers to assess the criteria's weights and chosen the best alternative [17].It helps to determine the best supplier (alternative) based on both qualitative and quantitative evaluations. It can also be used to help in complicated multi-criteria decisions. The main benefits of AHP are that its result is shown in a numerical format, ranking criteria is simplified, and evaluation bias and opinion are reduced; therefore, AHP was chosen to strive for suitable criteria and its associated weight to help improve the part delivery evaluation criteria to ensure the standardized and trustworthy evaluation process.

Due to the high number of purchased parts, the suppliers are classified into three groups, according to the importance of their supplied parts.

Group A (48 suppliers): These suppliers produce large and specialized parts such as engines, cushions and doors. These suppliers are difficult to replace, and their supplied parts are not regularly kept in stock. In addition, these suppliers always show on-time delivery performance.

Group B (130 suppliers): These supplies produce unspecialized parts such as headlights, brake lights, batteries and radios. These suppliers can be replaced easily. In addition, these suppliers send parts to the company according to the given cycle time.

Group C (25 suppliers): These suppliers produce small parts, which are ordered once a week, depending on the needs of the company. These parts are constantly kept in stock, which include the parts like screw, nut, bolt, tape and clip.

In each group, AHP is applied to specify the main criteria and their associated weights to be used in the part delivery performance evaluation so that the aforementioned problem can be effectively resolved. The steps can be described as follows.

4.1 Primary Criteria Selection

Numerous delivery problems and criteria for part delivery performance evaluation used in the automotive industry were collected. 3 main stakeholders as well as 12 evaluators were invited to participate in the brainstorming meeting. The criteria which deemed to be the most influences to the delivery performance and had the goal of improving delivery performance were selected. The results of the primary criteria selection in the part delivery evaluation consisted of 5 factors (on-time delivery, delivery cooperation, delivery part accuracy, package and delivery's documentation).

4.2 Sub-criteria selection

After the primary criteria were selected, once again the sub-criteria were chosen through stakeholder brainstorming. After brainstorming, the stakeholders vote to conclude the selected sub-criteria. It is noted that the sub-criteria must be related to the primary criteria. The results can then be exemplified as shown in Table 1.

Table 1.Results of selected sub-criteria

Criteria	Sub Criteria
A. On - time delivery	A.1 Line stop A.2 Pending Car (not complete) A.3 Change production schedule A.4 Not on time delivery but not effect to production line
B. Delivery cooperation	B.1 ASN confirm B.2 Information such as delivery delay information B.3 Delivery performance during emergency order B.4 Delivery flexibility B.5 Follow the rule of the factory B.6 Reply countermeasure ,Car Report and Survey
C. Delivery part accuracy	C.1 Part Short delivery quantity C.2 Part Over delivery quantity C.3 Part NG (damage, wrong parts, mixing Parts)
D. Package	D.1 Size standard D.2 No damage D.3 Cleaning D.4 Safety D.5 Easy for receiving check D6 Enough to use for requirement
E. Delivery's Documentation	E.1 Delivery documents must complete (part tag, delivery tag& delivery note) E.2 Part Tag and delivery tag base on document standard E.3 Delivery note must easy for receiving check

4.3 Analytic Hierarchical Process

After the relevant primary and sub-criteria were selected, the next step is to apply AHP to determine suitable weights for them. The process can be explained as follows.

4.3.1 Establishing a Hierarchical Structure

Following the criteria identification, the hierarchical structure (Figure 1.) can be established by creating a hierarchy chart, in which the highest factor is the goal. In this case, the goal was the delivery performance of part suppliers. One level lower the goal is the primary criteria used in determining the best supplier [18], which includes 5 factors (i.e. on-time delivery, delivery cooperation, delivery part accuracy, package and delivery's documentation). The sub-criteria located under the primary criteria consisted of 22 sub-factors (Table 1).

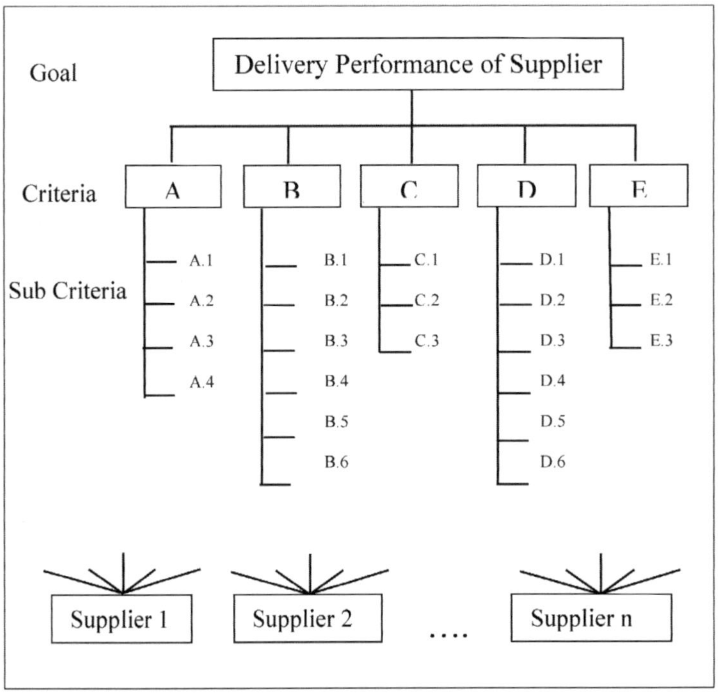

Figure 1. AHP based hierarchical model to evaluate delivery performance of suppliers.

Establishing a hierarchical structure shows the complicated relationships between the main criteria and sub-criteria. Creating the hierarchical structure leads to the ability to evaluate the comparison criteria within the same factor, which can affect the solution.

4.3.2 Ranking the Priority

This step is the most important step in the selection process because it ranks the importance of the evaluation criteria, which, if chosen incorrectly, will greatly affect the outcome. In this step, the details of factors are gathered and compared each factor in part supplier performance evaluations. The weight of each factor is then analysed with evaluator brainstorming results through scoring and evaluator comment. The scores are the outcome of comparing the importance of each factor in pairs. These pairwise comparisons proceed from the top of the hierarchical structure, starting with primary criteria and then moving through the sub-criteria.

In AHP, scores of 1-9 (Table2) are given based on importance of the pair of the factors to get satisfaction into the numbers, applying numerical values to non-numerical opinion.

Table 2. Measurement scales of AHP

Verbal judgment or Preference	Numerical rating
Extremely preferred	**9**
Very strongly preferred	**7**
Strongly preferred	**5**
Moderately preferred 3	**3**
Equally preferred 1	**1**
Intermediate values between two adjacent judgments (when compromise is needed)	**2,4,6 and 8**

Once the comparison result is obtained, a pairwise comparison matrix is created to compile the priority ranking of each criteria. In the 2nd level of the hierarchy, 5 critical factors dimensions are identified, i.e. delivery cooperation, delivery part accuracy, package and delivery's documentation. The decision first builds the pairwise comparison matrix for 5 factors (n=5) using intensity scales (1-9). The comparison judgment table (Table 3) shows the pairwise comparison matrix.

Table 3. Pairwise comparison matrix (criteria)

Factor	On Time	Cooperate	Accuracy	Package	Document
On Time	**1**	5	3	7	7
Cooperate	1/5	**1**	1/3	5	3
Accuracy	1/3	3	**1**	3	5
Package	1/7	1/5	1/3	**1**	1
Document	1/7	1/3	1/5	1	**1**
Total	1.819	9.533	4.867	17	17

From the pairwise comparison matrix, the next step is to calculate the priority or weight of the elements in the matrix by summing up the values in each column. Each value is then divided by the total of the column .The result of priority or weight of factors are shown in Table 4 (A). In Table 4 (A), the sum of each column must equal to 1 or 100%.

Table 4. (A) Priority matrix

Factor	On Time	Cooperate	Accuracy	Package	Document
On Time	**0.550**	0.524	0.616	0.412	0.412
Cooperate	0.110	**0.105**	0.068	0.294	0.176
Accuracy	0.183	0.315	**0.205**	0.176	0.294
Package	0.079	0.021	0.068	**0.059**	0.059
Document	0.079	0.035	0.041	0.059	**0.059**
Total	**1**	**1**	**1**	**1**	**1**

Table 4. (B) Priority matrix

Factor	Total	Weight	Percent	Rank
On Time	2.514	0.503	50.28%	1
Cooperate	0.754	0.151	15.08%	3
Accuracy	1.174	0.235	23.48%	2
Package	0.286	0.057	5.71%	4
Document	0.272	0.054	5.44%	5

The weight in Table IV (B) is the average of all factors in each row, for example the factor "Cooperate" weight is the sum of row factor cooperate (0.754) divided by the number of factors (n=5) giving the value of 0.151. From the analytical results (Table 5(B)), On-time delivery (0.503) is the most important criterion whereas other criteria obtain their weights as delivery part accuracy (0.235), delivery cooperate (0.151), package (0.057) and delivery's documentation (0.054).

4.3.3 Consistency Analysis

This step investigates the comparison results regarding their consistency relativity through calculating the consistency ratio (CR). CR can be calculated by dividing the Consistency Index (CI) by the Random Consistency Index (RI), which can be obtained from the matrix found above. If CR is less than or equal to 0.1, then the consistency is acceptable. If the CR greater than 0.1, then the consistency is unacceptable and the subjective judgment must be redone. The measure in Table 5 is calculated by summing the value of each row and then divide it by the identical weight.

Table 5. The value of each row

Factor	Measure
On Time	5.453
Cooperate	5.164
Accuracy	5.53
Package	5.108
Document	5.154
Total	26.409

The λ_{max} value is the total in Table V. divided the number of factors.

$$\lambda_{max} = 26.409/5 = 5.282 \quad (3)$$

To find the Consistency Ratio (CR), the random index (RI) when n = 5 is 1.12.

$$\begin{aligned} CR &= \lambda\ max\text{-}n / (n\text{-}1)\ RI \\ &= (5.282\text{-}5) / (5\text{-}1)\ (1.12) \\ &= 0.06 < 0.1 \end{aligned} \quad (4)$$

CR = 0.06 < 0.1, meaning the consistency is acceptable.

In the next level (3rd level) of decision making, the weights of the sub-criteria nested under the primary criteria have to be determined. The steps are similar to those explained previously. The result is shown in Table 6.

Table 6. Ranking of overall weight Criteria and Sub-Criteria

Criteria	Weight	Rank	Sub-Criteria	Weight	Rank
A. On-time delivery	0.503	1	A.1 Line stop	0.592	1
			A.2 Pending Car (not complete)	0.237	2
			A.3 Change production schedule	0.126	3
			A.4 Not on time delivery but not effect to production line	0.045	4
B. Delivery cooperate	0.151	3	B.1 Advance confirm order	0.19	2
			B.2 Information such as delivery delay information	0.328	1
			B.3 Delivery performance during emergency order	0.182	3
			B.4 Delivery flexibility	0.128	4
			B.5 Follow the rule of the factory	0.096	5
			B.6 Reply Countermeasure ,Car Report and Survey	0.076	6
C. Delivery part accuracy	0.235	2	C.1 Part Short delivery quantity	0.488	1
			C.2 Part Over delivery quantity	0.064	3
			C.3 Part NG (damage, wrong parts and mixing Parts)	0.448	2
D. Package	0.057	4	D.1 Size standard	0.164	3
			D.2 No damage	0.173	2
			D.3 Cleaning	0.053	6

Criteria	Weight	Rank	Sub-Criteria	Weight	Rank
			D.4 Safety	0.412	1
			D.5 Easy for receiving check	0.050	5
			D.6 Enough to use for requirement	0.148	4
E. Delivery's document	0.054	5	E.1 Delivery documents must complete	0.480	1
			E.2 Part tag and delivery tag Base on document standard	0.405	2
			E.3 Delivery note must easy for receiving check	0.115	3

After determining the weights for all criteria and sub-criteria in the delivery performance evaluation, these weights are set as a new evaluation standard. A formal training conducted by an in-house expert is provided to all evaluators so that the same evaluation baseline will be implemented in the company. To ensure that all evaluators understand and apply the given guideline correctly, the attribute MSA is conducted on each evaluator. As shown in Table 7, the percentage of repeatability and accuracy are equal to 100% meaning that all evaluators are competent to evaluate the delivery performance of the suppliers under the same standard.

Table 7. Attribute MSA

No. of Evaluator	Before Training		After Training	
	Appraiser Score (%)	Attribute Score (%)	Appraiser Score (%)	Attribute Score (%)
1	100	98	100	100
2	85	70	100	100
3	100	94	100	100
4	93	68	100	100
5	100	96	100	100
6	100	96	100	100
7	96	73	100	100
8	100	97	100	100
9	96	73	100	100
10	98	72	100	100
11	100	92	100	100
12	87	79	100	100
	Repeatability = 61%		Repeatability = 100%	
	Accuracy = 61%		Accuracy = 100%	

The test performance of 12 evaluators before training is shown in Figure 2. The overall effective repeatability and accuracy are equal to 61% and 61%, respectively. It was clear that all evaluator were failed by the testing criteria of the MSA which were set as 100%. Therefore, the company had to send all evaluators to the training program about the measurement for supplier evaluation. After completing the training course, the attribute MSA was re-conducted to these evaluators again and the testing results showed that all of them passed the given standard (Figure 3).

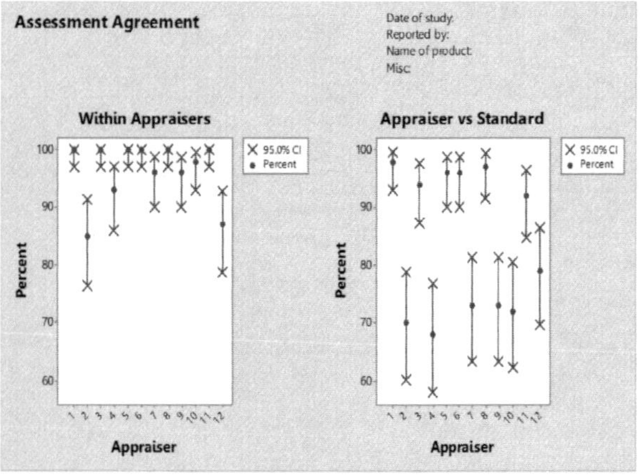

Figure 2. MSA Attribute test performance chart (Before training evaluator)

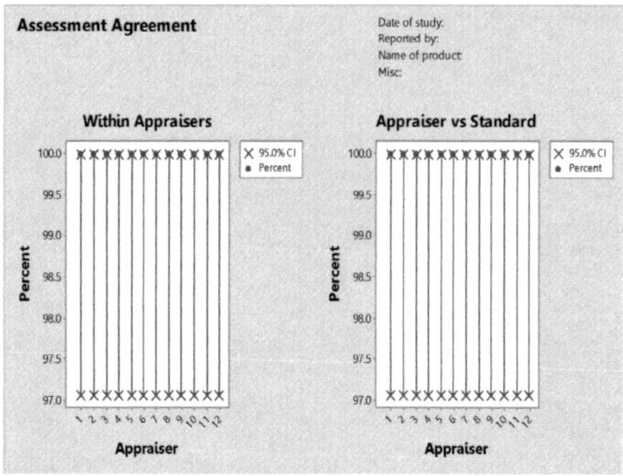

Figure 3. MSA Attribute test performance chart (After training evaluator)

5. Conclusion

Due to the competitive nature within the automotive industry, suppliers must develop their performances to meet the customer satisfaction criteria. Maintaining general quality, including capital, is necessary to help companies succeed and the best way for a company to do so is to use the best suppliers. The best suppliers will have the highest quality, best on-time delivery rate and facilitate the production process of the customer. Therefore, most customer companies give the highest priority to the suppliers and use part supplier performance evaluations.

In this study, the current evaluation results are ineffective to screen out between good and poor performance suppliers. The main reasons come from inappropriate criteria and their weightings are utilized. Therefore, the new criteria are established based on the experts in industry and the in-house evaluators. AHP is applied to find suitable weights for the primary and sub-criteria. The result shows that after the standard procedure is set and the evaluators are properly trained, the outputs of the part delivery supplier evaluation are much more accurate and trustworthy. After improving the criteria for delivery evaluation by using the AHP method and implementing the standardized method for evaluated supplier delivery performance, the evaluation results are much more trustworthy, reliable and easy-to-use for the evaluators.

References

[1] P. Humphreys, K.L. Mak and R. McIvor 1988 Procurement. Logistics Information Management **1**, 28-37.

[2] L. de Boer, E. Labro and P.Morlacchi 2001 A review of methods supporting supplier selectionEuropean *Journal of Purchasing & Supply Management*, **7**, 75-89.

[3] S.Chopra and P.Meindl 2007 Supply Chain Management: Strategy,Planning &Operations, 3rd ed. (Pearson Education Inc.)

[4] A. Arora, A.S. Arora and S.Palvia 2014 Social Media Index Valuation: Impact of Technological, Social, Economic and Ethical Dimensions *Journal of Promotion Management*, **20**,328-344.

[5] L.G. Vargas 1990 An overview of the analytic hierarchy process and its application, European *Journal of Operational Research*, **48(1),** 57-64.

[6] W. Li, W. Cui, Y. Chen and Y. Fu 2008 A group decision-making model for multi-criteria supplier selection in the presence of ordinal data, Service operations, Logistics, and Informatics, *IEEE International Conference*.

[7] A. Kaboli, M.-B. Aryanezhad, K. Shahanaghi and I. Niroomand 2007 A new method for plant location selection problem: a fuzzy-AHP approach, Systems, Man and Cybernetics, *IEEE International Conference*.

[8] S.A. Jozi, M.T. Shoshtary and A.R.K. Zadeh 2015 Environmental risk assessment of dams in construction phase using a multi-criteria decision making (MCDM) method, Human and Ecological Risk Assessment *An International Journal*, **21**, 1-16.

[9] S. Daichman, D. Greenberg, O. Pikovsky and J. Pliskin 2013 How to make a right decision in health care: Multi criteria decision analysis in the healthcare system, Digital Technologies (DT) *International Conference*.

[10] T.L. Saaty 1980 The analytic hierarchy process: planning, priority setting, resources (New York: McGraw).

[11] K. Wang and C.K. Wang, C. Hu 2005 Analytic hierarchy process with fuzzy scoring in evaluating multidisciplinary R&D projects in China, Engineering Management, IEEE Transactions on **52**,119-129.

[12] E.W.T.Ngai and E.W.C.Chan 2005 Evaluation of knowledge management tools using AHP,Expert Systems with Application **29**,889-899.

[13] A.N. Hag and G. Kannan, 2006 Fuzzy analytical hierarchy process for evaluating and selecting a vendor in a supply chain model *International journal of Advanced Manufacturing Technology***29**,826-835.

[14] G.D. Li, D. Yamaguchi and M. Nagai 2006 A grey-based decision-making approach to the supplier selection problem, *Mathematical and Computer Modelling*.

[15] Lee E.K., Ha.S, and Kim S.K. 2001 Supplier selection and management system considering relationships in supply chain management, IEEE Trans. On Engineering Management **48(3)**,307-318.

[16] D.Garg,S. Luthra and A. Haleem 2014 Ranking of performance measures of GSCM towards sustainability : Using analytic hierarchy process *International Science Index, Mechanical and Mechatronics Engineering***8**(3)2014.

[17] E.Loken 2007 Use of multi-criteria decision analysis methods for energy planning problems,Renewableabd Sustainable Energy Reviews,**11**,1584-1595.

[18] Dey P. K., Bhattacharya A. and Ho W. 2015 Strategic supplier performance evaluation: A case-based action research of a UK manufacturing organisation, *International Journal of Production Economics***166**, 192-214.

Numerical solution of optimal departure frequency of Taipei TMS

Lih-jier Young[1] and Chin-Hsin Chiu[2]

[1] Professor, Chung Hua University, Taiwan, R.O.C.

[2] Ph.D. Student, Chung Hua University, Taiwan, R.O.C.

E-mail: young@chu.edu.tw

Abstract. Route Number 5 (Bannan Line) of Taipei Mass Rapid Transit (MRT) is the most popular line in the Taipei Metro System especially during rush hours periods. It has been estimated there are more than 8,000 passengers on the ticket platform during 18:00~19:00 at Taipei main station. The purpose of this research is to predict a specific departure frequency of passengers per train. Monte Carlo Simulation will be used to optimize departure frequency according to the passenger information provided by 22 stations, i.e., 22 random variables of route number 5. It is worth mentioning that we used 30,000 iterations to get the different samples of the optimization departure frequency, i.e., 10 trains/hr which matches the practical situation.

1. Introduction

The mass rapid transit (MRT) system is the most important, efficient, convenient transportation between modern metropolitan and urban areas. Because the rail is always underground or elevated, MRT offers the fastest way from downtown to any other place [1]. MRT is therefore the most important mode of modern city transportation. It can carry more than 1,000 passengers on each train to reach their destination get elsewhere in even less than 10 minutes. However, during the 18:00~19:00 rush hour more than 8,000 passengers crowd into the platform of Route Number 5 (Bannan Line) of Taipei main station [2]. Maintaining optimization, i.e., get the optimization departure frequency to have these passengers on their way quickly and of course maintain the most benefit of the Taipei Metro System is the main goal of this research [3].

2. Monte Carlo Simulation

Due to the reasons of nonparametric statistics, full valuation, probabilistic iteration and easy operating, the Monte Carlo Simulation is chosen to model the complex situation of the train Taipei MRT system during the rush hours. Monte Carlo

Simulation is a very popular method of estimating a large amount of calculation. It is used in many research fields, such as statistical engineering, bio-medicine, psychology, atmospheric science, and management, but this application is specific to mass rapid transit transfer system. A lot of papers outline the risk estimate by MCS such as [4-7]. Probability estimate is based on the assumption of normal distribution. Gross [8], estimate the probability by MCS. Remaining life assessment of low pressure turbine rotor has been recommended based on MCS [9]. With the advent of computers, MCS has become a very common numerical method. Mark Demaria [10] used MCS to estimate Tropical Cyclone Wind Speed Probabilities. It is a random event process, similar to the scientific experimental process [11]. It is also used to model error estimation in order to estimate social risk [12].

Table 1 Numbers of Departure passenger of each station of Bannan Line during 18:00~19:00 per weekday.

	Station	Departure Passengers
a	Yongning	1607
b	Tucheng	860
c	Haishian	2951
d	Far East Hospital	2080
e	Futhong	4271
f	Banqiao	4956
g	Xinpu	6074
h	Jiang zihui	3863
i	Longshan Temple	3244
j	Ximen	6858
k	Taipei Main Station	8746
l	Shandao Temple	2541
m	Zhongxiao Xinsheng	2377
n	Fuxing	4784
o	Zhongxiao Dunhua	4701
p	S.Y.S. Memorial Hall	2603
q	Taipei City Hall	5015
r	Yongchun	2449
s	Houshanpi	2537
t	Kunyang	1502
u	Nangang	1103
v	Taipei Nangang Exhibition Center	1977

To analyze the optimization frequency of the Bannan Line of Taipei MTR from inputting each of the 22 independent parameters from each correspondence station the final income of each train plays a important role of this research. Table 1 shows the arrival of passengers for 22 stations of the Bannan Line during the time 18:00~19:00 of one weekday. In this paper, the Random Number Generator is made by Linear Congruential Generator (LCG) the equation is

$$Y_{i+1} = [AY_i + C] \quad MOD \quad M \qquad \mathrm{i} = 0,1,2,........ \tag{1}$$

where Y_0 is the seed value, A is the constant multiplier, C is the increment, and M is the modulus. The value of Y_0 is preset. The sequence of all other Y_i's will be obtained according to eq. (1) and the normalized value (Pseudo-Random Number) R_0, R_1, R_2, …., R_i will be assigned from eq. (2)

$$R_i = \frac{Y_i}{M} \ . \tag{2}$$

Performing each Monte Carlo iteration, a random number is generated for each random variable of those 22 stations as described above. The 22 simulations are processed deterministically for a few chosen standard deviations of the above mentioned 22 random variables. It is worth mentioning we repeated the iteration 30,000 times to get the sample of the profile of the Bannan Line with different departure frequencies.

3. Simulation the income of each train

To simulate the income of each train we have to set up a model to get the beginning and the end point of each passenger journey. It is reasonable to assume the alighting rate can be obtained as per the equations below:

$$GOR_s = \left(\frac{DP_s - \frac{DP_s}{n}}{\sum_{s=1}^{22} DP_s}\right)100\% \qquad s = 1,2,3,......,11 \quad \text{(before station k)} \tag{3}$$

$$GOR_s = \left(\frac{\frac{DP_s}{n}}{\sum_{s=1}^{22} DP_s}\right)100\% \qquad s = 12,13,14,......,22 \quad \text{(after station k)} \tag{4}$$

where GOR_s is the alighting rate of each station,

DP_s is the number of the departing passengers from of each station,

n is a specific number corresponding to each station and described in Table 2.

Table 2 Values of n at each station where the alphabet corresponds to Table 1.

	a	*b*	*c*	*d*	*e*	*f*	*g*	*h*	*i*	*j*	*k*	*l*	*m*	*n*	*o*	*p*	*q*	*r*	*s*	*t*	*u*	v
n	1	1.1	1.2	1.3	1.4	1.5	1.6	1.7	1.8	1.9	2	1.9	1.8	1.7	1.6	1.5	1.4	1.3	1.2	1.1	1.1	1

It is worthy of mention that the value of n for each station is assumed to have linearly increase from station a to k and then to decrease linearly from station k to v. It is reasonable that $n = 2$ at station k because station k is the central station of the line, i.e., the number of passenger for each journey is equal.

Table 3 Ticket Price of Bannan Line where the alphabet corresponds to Table 1.

Station	*a*	*b*	*c*	*d*	*e*	*f*	*g*	*h*	*i*	*j*	*k*	*l*	*m*	*n*	*o*	*p*	*q*	*r*	*s*	*t*	*u*	*v*
a	0																					
b	20																					
c	20	20																				
d	20	20	20																			
e	25	20	20	20																		
f	25	25	20	20	20																	
g	25	25	20	20	20	20																
h	30	25	25	20	20	20	20															
i	35	30	30	25	25	25	20	20														
j	35	35	30	30	25	25	25	20	20													
k	35	35	35	30	30	25	25	25	20	20												
l	40	35	35	30	30	30	25	25	20	20	20											
m	40	40	35	35	30	30	30	25	20	20	20	20										
n	40	40	40	35	35	30	30	30	25	20	20	20	20									
o	45	40	40	35	35	35	30	30	25	20	20	20	20	20								
p	45	45	40	35	35	35	30	30	25	25	20	20	20	20	20							
q	45	45	40	40	35	35	35	30	25	25	20	20	20	20	20	20						
r	50	45	45	40	40	35	35	35	30	25	25	25	20	20	20	20	20					
s	50	45	45	40	40	40	35	35	30	30	25	25	25	20	20	20	20	20				
t	50	50	45	45	40	40	40	35	30	30	25	25	25	25	20	20	20	20	20			
u	55	50	50	45	45	45	40	40	35	30	30	30	25	25	25	25	20	20	20	20		
v	55	55	50	50	45	45	45	40	35	35	30	30	30	25	25	25	25	20	20	20	20	0

After we have GOR_s of each station and consult the transportation fee listed in, we can than calculate the total income of each train by summation all incomes of each 22 stations. The ticket price of Bannan Line is shown on Table 3.

4. Results

The results of five different departure frequencies, i.e., 1, 2, 3, 10 and 20 trains each hour are shown in Tables 4 to 8. It is obvious that the income of each train can be calculated by the equations (5) and (6) below.

$$\mathrm{I} = \sum_{i=1}^{21} (\mathrm{Inc}) \tag{5}$$

$$\mathrm{Inc} = \sum_{1}^{\mathrm{Num}} \mathrm{Pri} \tag{6}$$

I: Total income of the train,
Inc: Income of each station,
Num: Number of passengers who board the train at on a specific station,
Pri: Ticket price that of each passenger needs to pay.

It is obvious from Table 4 where the departure frequency is 1 train/hr that some passengers cannot board the train at stations f, j, k, n, o, and q for the reason that the train's upper loading limit is 1,936 passengers (highlighted with yellow background). Therefore, the total income is NTD\$584,972 which is less than others. However, if we double the frequency, it can be seen from Table 5 that only if the passengers board the train at stations other than station k, then the total income will increase to NTD\$677,840 (338920×2). If the frequency increase three times, as the original Table 6 shows, the passengers of every station can get on the train and the total income increases to NTD\$691,308 (230436×3). Even if everyone can board the train, it will likely be crowded, especially at stations j and k. For the sake of passenger comfort level, we increase the frequency ten times from the original. It is very clear that even if the income is NTD\$691,330 (69133×10) which maintains the same income as Table 7, the comfort level is increased along with the willingness of passengers to take the line. Table 8 shows the result of 20 trains/hr with the income NTD\$681,040 (34052×20).

Table 4 Passengers of each station and total income of each train with departure frequency 1 train/hr where the alphabet corresponds to Table 1.

1 train/hr, passangers and income

	a	*b*	*c*	*d*	*e*	*f*	*g*	*h*	*i*	*j*	*k*	*l*	*m*	*n*	*o*	*p*	*q*	*r*	*s*	*t*	*u*	*v*	Rev
GOR	0	0.2	1.3	1.2	3.1	4.2	5.8	4	3.7	9.2	11.1	3.4	3.4	7.2	7.5	4.4	9.1	4.7	0.1	3.5	2.6	0	
Ticket Price	0	20	20	20	25	25	25	30	35	35	35	40	40	40	45	45	45	50	50	50	55	55	
Passangers	1620	410	930	1020	1260	1936	1860	1010	1230	1936	1936	1910	1780	1936	1936	1180	1936	570	330	210	120	0	
a	65	421	389	1256	1701	2349	1944	2098	5216	6294	2203	2203	4666	5468	3208	6634	3807	81	2835	2317	0	0	55153
b		107	98	318	431	595	492	531	1320	1593	558	558	1181	1384	812	1679	964	21	718	586	0	0	13942
c			223	721	977	1349	1116	1204	2995	3613	1265	1265	2678	3139	1841	3808	2186	47	1628	1330	0	0	31383
d				791	1071	1479	1224	1321	3284	3963	1387	1387	2938	3443	2020	4177	2397	51	1785	1459	0	0	34175
e					1323	1827	1512	1632	4057	4895	1714	1714	3629	4253	2495	5160	2961	63	2205	1802	0	0	41240
f						2807	2323	2507	6234	7521	2633	2633	5576	6534	3833	7928	4550	97	3388	2768	0	0	61332
g							2232	2409	5989	7226	2530	2530	5357	6278	3683	7617	4371	93	3255	2660	0	0	56228
h								1308	3252	3924	1374	1374	2909	3409	2000	4136	2374	51	1768	1444	0	0	29320
i									3961	4779	1673	1673	3542	4151	2435	5037	2891	62	2153	1759	0	0	34114
j										7521	2633	2633	5576	6534	3833	7928	4550	97	3388	2768	0	0	47461
k											2633	2633	5576	6534	3833	7928	4550	97	3388	2768	0	0	39940
l												2598	5501	6446	3782	7821	4489	96	3343	2731	0	0	36806
m													5126	6008	3524	7289	4183	89	3115	2545	0	0	31880
n														6534	3833	7928	4550	97	3388	2768	0	0	29098
o															3833	7928	4550	97	3388	2768	0	0	22564
p																4832	931	20	693	566	0	0	7042
q																	4550	97	3388	2768	0	0	10803
r																		29	998	815	0	0	1841
s																			193	157	0	0	350
t																				300	0	0	300
u																					0	0	0
v																							

Total Income= 584972

Note: 1936 is the upper limit of passangers each train

Table 5 Passengers of each station and total income of each train with departure frequency 2 trains/hr where the alphabet corresponds to Table 1.

2 trains/hr, passangers and income

	a	*b*	*c*	*d*	*e*	*f*	*g*	*h*	*i*	*j*	*k*	*l*	*m*	*n*	*o*	*p*	*q*	*r*	*s*	*t*	*u*	*v*	Rev
GOR	0	0.2	1.3	1.2	3.1	4.2	5.8	4	3.7	9.2	11.1	3.4	3.4	7.2	7.5	4.4	9.1	4.7	0.1	3.5	2.6	0	
Ticket Price	0	20	20	20	25	25	25	30	35	35	35	40	40	40	45	45	45	50	50	50	55	55	
Passangers	810	205	465	510	630	1020	930	505	615	1770	1936	955	890	1030	1100	590	1220	285	165	105	60	0	
a	32	211	194	628	851	1175	972	1049	2608	3147	1102	1102	2333	2734	1604	3317	1904	41	1418	1158	0	0	27576
b		53	49	159	215	297	246	265	660	796	279	279	590	692	406	839	482	10	359	293	0	0	6971
c			112	360	488	674	558	602	1497	1807	632	632	1339	1569	921	1904	1093	23	814	665	0	0	15691
d				395	536	740	612	660	1642	1981	694	694	1469	1721	1010	2088	1199	26	893	729	0	0	17088
e					662	914	756	816	2029	2448	857	857	1814	2126	1247	2580	1481	32	1103	901	0	0	20620
f						1479	1224	1321	3284	3963	1387	1387	2938	3443	2020	4177	2397	51	1785	1459	0	0	32314
g							1116	1204	2995	3613	1265	1265	2678	3139	1841	3808	2186	47	1628	1330	0	0	28114
h								654	1626	1962	687	687	1454	1704	1000	2068	1187	25	884	722	0	0	14660
i									1980	2389	836	836	1771	2076	1218	2518	1445	31	1076	879	0	0	17057
j										6876	2407	2407	5098	5974	3505	7248	4160	89	3098	2531	0	0	43392
k											2633	2633	5576	6534	3833	7928	4550	97	3388	2768	0	0	39940
l												1299	2750	3223	1891	3911	2244	48	1671	1366	0	0	18403
m													2563	3004	1762	3645	2092	45	1558	1273	0	0	15940
n														3476	2039	4218	2421	52	1803	1473	0	0	15481
o															2178	4505	2585	55	1925	1573	0	0	12821
p																2416	931	20	693	566	0	0	4626
q																	2867	61	2135	1745	0	0	6808
r																		14	499	408	0	0	921
s																			193	157	0	0	350
t																				150	0	0	150
u																					0	0	0
v																							

Total Income= 338920

Note: 1936 is the upper limit of passangers each train

Table 6 Passengers of each station and total income of each train with departure frequency 3 trains/hr where the alphabet corresponds to Table 1.

3 trains/hr, passangers and income

	a	b	c	d	e	f	g	h	i	j	k	l	m	n	o	p	q	r	s	t	u	v	Rev
GOR	0	0.2	1.3	1.2	3.1	4.2	5.8	4	3.7	9.2	11.1	3.4	3.4	7.2	7.5	4.4	9.1	4.7	0.1	3.5	2.6	0	
Ticket Pric	0	20	20	20	25	25	25	30	35	35	35	40	40	40	45	45	45	50	50	50	55	55	
Passangers	538	136	318	336	421	680	628	346	409	1189	1426	637	595	689	730	396	815	199	110	55	37		
a	22	140	129	417	565	780	646	697	1732	2090	732	732	1549	1816	1065	2203	1264	27	942	769	0	0	18316
b		35	33	105	143	197	163	176	438	528	185	185	392	459	269	557	320	7	238	194	0	0	4625
c			76	246	334	461	382	412	1024	1235	432	432	916	1073	630	1302	747	16	557	455	0	0	10731
d				260	353	487	403	435	1082	1305	457	457	968	1134	665	1376	790	17	588	480	0	0	11258
e					442	610	505	545	1356	1636	573	573	1212	1421	834	1724	989	21	737	602	0	0	13779
f						986	816	881	2190	2642	925	925	1958	2295	1346	2785	1598	34	1190	972	0	0	21542
g							754	813	2022	2440	854	854	1809	2120	1243	2572	1476	31	1099	898	0	0	18984
h								448	1114	1344	471	471	996	1168	685	1417	813	17	606	495	0	0	10044
i									1317	1589	556	556	1178	1380	810	1675	961	20	716	585	0	0	11344
j										4619	1617	1617	3424	4013	2354	4869	2794	59	2081	1700	0	0	29148
k											1939	1939	4107	4813	2823	5839	3351	71	2496	2039	0	0	29418
l												866	1835	2150	1261	2609	1497	32	1115	911	0	0	12275
m													1714	2008	1178	2437	1398	30	1041	851	0	0	10656
n														2325	1364	2821	1619	34	1206	985	0	0	10356
o															1445	2989	1716	37	1278	1044	0	0	8508
p																1622	931	20	693	566	0	0	3831
q																	1915	41	1426	1165	0	0	4548
r																		10	348	285	0	0	643
s																			193	157	0	0	350
t																				79	0	0	79
u																					0	0	0
v																							

Total Income = 230436

Table 7 Passengers of each station and total income of each train with departure frequency 10 trains/hr where the alphabet corresponds to Table 1.

10 trains/h, passangers and income

	a	*b*	*c*	*d*	*e*	*f*	*g*	*h*	*i*	*j*	*k*	*l*	*m*	*n*	*o*	*p*	*q*	*r*	*s*	*t*	*u*	*v*	Rev
GOR	0	0.2	1.3	1.2	3.1	4.2	5.8	4	3.7	9.2	11	3.4	3.4	7.2	7.5	4.4	9.1	5	0.1	3.5	2.6	0	
Ticket Pric	0	20	20	20	25	25	25	30	35	35	35	40	40	40	45	45	45	50	50	50	55	55	
Passangers	162	41	93	102	126	204	186	101	123	354	427	191	178	206	220	118	244	57	33	21	12	0	
a	6	42	39	126	170	235	194	210	522	629	220	220	467	547	321	663	381	8	284	232	0	0	5515
b		11	10	32	43	59	49	53	132	159	56	56	118	138	81	168	96	2	72	59	0	0	1394
c			22	72	98	135	112	120	299	361	126	126	268	314	184	381	219	5	163	133	0	0	3138
d				79	107	148	122	132	328	396	139	139	294	344	202	418	240	5	179	146	0	0	3418
e					132	183	151	163	406	490	171	171	363	425	249	516	516	6	221	180	0	0	4344
f						296	245	264	657	793	277	277	588	689	404	835	479	10	357	292	0	0	6463
g							223	241	599	723	253	253	536	628	368	762	437	9	326	266	0	0	5623
h								131	325	392	137	137	291	341	200	414	237	5	177	144	0	0	2932
i									396	478	167	167	354	415	244	504	289	6	215	176	0	0	3411
j										1375	481	481	1020	1195	701	1450	832	18	620	506	0	0	8678
k											581	581	1230	1441	845	1749	1003	21	747	611	0	0	8809
l												260	550	645	378	782	449	10	334	273	0	0	3681
m													513	601	352	729	418	9	312	312	0	0	3245
n														695	408	844	484	10	361	295	0	0	3096
o															436	901	517	11	385	315	0	0	2564
p																483	277	6	207	169	0	0	1142
q																	573	12	427	349	0	0	1362
r																		3	100	82	0	0	184
s																			58	47	0	0	105
t																				30	0	0	30
u																					0	0	0
v																						0	0

Total Income= 69133

Table 8 Passengers of each station and total income of each train with departure frequency 20 trains/hr where the alphabet corresponds to Table 1.

20 trains/hr, passangers and income

	a	*b*	*c*	*d*	*e*	*f*	*g*	*h*	*i*	*j*	*k*	*l*	*m*	*n*	*o*	*p*	*q*	*r*	*s*	*t*	*u*	*v*	Rev
GOR	0	0	1	1	3	4.2	5.8	4	3.7	9.2	11.1	3.4	3.4	7.2	7.5	4.4	9.1	5	0.1	3.5	3	0	
Ticket Pric	0	20	20	20	25	25	25	30	35	35	35	40	40	40	45	45	45	50	50	50	55	55	
Passangers	81	20	46	51	63	102	93	50	61	177	213	95	89	103	110	29	122	28	16	10	6	0	
a	3	21	19	63	85	117	97	105	261	315	110	110	233	273	160	332	190	4	142	116	0	0	2758
b		5	5	16	21	29	24	26	64	78	27	27	58	68	40	82	47	1	35	29	0	0	680
c			11	36	48	67	55	60	148	179	63	63	132	155	91	188	108	2	81	66	0	0	1552
d				40	54	74	61	66	164	198	69	69	147	172	101	209	120	3	89	73	0	0	1709
e					66	91	76	82	203	245	86	86	181	213	125	258	148	3	110	90	0	0	2062
f						148	122	132	328	396	139	139	294	344	202	418	240	5	179	146	0	0	3231
g							112	120	299	361	126	126	268	314	184	381	219	5	163	133	0	0	2811
h								65	161	194	68	68	144	169	99	205	118	3	88	72	0	0	1452
i									196	237	83	83	176	206	121	250	143	3	107	87	0	0	1692
j										688	241	241	510	597	350	725	416	9	310	253	0	0	4339
k											290	290	613	719	422	872	501	11	373	305	0	0	4394
l												129	274	321	188	389	223	5	166	136	0	0	1831
m													256	300	176	364	209	4	156	127	0	0	1594
n														348	204	422	242	5	180	147	0	0	1548
o															218	450	259	6	193	157	0	0	1282
p																119	68	1	51	41	0	0	281
q																	287	6	214	174	0	0	681
r																		1	49	40	0	0	90
s																			28	23	0	0	51
t																				14	0	0	14
u																					0	0	0
v																							0

Tota Income = 34052

It can be seen from Fig. 1 that the total income almost the same after the frequency increases up to 3 trains/hr and there is a maximum income value at the frequency of 10 trains/hr.

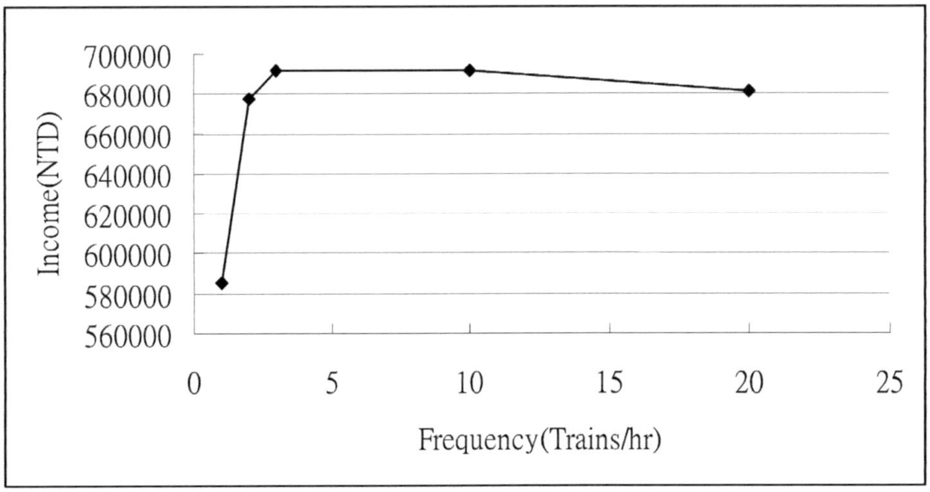

Figure 1 Income of different departing frequencies.

Figs 2-6 show the distribution and statistical data of 30,000 iterations of each different frequency. Each of them is normal standard distribution. Table 9 gives statistical data of the different departure frequency of each train.

According to the results shown above, it is wise to choose the departure frequency 10 trains/hr. The reason why we did not didn't pick up 20 trains/hr is that there is still an electricity cost of NTD$50,000 for each train. After consulting

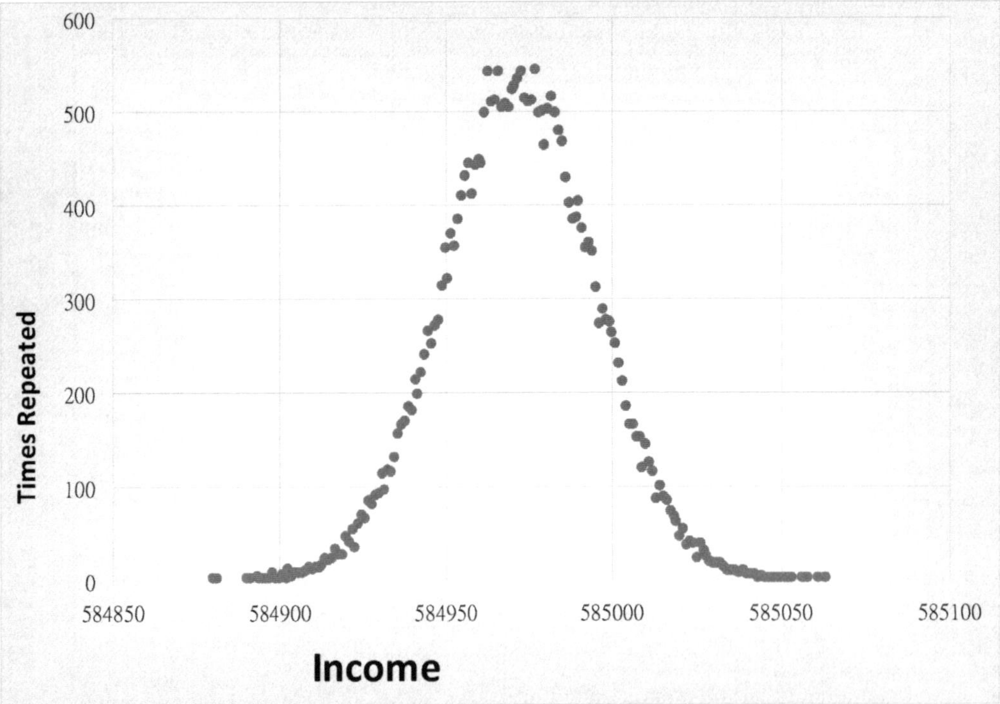

Figure 2 Distribution of 30,000 iterations of departing frequency 1 train/hr.

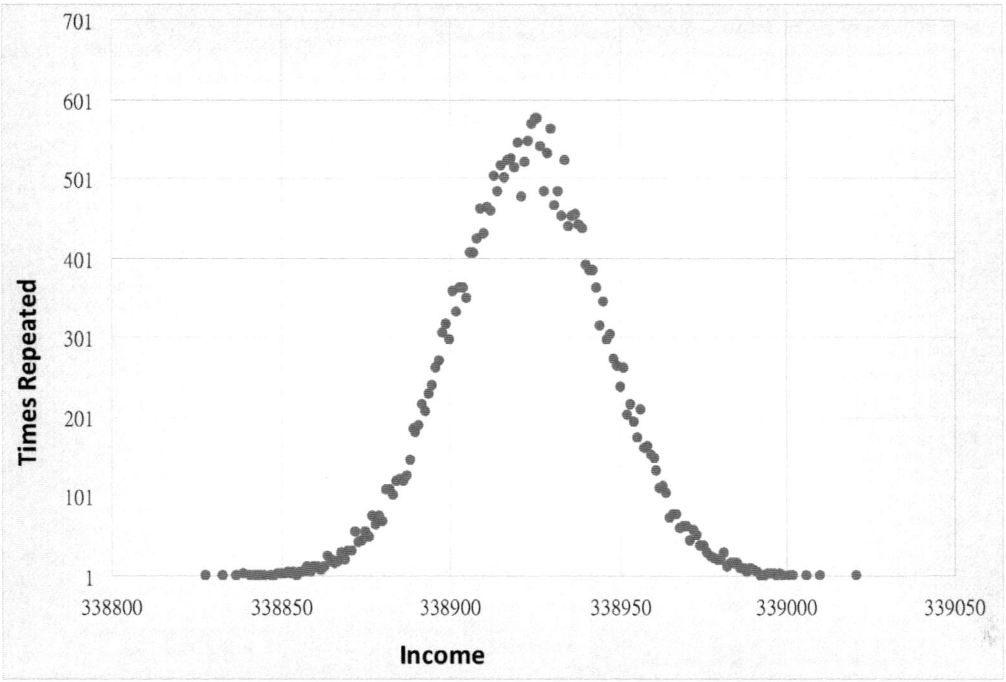

Figure 3 Distribution of 30,000 iterations of departing frequency 2 trains/hr.

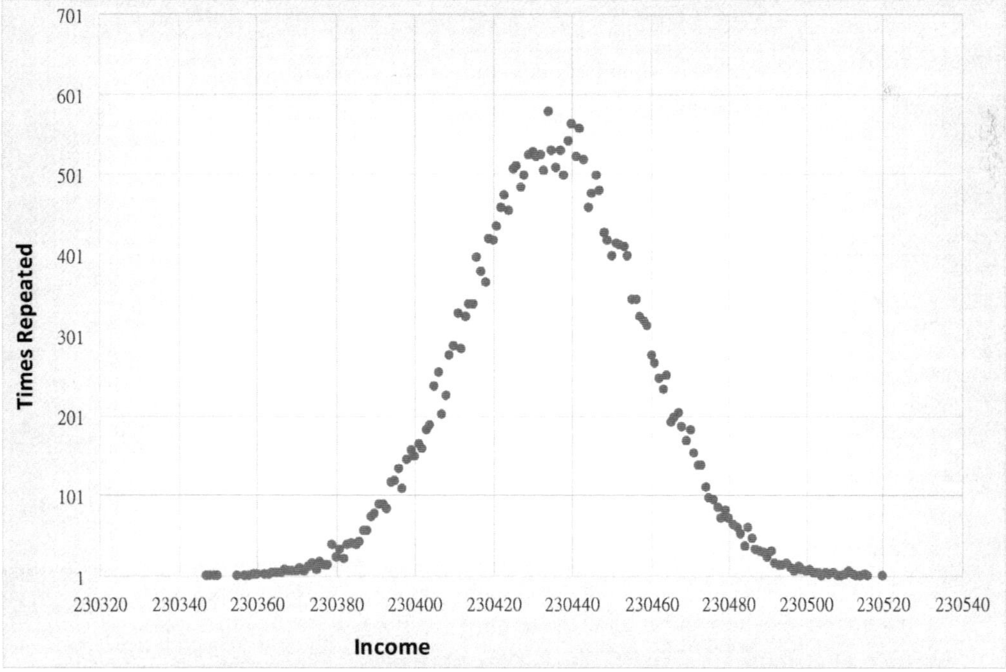

Figure 4 Distribution of 30,000 iterations of departing frequency 3 trains/hr.

the departure time table for the Bannan Line we found that the departure frequency is 11 trains/hr which matches the results of our research.

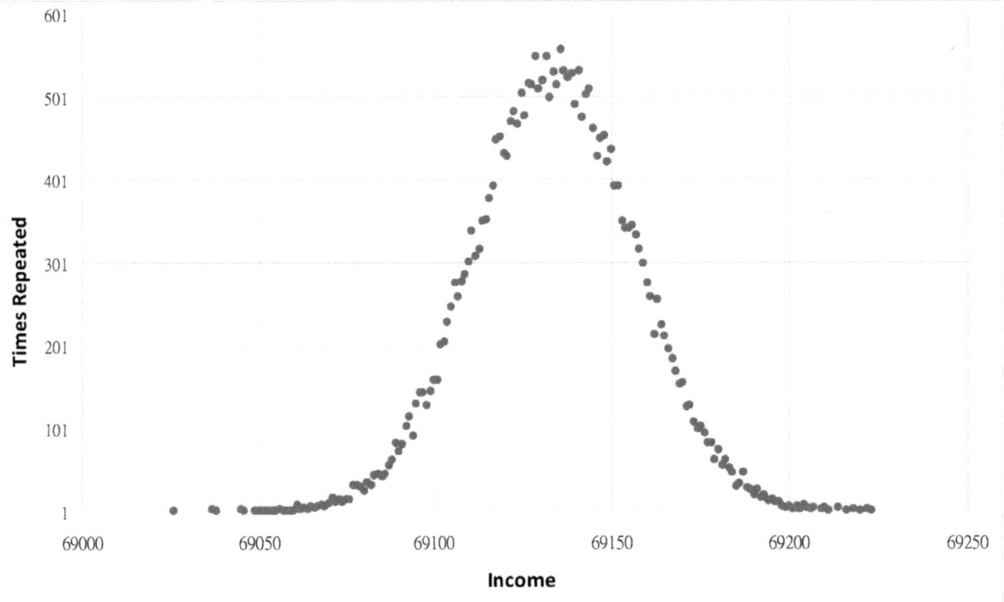

Figure 5 Distribution of 30,000 iterations of departing frequency 10 trains/hr.

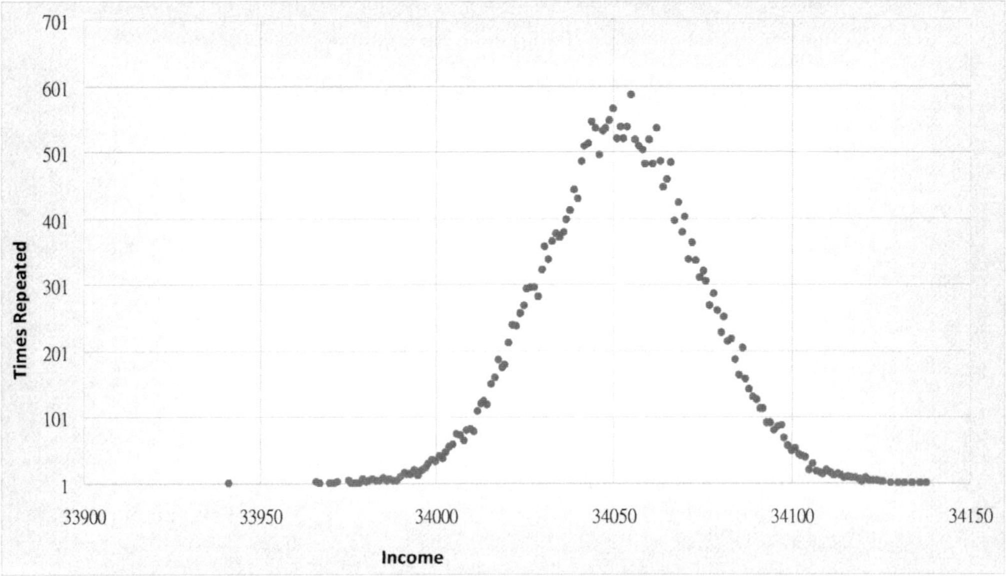

Figure 6 Distribution of 30,000 iterations of departing frequency 20 trains/hr.

Table 9: Statistical data of different departure frequencies.

Statistical Data					
departure frequency	1	2	3	4	5
mean	584972	338923	230435	69134	34052
standard error	0.13	0.129	0.13	0.13	0.13
median	584972	338923	230435	69134	34052
mode	584977	338926	230434	69133	34017
standard deviation	22.58	22.32	22	22.51	22
variance	509.8	498.34	500	507	496
kurtosis	0.018	0.0411	-0.02	0.05	0.001
skewness	-0.015	-0.013	-0.01	-0.02	-0.011
range	183	193	173	197	197
minimum	584880	338828	230347	69026	33941
maximum	585063	339021	230520	69223	34138
numbers of iteration	30000	30000	30000	30000	30000

5. Conclusions

This research focuses on the MRT of the Bannan line, which is the most popular line of the 5 lines of the MRT system. Based on the total entrance of passengers into of each station during the time of 18:00~19:00 on one weekday, we set up a procedure to obtain the optimization of departure frequency of trains by using Monte Carlo Simulation for 60,000 iterations. The frequency is 10 trains/hr which is almost the same as the real situation, i.e., 11 trains/hr. We can apply this research to other lines of the MRT and even other cities. However, there are still a few points that need to be discussed.

a. The given data varies according to the number assigned to the variable n. It is just an assumption for the specific number of each station n to calculate the number of passenger of different journeys. The results will be varied according to the different ways this number.
b. The number of passenger alighting the train is another assumption in this research. It is based on the GOR_s number which is the alighting rate calculated from the number of the departing passengers from each station. The results will be different if this assumption is changed.
c. Cost is an important factor of the total profit of the Bannan Line. We made no mention of any cost of this research owing to the reason that the comfort level is another factor of passengers' willingness to take the MRT.

6. References

[1] Lo, S. C., & Chang, W. J. (2012). Design of real-time fuzzy bus holding system for the mass rapid transit transfer system. *Expert Systems with Applications*, 39, 1718-1724.

[2] Caprara, A., Kroon, L., Monaci, M., Peeters, M., Toth, P. (2006). Passenger Railway Optimization. *ARRIVAL-TR-0035*.

[3] Ke, B. R., Lin, C. L., & Lai, C. W. (2011). Optimization of train-speed trajectory and control for mass rapid transit systems. *Control engineering Practice*, 19. 675-687.

[4] Kahneman, D., Slovic, P., & Tversky, A. (1982). Judgment under Uncertainty Heuristics Biases, Cambridge University Press.

[5] Schumacher, E. H., Seymour, T. L., Glass, J. M., Fencsik, D. E., Lauber, E. J., Kieras, D. E., & Meyer, D. E. (2001). Virtually Perfect Time Shearing in Dual-Task Performance: Unlocking the Central Cognitive Bottleneck. *Psychological Science*, 12, 2, 101-108.

[6] Rahman, A., Hosokawa, S., Oono, Y., Amakawa, T. Goto, N., & Tsurumi, S. (2002). Auxin and Ethylene Response Interactions during Arabidopsis Root Hair Development Dissected by Auxin Influx Modulators. *Plant Physiology*, 130, 1-10.

[7] Knaff, J. A., Cram, T. A., Schumacher, A. B., Kossin, J. A. P., & Demaria, M. (2007). Objective Identification of Annular Hurricanes. *American Meteorological Society*, 23, 17-28.

[8] Gross, B. O., Pgunwuyi, F., Moshary, S., Ahmed, & Cairns, B. (2004). Aerosol Retrieval Over Urban Areas Using Spatial Regression Between V/NIR and MIR Hyperion Channels, San Diego, Cal., Proc. SPIE, vol. 5547, 111, doi:10.1117/12.560097.

[9] Young, L. J., Yeh, B. H., & Young, P. (2014). Remaining Life Assessment of Low Pressure Turbine Rotor Using Monte Carlo Simulation. *America Society of Testing Materials*, 42, 3, 1-9.

[10] Mark Demaria, John A. Knaff Richard Knabb, Charles R. Sampson, Robert T. Demaria, 2009, A New Method for Estimating Tropical Cyclone Wind Speed Probabilities. *American Meteorological Society*, 24, 1573-1591.

[11] Vořechovský, M., (2012). Correlation Control in Small Sample Monte Carlo Type Simulation II: Analysis of Estimation Formulas, Random Correlation and Perfect Uncorrelatedness. *Probabilistic Engineering Mechanics,* 29, 105-120.

[12] David F., Andersen, J. A. M., Vennix, G. Richardson, P., & Rouwette, E. A. J. A. (2007). Group Model Building: Problem Structing, Policy Simulation and Decision Support. *The Journal of the Operational Research Society*, 58(5), 691-694.

IOP Conf. Series: Materials Science and Engineering **131** (2016) 012025 doi:10.1088/1757-899X/131/1/012025

Harmonic, Intermodulation and Cross-Modulation Distortion in Directly Modulated Quantum Cascade Lasers

J F Webb[1], K S C Yong[2] and M K Haldar[2]
[1] Department of Mechanical, Materials and Manufacturing Engineering, The University of Nottingham Malaysia Campus, Jalan Broga, 43500 Semenyih, Selangor Darul Ehsan, Malaysia
[2] Faculty of Engineering, Computing and Science, Swinburne University of Technology, Sarawak Campus, Jalan Simpang Tiga, 93350 Kuching, Sarawak, Malaysia
E-mail: jeffwebb@physics.org

Abstract. Using a simplified rate equation model, expressions for harmonic, intermodulation and cross-modulation distortion for a directly modulated quantum cascade laser can be derived. This paper shows how such derivations can be done and discusses some implications for quantum cascade lasers. It is important to understand such distortion, especially for applcations in communication systems.

1. Introduction

With quantum cascade lasers (QCLs) showing great potential in communication systems and the ubiquity of RF communication systems, QCLs can be utilized in these systems through transmission of RF over light carriers. Example of such system is community antenna television (CATV) system. These systems contribute to the emergent of sub-discipline of microwave photonics. Direct intensity modulation of the laser light is the simplest use of light carriers in the RF communication system. Such modulation is obtained by modulating the drive current of a laser with RF/microwave carriers. However, this modulation scheme produces harmonic, intermodulation and cross-modulation distortion. Analysis of these distortion is important for data transmission using amplitude as they will limit the available bandwidth in large signal modulation. Second and third harmonic distortions can be eliminated when the subcarriers are restricted to an octave bandwidth. The intermodulation and cross-modulation distortion, however, falls within the octave bandwidth. Cross-modulation distortion is known as intelligible crosstalk, in contrast with unintelligible crosstalk resulted by intermodulation distortion. It occurs when the modulation of one carrier is unintentionally transferred to another carrier. Hence, as will be done here, it is important to study nonlinear distortion in a direct intensity modulated QCL.

To the best knowledge of our knowledge, no previous work on harmonic, intermodulation and cross-modulation distortions for quantum cascade lasers has been reported. But harmonic and intermodulation distortions for interband laser have been widely investigated. Such investigation is done either using perturbation techniques [1–9] to linearise the rate equations or by a harmonic-balance method [10, 11]

2. Analytical Expressions for Harmonic, Intermodulation and Cross-Modulation Distortion

By the assumptions given by [12], the single-mode, two-level simplified rate equations [13] are given as follows:

$$\frac{dN_3}{dt} = \frac{I}{e} - \frac{N_3}{\tau_3} - G\left(N_3 - N_2\right)P \tag{1}$$

$$\frac{dN_2}{dt} = \frac{N_3}{\tau_3'} - \frac{N_2}{\tau_{21}} + G\left(N_3 - N_2\right)P \tag{2}$$

$$\frac{dP}{dt} = NG\left(N_3 - N_2\right)P + N\beta\frac{N_3}{\tau_{\text{sp}}}\frac{P}{\tau_{\text{p}}} \tag{3}$$

where N_2 and N_3 are the electron numbers in level 2 and level 3 respectively; G is the gain coefficient per stage; e is the magnitude of electric charge; τ_{21} is the lifetime representing the transitions from level 2 to 1 respectively; τ_{sp} is the spontaneous lifetime between level 3 and 2; τ_{p} is the photon lifetime; P is the photon number in the cavity; N is the number of stages; β is the coupling of spontaneous emission into the lasing mode; I is the injected current and t is time. τ_3 is given by $1/\tau_3 = 1/\tau_{31} + 1/\tau_{32} + 1/\tau_{\text{sp}}$ where τ_{31} and τ_{32} are the lifetimes representing the transitions from level 3 to 1 and 3 to 2 respectively, and τ_3' is given by $1/\tau_3' = 1/\tau_{32} + 1/\tau_{\text{sp}}$.

2.1. Perturbation analysis

The gain term at the rate equations makes the system nonlinear. The rate equations are linearized for small perturbations by writing various quantities as a perturbation series in powers of a perturbation parameter, ξ. The perturbation series for I and N_3 are given by:

$$I = I^{\{0\}} + \xi_{\omega_1} I^{\{\omega_1\}} + \xi_{\omega_2} I^{\{\omega_2\}} + \xi_{\omega_3} I^{\{\omega_3\}} \tag{4}$$

$$\begin{aligned} N_3 = {} & N_3^{\{0\}} + \xi_{\omega_1} N_3^{\{\omega_1\}} + \xi_{\omega_2} N_3^{\{\omega_2\}} + \xi_{\omega_3} N_3^{\{\omega_3\}} + \xi_{\omega_1}^{-1} N_3^{\{\omega_1\}*} + \xi_{\omega_2}^{-1} N_3^{\{\omega_2\}*} \\ & + \xi_{\omega_1}^2 N_3^{\{2\omega_1\}} + \xi_{\omega_1+\omega_2} N_3^{\{\omega_1+\omega_2\}} + \xi_{\omega_1+\omega_3} N_3^{\{\omega_1+\omega_3\}} + \xi_{\omega_1-\omega_2} N_3^{\{\omega_1-\omega_2\}} \\ & + \xi_{\omega_3-\omega_2} N_3^{\{\omega_3-\omega_2\}} + \xi_{2\omega_1-\omega_2} N_3^{\{2\omega_1-\omega_2\}} + \xi_{\omega_1+\omega_3-\omega_2} N_3^{\{\omega_1+\omega_3-\omega_2\}} \\ & + \xi_{\omega_1+\Delta\omega} N_3^{\{\omega_1+\omega_2-\omega_1\}} \end{aligned} \tag{5}$$

with expressions similar to N_3 for N_2 and P. Here, $I^{\{0\}}$ is the DC bias current which is higher than the threshold current, I_{th}. Superscripts of ξ denotes the powers while the subscripts denotes the angular frequency of the subcarriers. Those in braces of the quantities I, N_3, N_2 and P also denote the angular frequency of the subcarriers. Three sinusoidal drive currents of angular frequencies of ω_1, ω_2 and ω_3 are considered for obtaining three-carrier, third order intermodulation distortion of angular frequency $\omega_1 + \omega_3 - \omega_2$.

After equating coefficients with the same power and angular frequencies and some algebra, the equations are then written in matrix form as follows:

$$\left[D\left(\omega\right)\right]\left[U\left(\omega\right)\right] = \left[F\left(\omega\right)\right] \tag{6}$$

where ω represents $\omega_{1,2,3}$, $2\omega_{1,2,3}$, $3\omega_{1,2,3}$, $\omega_1+\omega_2$, $\omega_1+\omega_3$, $\omega_1-\omega_2$, $\omega_3-\omega_2$, $2\omega_1-\omega_2$, $\omega_1+\omega_3-\omega_2$, $\omega_1+\omega_2-\omega_1$,

$$\left[D\left(\omega\right)\right] = \begin{bmatrix} j\omega + \tau_3^{-1} + GP^{\{0\}} & -GP^{\{0\}} & \left(\tau_{\text{p}}N\right)^{-1} \\ -\tau_3'^{-1} - GP^{\{0\}} & j\omega + \tau_{21}^{-1} + GP^{\{0\}} & -\left(\tau_{\text{p}}N\right)^{-1} \\ -NGP^{\{0\}} - \beta N\tau_{\text{sp}}^{-1} & NGP^{\{0\}} & j\omega \end{bmatrix} \tag{7}$$

and $[U(\omega)] = \left[N_3^{\{\omega\}} N_2^{\{\omega\}} P^{\{\omega\}}\right]^{\mathrm{T}}$ is the column matrix of the unknown quantities and T indicates transpose operation. $[F(\omega)]$ is the column matrix of the forcing terms and is given in Table 1 and 2. The forcing term matrix is given as the transpose of row matrix to save space. Here, the superscripts in braces of the forcing terms denote the frequency of the phasors and not the order.

Table 1. Forcing matrix for fundamental, harmonic, intermodulation and cross-modulation terms

ω	$[F(\omega)]^{\mathrm{T}}$
$\omega_{1,2,3}$	$\left[I^{\{\omega_{1,2,3}\}}/e \quad 0 \quad 0\right]$
$2\omega_{1,2,3}$	$\left[-S^{\{2\omega_{1,2,3}\}} \quad S^{\{2\omega_{1,2,3}\}} \quad NS^{\{2\omega_{1,2,3}\}}\right]$
$3\omega_{1,2,3}$	$\left[-S^{\{3\omega_{1,2,3}\}} \quad S^{\{3\omega_{1,2,3}\}} \quad NS^{\{3\omega_{1,2,3}\}}\right]$
$\omega_1+\omega_2$	$\left[-S^{\{\omega_1+\omega_2\}} \quad S^{\{\omega_1+\omega_2\}} \quad NS^{\{\omega_1+\omega_2\}}\right]$
$\omega_1+\omega_3$	Exchange ω_2 with ω_3 above.
$\omega_1-\omega_2$	$\left[-S^{\{\omega_1-\omega_2\}} \quad S^{\{\omega_1-\omega_2\}} \quad NS^{\{\omega_1-\omega_2\}}\right]$
$\omega_3-\omega_2$	Exchange ω_1 with ω_3 above.
$2\omega_1-\omega_2$	$\left[-S^{\{2\omega_1-\omega_2\}} \quad S^{\{2\omega_1-\omega_2\}} \quad NS^{\{2\omega_1-\omega_2\}}\right]$
$\omega_1+\omega_3-\omega_2$	$\left[-S^{\{\omega_1+\omega_3-\omega_2\}} \quad S^{\{\omega_1+\omega_3-\omega_2\}} \quad NS^{\{\omega_1+\omega_3-\omega_2\}}\right]$
$\omega_1+\omega_2-\omega_1$	$\left[-S^{\{\omega_1+\omega_2-\omega_1\}} \quad S^{\{\omega_1+\omega_2-\omega_1\}} \quad NS^{\{\omega_1+\omega_2-\omega_1\}}\right]$

The forcing terms in the matrix are defined in Table 2.

Note that the forcing terms for the nonlinear distortions of QCLs have a strong resemblance with those for interband lasers, but with difference in the electron number terms. In QCLs, the forcing terms consist of electron number differences instead of absolute electron number as in interband lasers. The forcing terms for the higher order distortions are given in terms of lower order distortions and are found once the lower order distortions are known. The second order intermodulation distortions of frequency $\omega_1+\omega_2$, $\omega_1+\omega_3$, $\omega_1-\omega_2$, and $\omega_3-\omega_2$ are given here as they are needed for finding the two-carrier, third order intermodulation distortion, three-carrier, third order intermodulation distortion and cross-modulation distortion. These second order intermodulation distortion will not be a hindrance in a communication system as they will be filtered out with a RF bandpass filter following the optical receiver.

2.2. Fundamental response

The optical modulation depth (OMD), m_{d} is defined as [14]

$$m_{\mathrm{d}} = \frac{P^{\{\omega_{1,2,3}\}}}{P^{\{0\}}} = \frac{\left|I^{\{\omega_{1,2,3}\}}\right|}{I^{\{0\}} - I_{\mathrm{th}}} \tag{8}$$

Table 2. Definition of forcing terms in the matrix

ω	$S^{\{\omega\}}$
$2\omega_{1,2,3}$	$0.5G\left(N_3^{\{\omega_{1,2,3}\}} - N_2^{\{\omega_{1,2,3}\}}\right)P^{\{\omega_{1,2,3}\}}$
$3\omega_{1,2,3}$	$0.5G\left[\left(N_3^{\{\omega_{1,2,3}\}} - N_2^{\{\omega_{1,2,3}\}}\right)P^{\{2\omega_{1,2,3}\}} + \left(N_3^{\{2\omega_{1,2,3}\}} - N_2^{\{2\omega_{1,2,3}\}}\right)P^{\{\omega_{1,2,3}\}}\right]$
$\omega_1+\omega_2$	$0.5G\left[\left(N_3^{\{\omega_1\}} - N_2^{\{\omega_1\}}\right)P^{\{\omega_2\}} + \left(N_3^{\{\omega_2\}} - N_2^{\{\omega_2\}}\right)P^{\{\omega_1\}}\right]$
$\omega_1+\omega_3$	Exchange ω_2 with ω_3 above.
$\omega_1-\omega_2$	$0.5G\left[\left(N_3^{\{\omega_1\}} - N_2^{\{\omega_1\}}\right)P^{\{\omega_2\}*} + \left(N_3^{\{\omega_2\}} - N_2^{\{\omega_2\}}\right)^*P^{\{\omega_1\}}\right]$
$\omega_3-\omega_2$	Exchange ω_1 with ω_3 above.
$2\omega_1-\omega_2$	$0.5G\Big[\left(N_3^{\{\omega_1\}} - N_2^{\{\omega_1\}}\right)P^{\{\omega_1-\omega_2\}} + \left(N_3^{\{\omega_1-\omega_2\}} - N_2^{\{\omega_1-\omega_2\}}\right)P^{\{\omega_1\}} + \left(N_3^{\{\omega_2\}} - N_2^{\{\omega_2\}}\right)^*P^{\{2\omega_1\}} + \left(N_3^{\{2\omega_1\}} - N_2^{\{2\omega_1\}}\right)P^{\{\omega_2\}*}\Big]$
$\omega_1+\omega_3-\omega_2$	$0.5G\Big[\left(N_3^{\{\omega_1\}} - N_2^{\{\omega_1\}}\right)P^{\{\omega_3-\omega_2\}} + \left(N_3^{\{\omega_3-\omega_2\}} - N_2^{\{\omega_3-\omega_2\}}\right)P^{\{\omega_1\}} + \left(N_3^{\{\omega_2\}} - N_2^{\{\omega_2\}}\right)^*P^{\{\omega_1-\omega_3\}} + \left(N_3^{\{\omega_1-\omega_3\}} - N_2^{\{\omega_1-\omega_3\}}\right)P^{\{\omega_2\}*} + \left(N_3^{\{\omega_3\}} - N_2^{\{\omega_3\}}\right)P^{\{\omega_1-\omega_2\}} + \left(N_3^{\{\omega_1-\omega_2\}} - N_2^{\{\omega_1-\omega_2\}}\right)P^{\{\omega_3\}}\Big]$
$\omega_1+\omega_2-\omega_1$	$0.5G\Big[\left(N_3^{\{\omega_1\}} - N_2^{\{\omega_1\}}\right)^*P^{\{\omega_1+\omega_2\}} + \left(N_3^{\{\omega_1+\omega_2\}} - N_2^{\{\omega_1+\omega_2\}}\right)P^{\{\omega_1\}*} + \left(N_3^{\{\omega_2-\omega_1\}} - N_2^{\{\omega_2-\omega_1\}}\right)P^{\{\omega_1\}} + \left(N_3^{\{\omega_1\}} - N_2^{\{\omega_1\}}\right)P^{\{\omega_2-\omega_1\}}\Big]$

as the light current characteristic of QCLs is linear at above threshold. Here, $P^{\{\omega_{1,2,3}\}}$ is the peak value of the sinusoidal light intensity corresponding to sinusoidal peak current, $I^{\{\omega_{1,2,3}\}}$; $P^{\{0\}}$ is the DC value at the bias current, $I^{\{0\}}$, and I_{th} is the threshold current. Small OMD ranging from 1% to 10% is considered here.

Using Cramer's rule, (6) is solved for the expressions of the unknown quantities. The denominator for all the expressions is the determinant of $[D(\omega)]$ and is given by

$$\begin{aligned}|D(\omega)| = &- j\omega^3 - \left(\frac{1}{\tau_{21}} + \frac{1}{\tau_3} + 2GP^{\{0\}}\right)\omega^2 \\ &+ \left(\left(\frac{2}{\tau_{\text{p}}} + \frac{1}{\tau_{31}} + \frac{1}{\tau_{21}}\right)GP^{\{0\}} + \frac{1}{\tau_3\tau_{21}} + \frac{\beta}{\tau_{\text{p}}\tau_{\text{sp}}}\right)j\omega \\ &+ \left(\left(\frac{1}{\tau_{31}} + \frac{1}{\tau_{21}}\right)\frac{GP^{\{0\}}}{\tau_{\text{p}}} + \frac{\beta}{\tau_{21}\tau_{\text{p}}\tau_{\text{sp}}}\right)\end{aligned} \tag{9}$$

The expression of the fundamental component of the photon number solved from (6) for $\omega = \omega_{1,2,3}$ is given as

$$P^{\{\omega_{1,2,3}\}} = m_{\text{d}}\frac{M_P(\omega_{1,2,3})}{|D(\omega_{1,2,3})|} \tag{10}$$

where

$$M_P(\omega_{1,2,3}) = \frac{I^{\{0\}} - I_{\text{th}}}{e}\left[\left(NGP^{\{0\}} + \frac{\beta N}{\tau_{\text{sp}}}\right) j\omega_{1,2,3} - \left(\frac{1}{\tau_3'} - \frac{1}{\tau_{21}}\right) NGP^{\{0\}} + \left(\frac{1}{\tau_{21}} + GP^{\{0\}}\right)\frac{\beta N}{\tau_{\text{sp}}}\right] \tag{11}$$

The normalized modulation response has been reported by [13]. As the magnitude of the distortions is of interest, the un-normalized modulation response is considered in this chapter.

Fundamental components of N_3 and N_2 are needed for obtaining solution for higher order distortions. These components are solved from (6) for $\omega = \omega_{1,2,3}$ and are given as

$$N_3^{\{\omega_{1,2,3}\}} = m_{\text{d}} \frac{M_{N_3}(\omega_{1,2,3})}{|D(\omega_{1,2,3})|}; \quad N_2^{\{\omega_{1,2,3}\}} = m_{\text{d}} \frac{M_{N_2}(\omega_{1,2,3})}{|D(\omega_{1,2,3})|} \tag{12}$$

where

$$M_{N_3}(\omega_{1,2,3}) = \frac{I^{\{0\}} - I_{\text{th}}}{e}\left[-\omega_{1,2,3}^2 + \left(\frac{1}{\tau_{21}} + GP^{\{0\}}\right) j\omega_{1,2,3} + \frac{GP^{\{0\}}}{\tau_{\text{p}}}\right] \tag{13}$$

$$M_{N_2}(\omega_{1,2,3}) = \frac{I^{\{0\}} - I_{\text{th}}}{e}\left[\left(\frac{1}{\tau_3'} + GP^{\{0\}}\right) j\omega_{1,2,3} + \frac{GP^{\{0\}}}{\tau_{\text{p}}} + \frac{\beta}{\tau_{\text{p}}\tau_{\text{sp}}}\right] \tag{14}$$

2.3. Second harmonic ($2\omega_{1,2,3}$) and third harmonic ($3\omega_{1,2,3}$) response

Using Cramer's rule, expressions for second and third harmonic are found by solving (6) for $\omega = 2\omega_{1,2,3}$ and $\omega = 3\omega_{1,2,3}$ respectively. The harmonic expressions are given in terms of ω as

$$N_3^{\{\omega\}} = \frac{M_{N_3}(\omega)}{|D(\omega)|}; \quad N_2^{\{\omega\}} = \frac{M_{N_2}(\omega)}{|D(\omega)|} \tag{15}$$

$$P^{\{\omega\}} = \frac{M_P(\omega)}{|D(\omega)|} \tag{16}$$

where

$$M_{N_3}(\omega) = \left[\omega^2 - \left(\frac{1}{\tau_{21}} + \frac{1}{\tau_{\text{p}}}\right) j\omega + \frac{1}{\tau_{21}\tau_{\text{p}}}\right] S^{\{\omega\}} \tag{17}$$

$$M_{N_2}(\omega) = \left[-\omega^2 + \left(\frac{1}{\tau_{31}} + \frac{1}{\tau_{\text{p}}}\right) j\omega + \frac{1}{\tau_{31}\tau_{\text{p}}}\right] S_{\{\omega\}} \tag{18}$$

$$M_P(\omega) = \left[-\omega^2 + \left(\frac{1}{\tau_{21}} + \frac{1}{\tau_3} - \frac{\beta}{\tau_{\text{sp}}}\right) j\omega + \frac{1}{\tau_3\tau_{21}} - \frac{\beta}{\tau_{21}\tau_{\text{sp}}}\right] NS^{\{\omega\}} \tag{19}$$

Second and third harmonics are found by substituting the appropriate angular frequency and forcing terms into (16). Note that second harmonic can be obtained once the fundamental components of N_3, N_2 and P are found and third harmonic can be obtained once both fundamental component and second harmonic are found. Fundamental component is found using (12) and (10) and second harmonic is found using (15) and (16). Only third harmonic of P is of interest and the third harmonic of N_3 and N_2 need not to be found.

IOP Conf. Series: Materials Science and Engineering **131** (2016) 012025 doi:10.1088/1757-899X/131/1/012025

2.4. Two-carrier, third order intermodulation distortion $(2\omega_1 - \omega_2)$

The expressions for two-carrier, third order intermodulation distortion (IMD2) are obtained by solving (6) for $\omega = 2\omega_1 - \omega_2$. Using Cramer's rule, the expression for P are obtained as given in (16) with $\omega = 2\omega_1 - \omega_2$. The solution for this distortion can be obtained once the fundamental component, second harmonic and second order intermodulation of N_3, N_2 and P are found. The fundamental component is found using (12) and (10) and second harmonic and second order intermodulation are found using (15) and (16) with their respective angular frequencies.

Finally, note that expressions for three-carrier, third order intermodulation distortion can be derived in a similar manner.

3. Conclusion

Expression for harmonic and intermodulation distortion in QCLs have been derived and discussed. In future work these expressions will be analysed further including graphs for some typical parameter values.

References

[1] Lau K Y and Yariv A 1984 *Applied Physics Letters* **45** 1034–1036
[2] Darcie T E, Tucker R S and Sullivan G J 1985 *Electronics Letters* **21** 655–666
[3] Iannone P and Darcie T E 1987 *Electronics Letters* **23** 1361–1362
[4] Way W I 1987 *Journal of Lightwave Technology* **LT-5** 305–315
[5] Darcie T E 1987 *Journal of Lightwave Technology* **LT-5** 1103–1110
[6] Haldar M K, Kooi P S, Mendis F V C and Guan Y L 1992 *Journal of Applied Physics* **71** 1102–1108
[7] Helms J 1992 *Journal of Lightwave Technology* **10** 1901–1906
[8] Wang J, Haldar M K and Mendis F V C 1993 *Electronics Letters* **29** 1341–1343
[9] Chen J, Ram R J and Helkey R 1999 *IEEE Journal of Quantum Electronics* **35** 1231–1237
[10] Iezekiel S, Snowden C M and Howes M J 1989 *Electronics Letters* **25** 529–530
[11] Odermatt S, Witzigmann B and Schmithüsen B 2006 *Optical and Quantum Electronics* **38** 1039–1044
[12] Webb J F and Haldar M K 2012 *J. App. Phys.* **111** 093116–1–093116–6
[13] Yong K S C, Haldar M K and Webb J F 2014 *Optical and Quantum Electronics* **46** 659–667
[14] Agrawal G P 2002 *Fiber-optic Communication System* (Singapore: Wiley-Interscience)